THE SOLAR HOME BOOK

art direction by ***Linda Goodman*** *illustrations by* ***Edward A. Wong***
pencil drawings by Rachel Dutton

The Solar Home Book

heating, cooling and designing with the sun

by Bruce Anderson with Michael Riordan

Brick House Publishing Co., Inc. Andover, Massachusetts

Acknowledgements

Linda Goodman: Art direction
Edward A. Wong: Book design, illustrations
Ned Williams: Cover design
Marshall Henrichs: Cover photo
Rachel Dutton: Pencil drawings
R. G. Beukers: Copy editing, proofreading
Lynn Nelson: Reviewing
Trudy Smith: Typesetting
Robert Cooney: Paste-up
Jean Jacobson: Copy editing
R. Schorr Berman: Final editing

This book was edited and produced by Cheshire Books.
Published by: Brick House Publishing Co.
3 Main St.
Andover, MA 01810

Distributed in Canada by: Firefly Books
2 Essex Avenue #5
Thornhill, Ontario

Printed in the United States of America
Library of Congress Catalog Card Number 76-29494
ISBN: 0-917352-01-7
11 12 13 14 15 16 17 18 19 20

The authors also thank the following:

The American Society of Heating, Refrigerating, and Air-conditioning Engineers, Inc., for permission to adapt and reprint tables from *ASHRAE Guide and Data Book,* 1970; *ASHRAE Handbook and Product Directory,* 1974; *ASHRAE Handbook of Fundamentals,* 1967 and 1972; and the Symposium on *Solar Energy Applications,* 1974.

Pacific Gas and Electric Co. for permission to reprint the infrared photographs on page 73.

Revere Copper and Brass Co. for permission to reprint the graphs on p. 179.

Rodale Press, Emmaus, Pa., for permission to adapt and reprint material from *Low Cost Energy-Efficient Shelter,* ed. by Eugene Eccli. Copyright © 1976 by Eugene Eccli.

John Wiley & Sons, Inc., for permission to reprint material from *Architectural Graphic Standards,* by C. G. Ramsey and H. R. Sleeper. Copyright © 1970 by John Wiley & Sons, Inc., and for permission to adapt material from *Solar Energy Thermal Processes,* by John A. Duffie and William A. Beckman. Copyright © 1974 by John Wiley & Sons, Inc.

Author's Note

The Solar Home Book has benefitted from the efforts of many fine people. Its first vague stirrings occurred in the autumn of 1972, when I wrote my Master's thesis, *Solar Energy and Shelter Design,* for the School of Architecture at the Massachusetts Institute of Technology. In 1974 and early 1975, I expanded this thesis into the manuscript for a full-length book, *Solar Energy in Building Design.* This project drew upon the valuable contributions of my associates at Total Environmental Action (TEA). Douglas Mahone deserves special recognition for working as my unofficial editor during those trying times and for writing a chapter on solar water heaters. But much to our dismay, our publisher quietly closed up shop not a week after the manuscript was in their hands.

In the Fall of 1975, Richard Katzenberg and Michael Riordan offered to finance an independent publication of the manuscript. Under Michael's direction, a thoroughly revised and edited version began to emerge during the first few months of 1976. Flying the banner of Cheshire Books, a coterie of artists, writers, and other free spirits labored night and day in the stuffy attics and musty cellars of Menlo Park and Palo Alto to produce in six short months what established publishing companies take years to emulate. The end result of all these efforts rests in your hands—an informative, enjoyable, and easy-to-read introduction and guide to the uses of solar energy in the home.

Simultaneously, the McGraw Hill Book Company produced a hardbound textbook version in cooperation with TEA. *Solar Energy: Fundamentals in Building Design* is based on the same original manuscript but directed toward professional audiences.

My humblest thanks go to Richard Katzenberg and Michael Riordan for their confidence in the manuscript and their courage in financing this homespun version. The quality of Michael's editing and writing and the unity of thought he gave the book escape my powers of description. In some sense, the book is as much his as it is mine. And I am truly indebted to Linda Goodman, who transformed a work of science and technology into one also of art. Her unflinching insistence on quality and coherence in all aspects of the book made it the beautiful work it is. Edward A. Wong assisted her in designing a book format appropriate to the subject material and to the audience we were trying to reach. He also contributed the many fine illustrations that bring the text to life. I would like to thank Rachel Dutton for the superb pencil drawings that grace the first page of each chapter and Minna Resnick for her outstanding cover design. Expert copy-editing came from R. G. Beukers, who also supplied much of the humor in the text. Trudy Smith patiently and flawlessly set the type on her IBM Composer. And special thanks go to Lynn Nelson, who spent long hours in search of do-it-yourself projects for inclusion in Chapter 7. The writing therein is mainly hers, with contributions from Douglas Mahone and some judicious editing supplied by Michael. Finally, I would like to thank Tom Gage for the support and guidance he gave to this publishing effort.

The use of solar energy is at a crossroads. It can be used in ways that perpetuate the *status quo* and hasten the demise of an over-exploited environment, or it can be used in ways which will enrich our lives and bring us closer to our natural surroundings. I hope my book will help us to walk this latter path.

Bruce Anderson
Harrisville, New Hampshire
September 1976

Table of Contents

Foreword

For generations, Americans have viewed cheap and plentiful energy as their birthright. Coal, oil or gas have always been abundantly available to heat our homes, power our automobiles, and fuel our industries. But just as the supply of these fossil fuels begins to dwindle and we look to the atom for salvation, we are beginning to perceive the environmental havoc being wrought by our indiscriminate use of energy. Our urban and suburban skies are choked with smog; our rivers and shores are streaked with oil; even the food we eat and the water we drink are suspect. And while promising us temporary relief from energy starvation, nuclear power threatens a new round of pollution whose severity is still a matter of speculation.

The residential use of solar energy is one step toward reversing this trend. By using the sun to heat and cool our homes, we can begin to halt our growing dependence on energy sources that are polluting the environment and rising in cost. The twin crises of energy shortage and environmental degradation occur because we have relied on concentrated forms of energy imported from afar. We had little say in the method of energy production and accepted its byproducts just as we grasped for its benefits. But solar energy can be collected right in the home, and we can be far wiser in its distribution and use.

Unlike nuclear power, solar energy produces no lethal radiation or radioactive wastes. Its generation is not centralized and hence not open to sabotage or blackmail. Unlike oil, the sun doesn't blacken our beaches or darken our skies. Nor does it lend itself to foreign boycott or corporate intrigue. Unlike coal, the use of solar energy doesn't ravage our rural landscapes with strip mining or our urban atmospheres with soot and sulphurous fumes.

Universal solar heating and cooling could ease fuel shortages and environmental pollution substantially. Almost 15 percent of the energy consumed in the United States goes for home heating, cooling, and water heating. If the sun could provide two thirds of these needs, it would reduce the national consumption of non-renewable fuels by 10 percent and world consumption by more than 3 percent. National and global pollution would drop by similar amounts.

But solar energy has the drawback of being diffuse. Rather than being mined or drilled at a few scattered places, it falls thinly and fairly evenly across the globe. The sun respects no human boundaries and is available to all. Governments and industries accustomed to concentrated energy supplies are ill-equipped, by reason of economic constraints or philosophical prejudices, to harness this gentle source of energy. These institutions are far more interested in forms of energy that lend themselves to centralization and control. Hence the United States government spends billions for nuclear power while solar energy is just a subject for study—a future possibility, maybe, but not right now.

This book speaks to the men and women who cannot wait for a hesitant government to "announce" a new solar age. We can begin to fight energy shortages and environmental pollution in our own homes and surroundings. Solar heating and cooling are feasible *today*—not at some nebulous future date. The solar energy falling on the walls and roof of a home during winter is several times the amount of energy needed to heat it. All it takes to harness this abundant supply is the combination of ingenuity, economy and husbandry that has been the American ideal since the days of Franklin and Thoreau.

There are many simple but elegant ways of capturing the sun's energy. Homes can be designed to respond to local climates instead of isolating the inhabitants from the outdoors. Dwellings can partake of the natural energy flows of sunlight and wind that are normally excluded. And they can be built to retain the energy they do capture. Such simple, low-technology methods are cheaper and more reliable than the many complex, high-technology devices being developed to harness the sun's energy.

Anyone with good building skills and a knowledge of materials can take advantage of these simple methods. Millions of people in cities, towns, and villages across the country can design and build homes suited to local climates and needs. It is these individual homeowners and builders, designers and architects, contractors and tradespeople who will make solar heating and cooling a reality for all. This book has been written to guide them in their work.

Michael Riordan
Menlo Park, California
September, 1976

1 Introduction

Harnessing the Sun's Energy for Use in Home Heating and Cooling

Lashing it to the sacred Intihuatana stone, an Inca priest prevents the sun from "running away" at the winter solstice.

Now in houses with a south aspect, the sun's rays penetrate into the porticoes in winter, but in summer the path of the sun is right over our heads and above the roof, so that there is shade. If, then, this is the best arrangement, we should build the south side loftier to get the winter sun and the north side lower to keep out the cold winds.

Socrates, as quoted by
Xenophon in *Memorabilia*

Human shelter has often reflected an understanding of the sun's power, generosity and cruelty. Primitive shelters in tropical areas have broad thatched roofs that provide shade from the scorching midday sun and keep out frequent rains. The open walls of these structures allow cooling breezes to carry away accumulated heat and moisture. In the American southwest, Pueblo Indians built thick adobe walls and roofs that kept the interiors cool during the day by absorbing the sun's rays. By the time the cold desert night rolled around, however, the absorbed heat had penetrated the living quarters to warm the humble inhabitants. Their communal buildings faced south or southeast to absorb as much of the winter sun as possible.

Even the shelters of somewhat more civilized peoples have taken advantage of the sun. The entire Meso-American city of Teotihuacan, as large as ancient Rome at its height, was laid out on a grid facing 15° west of south. The axis of a Roman military camp was always oriented within 30° of true south. Early New England houses had masonry filled walls and compact layouts to minimize heat loss during frigid winter months. The kitchen, with its constantly burning wood stove, was located on the north side of the house to permit the other rooms to occupy the prime southern exposure. Only in the present century, with abundant supplies of cheap fossil fuels at our command, have we seriously ignored the sun when designing our buildings.

On a higher technical plane, the use of the sun's energy for home heating began in 1939, when the Massachusetts Institute of Technology built its first solar house. For the first time, *solar collectors* placed upon the roof gathered sunlight for interior heating. By 1960, more than a dozen structures had been built that utilized modern methods to harness the sun's energy. The most important of these experimental buildings are described in Chapter 2.

In the early 1970's, particularly since the Arab oil embargo, interest in solar energy kindled to a fiery blaze. By early 1976, hundreds of solar homes had been built and thousands were on the drawing boards. Hundreds of manufacturers were producing solar collectors where there had been few three years before. And in addition to heating homes, the sun's energy was being used to heat domestic water supplies and to power air-conditioners. Recent solar homes in all climatic areas of the United States are also surveyed in Chapter 2.

In spite of this fervor, many obstacles hamper the full-scale use of solar energy.

Measurement of Heat and Solar Energy

There are two basic types of measurement used to describe heat energy—quantity of heat and intensity of heat. We are most familiar with measurements of temperature in degrees Fahrenheit (°F), which refer to the intensity of heat. If a swimming pool has a temperature of 75°F, we know nothing about the quantity of heat in the pool. Intuitively, we know it takes a much larger quantity of heat to raise a pool to 75°F than to raise a kettle of water to 75°F, but the intensity of heat is the same in both.

In the English system of measurement, the unit of heat quantity is the British Thermal Unit, or Btu, the amount of heat needed to raise one pound of water one degree Fahrenheit. In the metric system, the unit of heat quantity is the calorie, or cal, the amount of heat required to raise one gram of water one degree Centigrade (°C). One Btu is equivalent to about 252 cal. It takes the same quantity of heat, 100 Btu or 25,200 cal, to heat 100 pounds of water 1°F as it does to heat 10 pounds of water 10°F.

Heat is one form of energy and sunlight is another—radiant energy. An important characteristic of energy is that it is never lost—energy may change from one form to an equivalent amount of another, but it never disappears. Consequently, we can describe the amount of solar energy striking a surface in terms of an equivalent amount of heat. We measure the solar energy striking a surface in a given time period in units of Btu/ft^2/hr or cal/cm^2/min. Outside the earth's atmosphere, for example, solar energy strikes at the average rate of 429 Btu/ft^2/hr or 1.94 cal/cm^2/min.

The radiant energy reaching us from the sun has a distribution of wavelengths (or colors). We describe these wavelengths in units of microns, or millionths of a meter. The wavelength distribution of solar energy striking the earth's atmosphere and reaching the ground are shown in the accompanying chart.

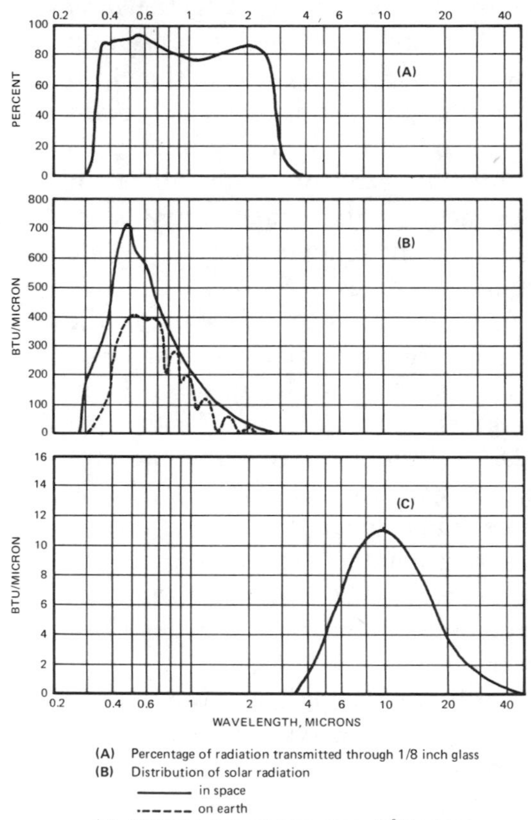

(A) Percentage of radiation transmitted through 1/8 inch glass
(B) Distribution of solar radiation
——— in space
- - - - - on earth
(C) Distribution of thermal radiation from a 95°F black body

About half of the solar radiation reaching the ground falls in the visible range, 0.4 to 0.7 microns. Most of the radiation in the ultraviolet range, with wavelengths below 0.4 microns, is absorbed in the upper atmosphere. A substantial portion of the infrared radiation, with wavelengths greater than 0.7 microns, reaches the earth's surface. A warm body emits even longer wave infrared radiation. Since glass transmits very little radiation at these longer wavelengths, it traps this thermal radiation.

Most of these stumbling blocks are nontechnical in nature. We have the technology to harness the sun for home heating and cooling. But initial costs of solar equipment remain burdensome to the average homeowner, and members of the financial community are skeptical about solar energy. Also, the effective use of the sun requires that homeowners assume new attitudes toward their surroundings. The exaggerated needs and wasteful habits born in an age of abundant fossil fuels are incompatible with the use of this gentle energy. We must become more attuned to our own environment if we are to make the best use of solar energy.

As we embark upon this new era of solar architecture, we should not forget the lessons of history. Though many of the old ways of using the sun require personal attention, they are reliable and cause little harm to the ecosphere. They have withstood the test of time through simplicity of design and sensitivity to real human needs. Only when we have made every effort to use the sun in simple ways should we introduce new technologies—and then only sparingly, patiently, and carefully.

Basic Heat Theory and Solar Phenomena

Most of the energy reaching us from the sun consists of visible light and *infrared* rays. These two forms of radiation are similar, differing only in their wavelengths. Upon striking an object, a portion of this radiation is absorbed and transformed into an equivalent amount of heat energy. This heat is simply the motion of the atoms and molecules in that object. It is stored in the material itself or *conducted* to surrounding materials, warming them in turn. Heat can also be carried off by air and water flowing past these warm materials in what we call *convection* heat flow.

That a material can be heated by the sun is obvious to anyone who has walked barefoot over a sun-baked pavement. What may not be so obvious is that the pavement also *radiates* some of the heat energy away in the form of infrared rays. You can feel this *thermal radiation* by putting your hand near an iron poker after it has been heated in a fireplace. It is this radiation of energy back into space that keeps the earth from overheating and saves us from frying to a crisp.

The amount of solar energy reaching the earth's surface is truly enormous. It frequently exceeds 200 Btu per hour on a square foot of surface, or enough to power a 60 watt light bulb if we could convert all the solar energy to electricity. But we're fortunate to convert even 10 percent, and solar electricity is still in its infancy. On the other hand, efficiencies of 50 percent are not unreasonable for the conversion of solar energy into heat for a house. And, as explained in Chapter 3, the energy falling on a house during winter is generally several times what is needed inside.

Glass is the "miracle" substance that makes modern solar heating possible. Glass transmits visible light but it absorbs thermal radiation. You can prove this to yourself by sitting in front of a blazing fire. Your face becomes unbearably hot if you sit too close. But what happens if you place a pane of glass in front of your face? You can still see the fire but your face is not nearly as hot as before. The longwave infrared rays which carry most of the fire's radiant energy are absorbed by the glass, while the shortwave visible rays penetrate to your eyes. In the same way, once sunlight passes through a window and is transformed into heat energy inside, this energy will not be radiated back outside. This phenomenon, known as the *greenhouse effect,* is responsible for that hot stuffy car that you left in the sun after locking the doors and rolling up the windows. Other transparent materials, particularly plastics, absorb this thermal radiation, but none quite so well as glass.

The basic principles of solar collection for home heating and cooling are embodied in the greenhouse. The sun's rays pass through glass or a transparent plastic and are absorbed in a dark surface. The heat produced cannot escape readily because thermal radiation and warm air currents are trapped by the glass or plastic. The accumulated solar heat is then transported to the living quarters or stored for later use.

There is often an overabundance of solar energy when it is not needed, and none at all when it is most in demand. Some means is required to *store* the collected solar heat for use at night or during extended periods of

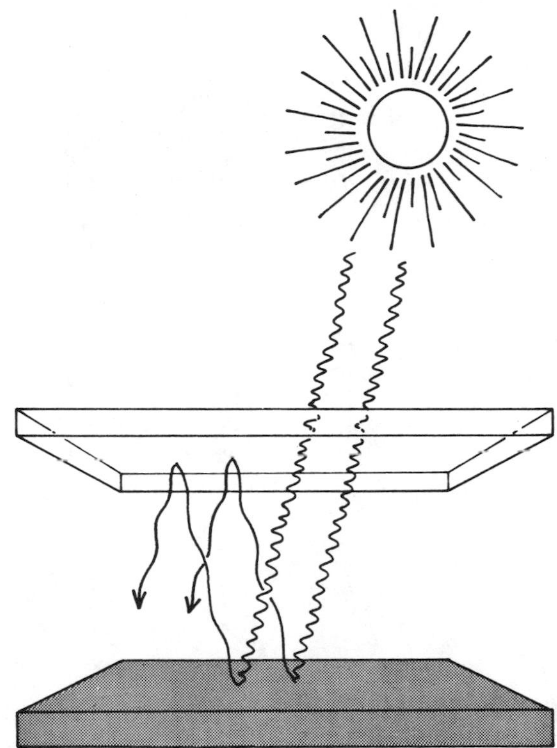

The greenhouse effect. Shortwave solar radiation penetrates the glass and warms the surface below. Longwave thermal radiation is absorbed by the glass, and some of this heat is kept inside.

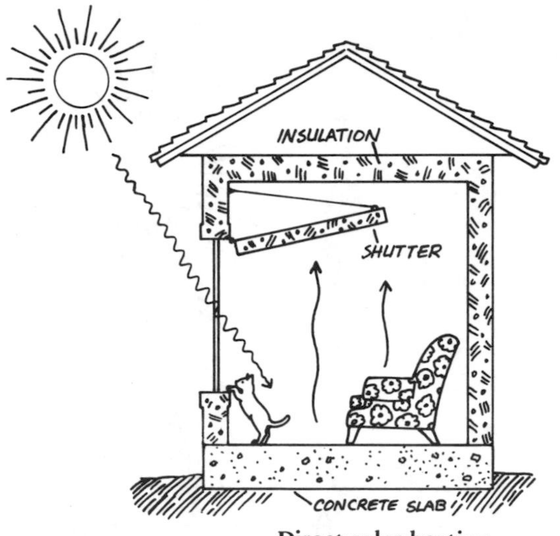

Direct solar heating.

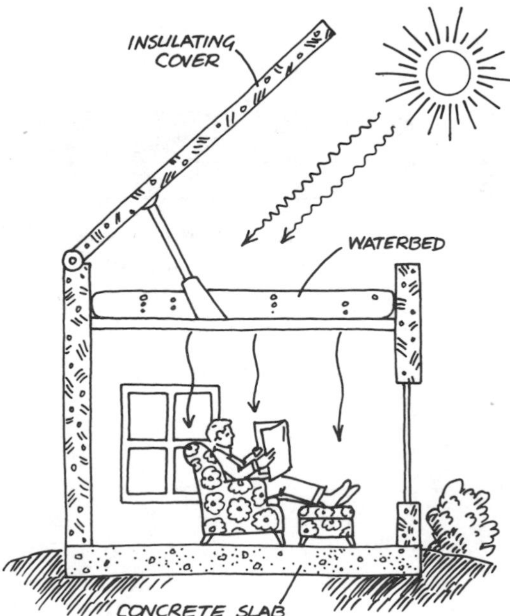

An integrated system for solar heating.

cloudiness. Any material absorbs heat as its temperature rises and releases heat as its temperature falls. The objects inside a house—the walls, floors, and furniture—can therefore serve as heat storage devices. Extra heat can be stored in insulated tanks of water or beds of gravel located within the rooms or in the cellar.

Solar Heating Methods

The growing variety of methods to trap solar radiation for home heating can be grouped into three broad categories—direct, integrated, and indirect. In *direct methods,* the sun's rays penetrate directly into the home. Massive internal structures, such as concrete floors and adobe walls, absorb the heat generated and release it when the sun is not shining. Insulating shutters limit the escape of this heat through the windows into the cold night air. The directly heated solar home acts as a solar collector, a heat storehouse, and a heat trap. Direct methods, which are discussed at length in Chapter 4, are the simplest methods of solar heating. They require at most a rearrangement of standard construction practices. They may also require more attention from homeowners than oil, gas, or electric heating systems. Nevertheless, the basic principles of direct solar heating are essential to the design of any solar home. It makes very little sense to solar heat a flimsy, drafty, uninsulated house.

Integrated systems, which use ingenious adaptations of the natural thermal properties of materials to collect and distribute heat, are discussed in Chapter 5. Heat from the sun is absorbed directly and stored in concrete walls, containers of water inside the house, or water bags on the roof. The heat flows to the rooms without the help of complicated ducts, piping, or pumps. In the example shown, thermal radiation brings heat to the rooms from warm bags of water on the roof. Such solar heating systems are often indistinguishable from the fabric of the home itself. They are also called *passive systems* because they require a minimum of mechanical power to distribute the heat. Although passive systems frequently require radical departures from standard building practice, their simplicity results in effective and reliable home heating.

Indirect systems for solar heating generally use rooftop solar collectors and separate heat storage devices. Heat moves from the collectors to storage or to the rooms by indirect routes, usually involving pipes or ducts. Pumps or fans are required to circulate liquids or air through the collector and back to the insulated heat storage container, often a tank of water or a pile of rocks. When the house needs heat, air or water from the house heating system is warmed by the stored heat and circulated to the rooms. Indirect heating systems, which are discussed in Chapter 6, are also called *active systems* because of their reliance on mechanical power to move the heat. They are much more complex than passive systems and consequently more prone to failure.

Indirect systems are very popular because they require little owner attention, are more readily applied to existing homes, and are more intriguing to a gadget-loving society.

Most indirect solar heating systems use an array of *flat-plate collectors* to gather solar energy. These collectors have one or more glass or plastic cover plates with a black *absorber* beneath them. The cover plates reduce the loss of energy through the front and insulation behind the absorber reduces the heat loss through the back. Heat from the absorber is conducted to a transfer fluid, either air or a liquid, which flows in contact with it and carries off the heat. In *concentrating collectors,* reflective surfaces concentrate the sun's rays onto a very small area. This solar energy is then absorbed by a black surface and converted to heat that is carried off by a fluid. Concentrating collectors can produce very high temperatures, but they usually require mechanical devices to track the sun across the sky. Both flat-plate and concentrating collectors are also discussed in Chapter 6.

Solar heating systems almost always require a backup, or *auxiliary,* heating system such as an oil furnace or wood stove. Rarely is it economically advisable to construct a heat storage unit capable of carrying a house through the longest periods of cold, sunless weather. Depending upon the climate, the house, and the solar heating system design, 50 to 90 percent of a house's heating needs can be readily supplied by the sun. Because the sun is used to modify the indoor climate and doesn't provide *all* its heating needs, we often speak of *solar tempered,* rather than solar heated, homes. This description also reflects our feeling that human shelter should be an extension of the natural environment rather than a separate entity.

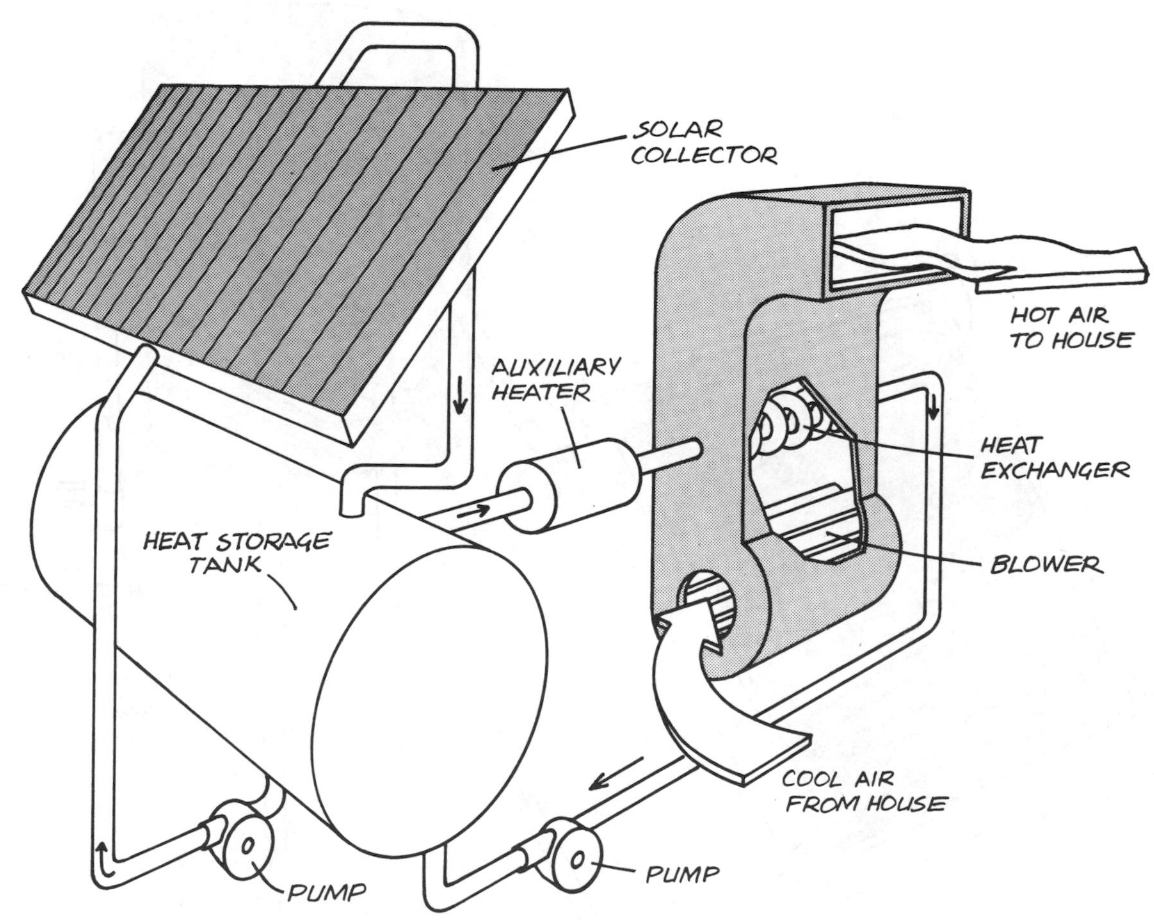

A typical indirect or active system for solar heating.

Cooling, Water Heating, and Electricity

Solar cooling of homes is possible with the absorption cooling method used in gas fired refrigerators. But present cooling equipment requires operating temperatures well above those required for efficient collection of solar energy. Much current research is being devoted to lowering these operating temperatures and to developing collectors that operate efficiently at higher temperatures. Because of the higher cost of such equipment, however, it will be several years before solar cooling systems are commercially viable.

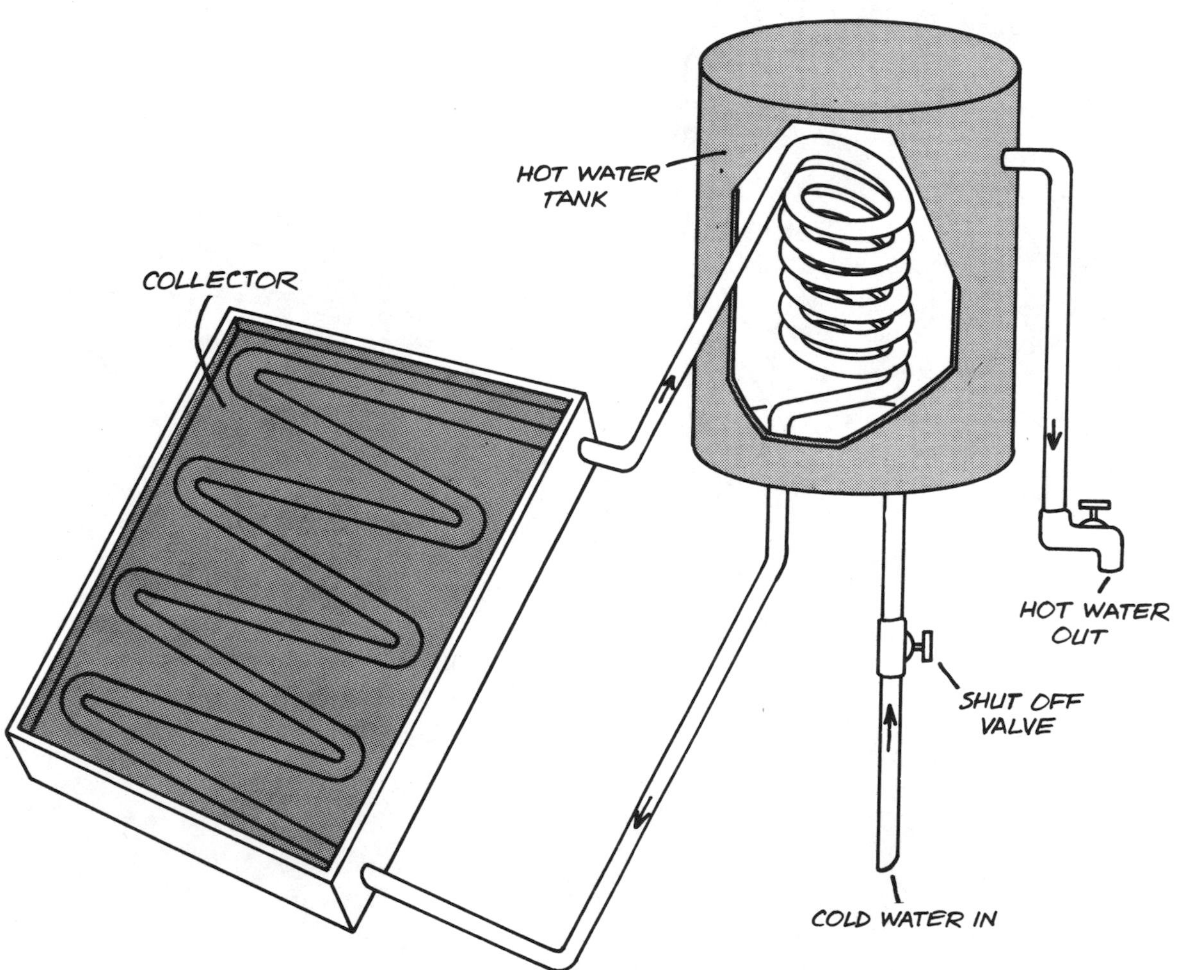

Solar hot water heater with heat exchanger inside the tank.

Some *nocturnal cooling* can be achieved by exposing the solar collection surfaces to the night sky in summer. Particularly in arid areas, where the nights are clear and cool, the collectors can radiate a significant amount of heat into the sky. In this manner, the storage medium can be cooled at night and used for cooling the next day. Nocturnal cooling is best suited to direct methods and integrated systems; it will be discussed further in Chapters 4 and 5.

Solar water heating units are used by millions in Australia, Israel and Japan, and are becoming commonplace in California and Florida. But because of available low-cost fuels and difficulties in protecting solar collectors from freezing, solar water heating has not been widespread in northern climates. With rising fuel costs and improved collectors, however, the solar heating of domestic hot water is gradually being adopted in colder climates. A solar water heater makes an excellent small scale project for the do-it-yourself enthusiast. It can be easily combined with an existing water heater which becomes the backup system. Solar water heaters are discussed in Chapter 7 together with greenhouses and a number of other small projects.

Solar energy can be converted to electricity through five basic methods:

1) using solar energy through *photosynthesis* to grow trees, plants, and algae, which are burned instead of coal in power plants;
2) using the solar heated upper levels and the cold lower depths of the ocean to operate a low temperature difference heat engine and drive a generator;
3) using concentrating solar collectors to heat fluids to drive generators in the same way that electricity is produced in coal-fired power plants;
4) using the solar driven wind to power a turbine or windmill;
5) using *photovoltaic* cells for the direct conversion of solar energy to electricity.

Only the last two offer any promise for generating electricity at the scale of a single house. The third alternative may eventually provide solar electricity for small communities. Unfortunately the centralized generation of electrical power is an established trend in the United States. The rest of this book suggests that many such trends should change.

Solar Home Design

Combining solar heating and cooling into a house design amenable to human needs and the constraints of climate is by no means an easy task. The variety of climates and personal tastes will dictate a veritable plethora of unique solar home designs. Such diversity makes it impossible to provide you with a few standard house designs applicable to your own situation. Solar house design is evolving so rapidly that it would be presumptuous to suggest that the best designs for all climates are immediately at hand.

Nevertheless, a few basic principles are beginning to emerge from more than thirty years of experience in building solar homes. This book is an attempt to present these principles in a form that appeals to the general reader. It is directed to those who would design and build a solar tempered home as well as to those who would hire an architect and contractor to do so. The book does not pretend to be a catalogue of designs. Rather its purpose is to teach the principles of solar home design. Many prime examples of solar tempered homes are described in order to show these principles in action. The rest is up to you and your unflagging enthusiasm for building a home that recognizes the power of the sun.

FURTHER READING

Brinkworth, B. J. *Solar Energy for Man.* Wiltshire, England: Compton Press, 1973.

Clark, Wilson. *Energy for Survival.* Garden City, New York: Anchor Books, 1975.

Daniels, Farrington. *Direct Use of the Sun's Energy.* New Haven: Yale University Press, 1964.

Halacy, D. S. *The Coming Age of Solar Energy.* New York: Harper and Row, 1964.

The solar driven wind powers our windmills.

2
Solar Architecture

A Survey of Important Solar Heated and Cooled Homes, Laboratories, and Office Buildings

A Hopi kachina gazes from the roof of the Acoma pueblo.

The physical architecture of the new society will require the utilization of sophisticated technologies of the present to bridge the gap of history to the styles of the past, when climate reigned supreme.

Wilson Clark,
Energy for Survival

Solar tempered homes have existed since Neolithic times, when people crawled from their caves, rubbed the darkness from their eyes, and piled or pounded together their first structures. Any shelter, no matter how flimsy, can capture sunlight in its skin and transfer some of the absorbed heat inside. It can also provide protection from undesirable solar radiation. Such moderation of climate is the essence of shelter.

Early cultures were adept at using the sun to temper their indoor climates. They chose designs, materials, and orientations that exploited this powerful energy source. But the penetration of solar heat through walls and roofs provided adequate warmth only in the mildest climates. For centuries, people in colder regions have had to bundle up or huddle around fires to keep warm in winter.

The use of transparent substances to enhance solar heat gains is a comparatively recent occurrence. With the advent of plate glass in the present century has come the possibility of heating a shelter mainly with the sun. Experimenters were groping in this direction late in the 1930's, when "solar houses" began to appear. The first structure to earn this name was a conventional wood frame house built near Chicago in 1938. The only unusual feature of this dwelling was its exceptionally large windows on the south side. In 1939, the first flat-plate solar collectors were used to heat another house in Cambridge, Massachusetts. Even more sophisticated technologies for trapping sunlight and transferring the heat inside have been tried since then. The diversity of engineering methods and architectural styles evolved in this search are difficult to summarize in a few pages, but the following survey of solar architecture should give you a good introduction to this work.

EARLY SOLAR BUILDINGS

Thirteen solar heated buildings completed before 1960 are described in the following pages. These homes, offices, and laboratories were selected for their instructive value and for the ample performance data available. Because most of them were experimental projects, they emphasize the engineering aspects of harnessing the sun. Existing architectural styles were only slightly modified to accomodate the solar apparatus in most of these buildings. These examples illustrate the possibilities, as well as the pitfalls, of heating a home with the sun.

MIT Solar House I

The first structure using flat-plate collectors for solar heating was built at MIT in

MIT Solar House I
Cambridge, Massachusetts (42°N)
MIT, 1939-41

Heated floor area:	500 ft^2
Collector - type:	water
area:	360 ft^2
tilt:	30° S
Storage - type:	water
volume:	17,400 gal
Heat distribution:	warm air
Auxiliary heating:	none
Percent solar heated:	100%

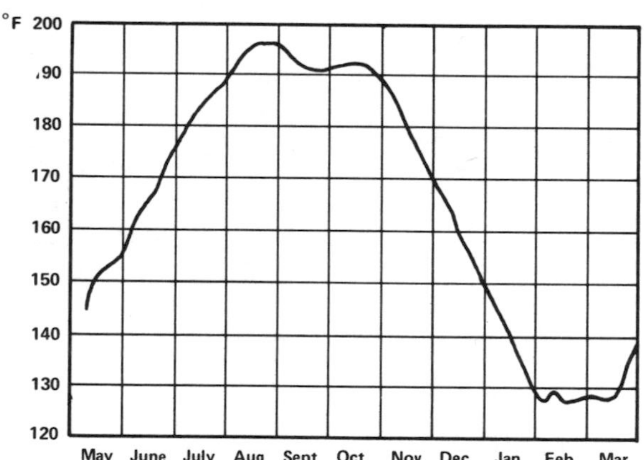

Storage tank temperatures in MIT I. Solar heat was captured in summer and stored for winter use.

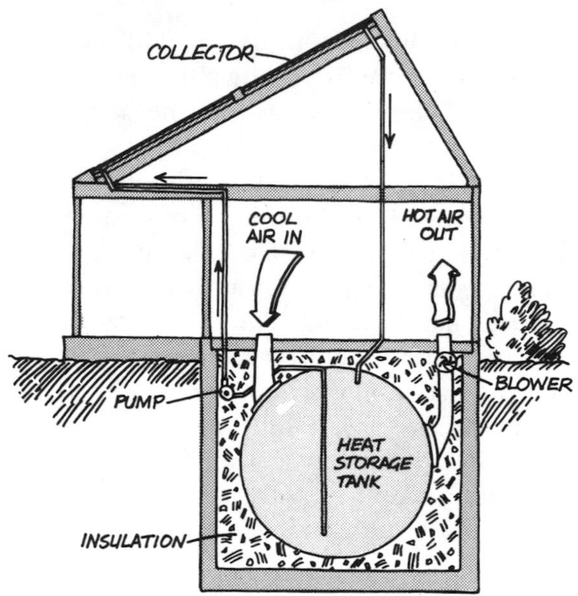

Solar heating system in the first MIT house.

1939. This unassuming two room laboratory building was used to develop methods for calculating the performance of these collectors, which were mounted on the south facing roof with a *tilt angle* of 30° to the horizontal. They were composed of blackened copper sheets sitting behind three glass plates in insulated boxes. Water from an enormous basement tank was pumped to the roof and flowed upwards through 3/8-inch copper tubes soldered at intervals of 6 inches to the back of the copper sheet. The solar heated water then returned to the storage tank. A blower sucked cool room air around the tank surface and distributed warm air to the rooms.

In addition to winter collection, heat was captured in the summer and stored in the well-insulated tank for winter use. No auxiliary heat was needed for two years since the storage water temperature never fell below 125°F. But the added expense of long term heat storage made it impractical at that time. Further, the tilt angle of the collectors was later judged to be too small for winter heat collection. Some problems encountered with this first solar heating system included waterlogged insulation traced to leaky pipes, and breakage of inner glass panes due to thermal expansion.

The director of the MIT research team, Hoyt C. Hottel, together with his assistant B. B. Woertz, derived equations for collector performance from the results of this experiment. Their publication, "The Performance of Flat-Plate Solar Heat Collectors," is still the basic guide for design. They found that the *collector efficiency* (the percent of solar radiation striking the glass and converted to useful heat) was a function of the number of glass panes and of the outdoor and collector temperatures. As expected, this efficiency fell sharply with a rise in the collector temperature. The method of bonding copper tubes to the copper surface was also discovered to be crucial to good performance.

Purdue House

Most of the solar houses fashionable in the 1940's were nothing more than conventional houses graced by large picture windows facing south. The sunlight that penetrated the glass was absorbed inside the house, contributing additional heat. In an experimental solar house built near Chicago in 1940, the savings in heating costs were reported to be 18 percent. Fuel savings as high as 30 percent were claimed for other solar houses.

Some of the most important work with these houses was done by F. W. Hutchinson of Purdue University. In the summer of 1945, he built two small, wood-frame houses on the Purdue campus. They were thermally and structurally identical, except that one house had a much larger window area. All windows in both houses were *double glazed* (two panes of glass) and most of the windows in the solar house faced south.

During the winter of 1945-46, both houses were left unheated (and unoccupied!). The only warmth was provided by the solar energy that penetrated through the walls, roofs, and windows. As long as the weather was clear and sunny, the average air temperature was markedly higher in the solar house. When the outside air temperature was 30°F on January 15, the temperature of the *unheated* solar house was 80°F, while that of the orthodox house was 50°F. But the temperature inside the solar house plummeted after the sun had set because of rapid heat loss through the large expanses of glass. By early morning, the solar house was often colder than the other. Through long periods of cloudy weather, the solar house remained about 5°F colder than the ordinary house.

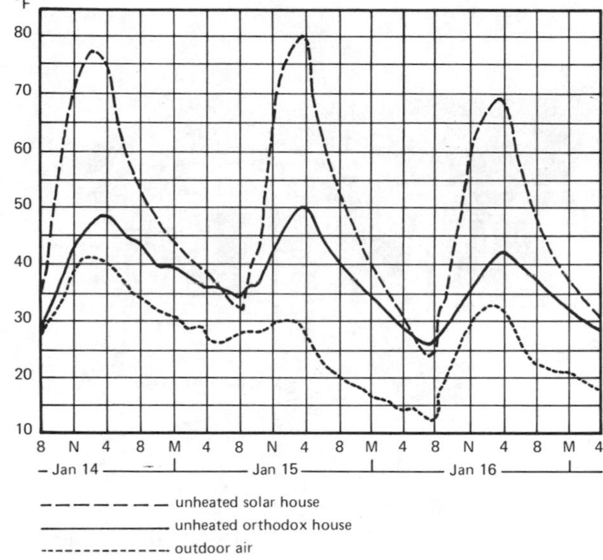

Daily air temperatures inside Purdue test house. Wide temperature fluctuations occurred in the unheated solar house because of its large south windows.

From an analysis based on this study, Hutchinson concluded that the solar heat gain through double glass is greater than the excess heat loss through it. But he also noted that effective control of this solar heat was critical. The pronounced temperature swings that occurred on sunny winter days could not be tolerated in any normal house. Large quantities of heat would have to be excluded—either by drawing shades or venting the rooms—if the house couldn't absorb this heat. Another problem with the solar house was that the larger heat losses through the glass (instead of insulated walls) required a larger and costlier auxiliary heating system. Marked improvements in the storage and

Purdue House
Lafayette, Indiana (40°N)
Hutchinson, 1945-47

Heated floor area:		830 ft^2
Collector -	type:	windows
	area:	112 ft^2
	tilt:	90° S
Storage -	type:	walls, floors
	volume:	unspecified
Heat distribution:		radiation and convection
Auxiliary heating:		none

Boulder House	
Boulder, Colorado (42°N)	
Löf, 1945-47	
Heated floor area:	1000 ft^2
Collector - type:	air
area:	463 ft^2
tilt:	27° S
Storage - type:	gravel
volume:	180 ft^3
Heat distribution:	warm air
Auxiliary heating:	gas furnace
Percent solar heated:	26%

The Boulder House.

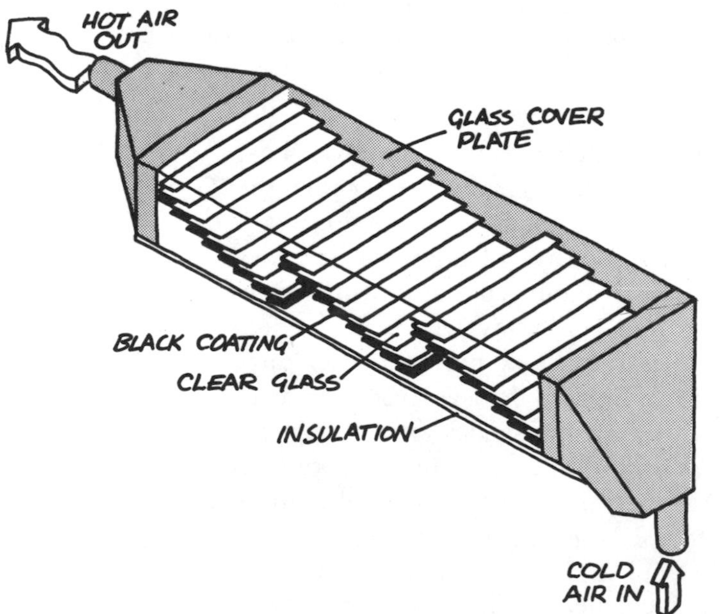

Overlapping-plate collector.

control of the solar heat would have been realized with a more massive building and the use of insulating shutters on the windows.

Boulder House

In 1945, Dr. George Löf, a pioneer of solar home heating, designed a collector and installed it on an existing 5-room bungalow in Boulder, Colorado. The primary objective in his design was the "maintenance of simplicity and economy in construction." The solar collector consisted of a sheet metal trough containing a series of overlapping glass plates arrayed in stairstep fashion. Each pane of glass was blackened on one third of its area. The glass was arranged so that each black surface was beneath two clear surfaces and separated from them by a ¼-inch air gap. One or two cover plates were supported on the top edges of the trough to form an air tight enclosure. The collectors were placed on the surface of the gently sloping roof and insulated from it.

Sunlight penetrated the transparent surfaces and was absorbed in the black areas, raising their temperatures to between 150°F and 250°F. Air entered at the lower end of the trough and exited from the upper end at a temperature close to that of the black areas. The efficiency of this overlapped plate collector ranged from 30 to 65 percent, depending upon the air velocity and number of glass plates. Warmed air from the roof-mounted collector was gathered in a duct at the roof ridge and transported to the gravel heat storage bed in the basement. This air returned to the collector after transferring most of its heat to the gravel. Room air was passed through the gravel bed to bring heat to the house as needed. Warm air also circulated to the rooms directly from the collector.

The primary significance of this project was its successful integration of a functioning solar system with a conventional gas furnace heating unit. In the first winter of operation, the solar unit supplied 26 percent of the total heat requirements of the house. Löf stated that 55 percent solar heating would have been achieved with a smoothly functioning system. Initial difficulties included glass breakage due to thermal expansion, heat losses from the ducts, and hot air leakage at the dampers.

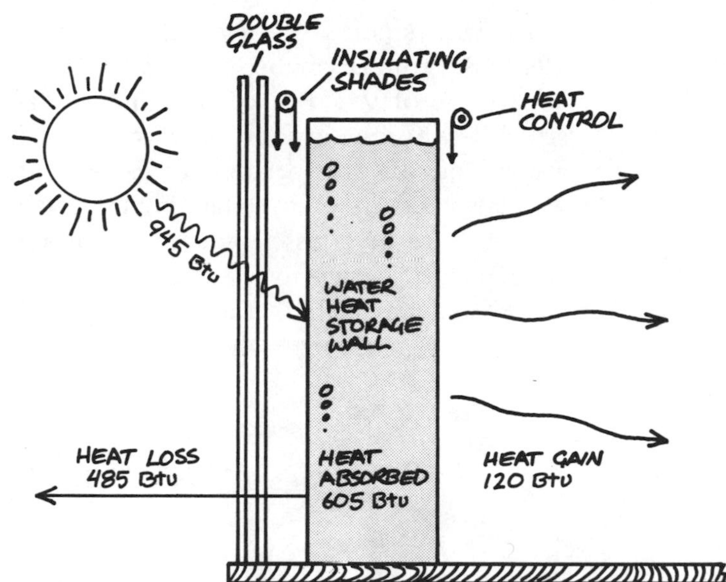

Heat storage wall in MIT Solar House II. Heat flows are in Btu per square foot per day.

MIT Solar House II	
Cambridge, Massachusetts (42°N)	
MIT, 1947-48	
Heated floor area:	392 ft²
Collector - type:	water
area:	224 ft²
tilt:	90° S
Storage - type:	water
volume:	830 gal
Heat distribution:	radiation and convection
Auxiliary heating:	electric
Percent solar heated:	38-48%

MIT Solar House II

Built in the summer of 1947, MIT's second house combined the collection, storage and room heating functions in a single unit—the south wall. Metal containers of water sitting behind vertical south-facing glass collected and stored the solar energy. Heat flowed to the rooms directly from the back sides of these containers. Such a heat storage wall provided some of the solar heat control advocated earlier by Hutchinson.

This laboratory building was oriented from east to west and was divided into seven adjacent cubicles to compare seven different wall configurations. Each cubicle had 56 square feet of floor area and 32 square feet of south windows. The first cubicle had no storage wall, and the storage walls in the other six were either 4 or 9 inches thick. The first six cubicles were faced with double glass, but the seventh was triple glazed. The sun's rays penetrated through the glass to the blackened south face of each storage wall. Most of this sunlight was absorbed, warming the water to temperatures as high as 125°F. The stored heat flowed to four of the cubicles by thermal radiation and to the remaining two cubicles by fan-driven air convection. Each cubicle had a pair of reflecting shades that were lowered at night or on cloudy days to conserve the stored heat.

The performance of these heat storage

walls was studied during the winter of 1947-48. From December through February, an average of 945 Btu of sunlight per day struck each square foot of glass, and an average of 605 Btu per day, or 64 percent, was absorbed in each square foot of wall. But outward heat losses reduced the overall collection efficiency to less than 20 percent. Quite understandably, the MIT engineers concluded that further research in this direction would be fruitless. Instead, they recommended that solar heat collection and storage be *separated,* so that nighttime heat losses could be eliminated and more than half the collected solar heat kept inside. However, modern insulation materials and techniques can do much better than the reflecting shades used in this house. One can readily isolate the heat storage wall from the outside when the sun isn't shining. With improved insulation, south wall collectors are more than adequate for solar heating, particularly when cost, durability, and maintenance are important.

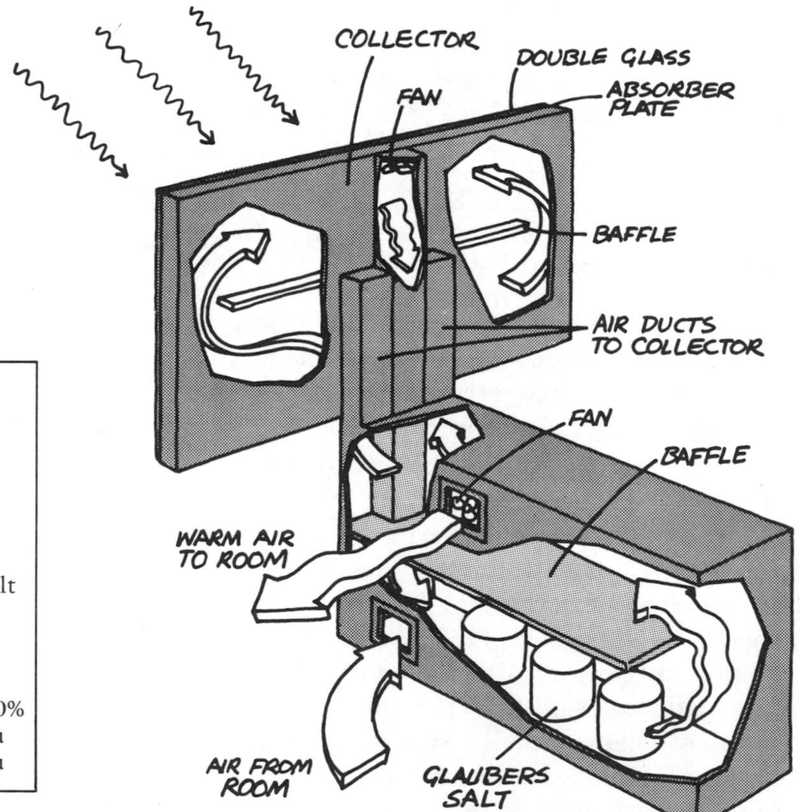

Solar heating system in the Dover House.

Dover House	
Dover, Massachusetts (42°N)	
Telkes, 1948-53	
Heated floor area:	1456 ft^2
Collector - type:	air
area:	720 ft^2
tilt:	90° S
Storage - type:	Glaubers salt
volume:	470 ft^3
Heat distribution:	warm air
Auxiliary heating:	electric
Percent solar heated:	about 80%
Winter heating needs:	47 MBtu
Solar heat supplied:	38 MBtu

Dover House

In early 1948, Dr. Maria Telkes and architect Eleanor Raymond completed a solar house for Amelia Peabody in Dover, Massachusetts. They tried to achieve 100 percent solar heating without a backup furnace. A five to seven day storage capacity was attained by using Glaubers salt ($Na_2SO_4 \cdot 10H_2O$) as the storage medium. The heat required to melt one pound of this salt is 104 Btu. One cubic foot of Glaubers salt, weighing 92 pounds, absorbs 9500 Btu when it melts at about 90° F. Over the temperature range from 80° F to 100° F, one cubic foot of this salt can store 10,700 Btu. By comparison, one cubic foot of water can store only 1250 Btu, and one cubic foot of solid rock only 720 Btu, for the same temperature rise.

Not only does Glaubers salt store almost nine times more heat per volume than water, but it can collect and store the heat at a relatively constant and moderate temperature. The primary drawback was that the salt stratified in its container, resulting in imperfect reversibility between liquid and solid states. After several cycles, the solution no longer stored large quantities of heat because of incomplete solidification. Dr. Telkes appears to have solved this problem in her recent work. (See the discussion of the University of Delaware solar house.)

The two-story Dover house had a vertical collector on the upper part of its south wall, plus 180 square feet of south-facing windows. Air flowed in a gap behind the blackened sheet metal absorber and carried the solar heat to the storage bins, where five-gallon steel cans of Glaubers salt absorbed it. The total volume of 470 cubic feet of salt could store 5 million Btu (5 MBtu), or enough heat to last for five to seven cold cloudy days. Unfortunately, the salt deteriorated in its thermal storage capacity, and the house was fully solar heated only during the first winter. The house was converted to standard heating in 1953, when it was remodeled and enlarged.

The Dover House in Dover, Massachusetts.

MIT Solar House III

The MIT solar research team remodeled their second solar house in 1948, converting it into a small home for a married student, his wife and baby. This house used a flat-plate collector on the roof and a tank of water in the attic to collect and store solar heat. Radiant ceiling panels distributed this heat to the rooms. Solar energy was also collected through 180 square feet of windows on the south wall. The performance of the house agreed closely with predictions based upon the results of the first MIT house. Through four consecutive heating seasons (the period from October through April), solar energy supplied about three quarters of the total heating needs.

Both the collector and storage tank were purposely undersized in order to require auxiliary heating. The collectors were double glazed, with 3/8-inch copper tubes soldered at 6-inch intervals to the back of the blackened copper absorber plates. Behind these plates were 4 inches of mineral wool insulation to reduce the heat loss from the collector to the inside. A tilt angle of 57°, or 15° greater than the local latitude (42°22′ for Cambridge), was used for optimum

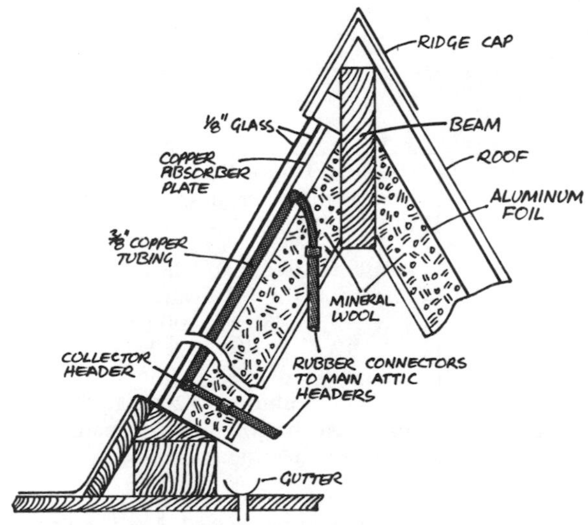

Flat-plate collector used in MIT III.

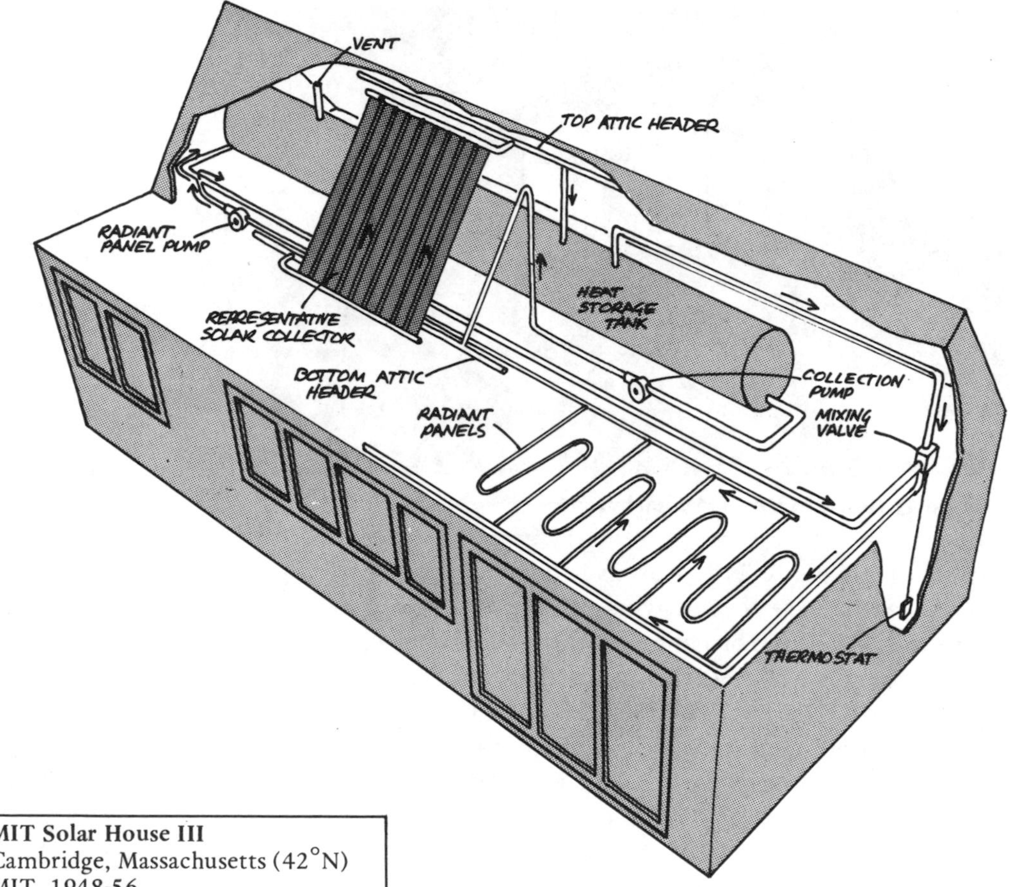

Solar heating system in MIT III. Heat was distributed to the rooms by radiant ceiling panels.

MIT Solar House III
Cambridge, Massachusetts (42°N)
MIT, 1948-56

Heated floor area:	608 ft^2
Collector - type:	water
area:	400 ft^2
tilt:	57° S
Storage - type:	water
volume:	1200 gal
Heat distribution:	radiant panels
Auxiliary heating:	electric
Percent solar heated:	69-82%
Winter heating needs:	35-40 MBtu
Solar heat supplied:	26-30 MBtu

collection of sunlight during the winter months. At this angle, the collector faced the midday sun squarely in early November and again in the middle of February. Due to remodeling constraints, the 30 foot long storage tank was placed in the attic. Its 1200 gallon capacity, or 3 gallons per square foot of collector, could store enough heat for two consecutive average winter days. Three electric heaters installed in the tank near its outlet provided auxiliary heat during longer cloudy periods.

The collector pump started when the collector temperature reached 5° F above the storage tank temperature, and circulation continued as long as this condition persisted. This method prevented circulation during momentary periods of sunshine and delivery of warm storage water to a cold collector. When the pump stopped, the water drained back into the storage tank, eliminating the need for an antifreeze solution to prevent freezing. The radiant ceiling panels operated in a conventional manner. When the thermostat called for heat, another pump began circulation through these panels, and the mixing valve mixed warm water from storage with return water from the panels as required by the demand.

The performance of this system was studied closely in the years from 1949 to 1953. Of the 219,000 Btu of solar radiation striking a square foot of the collector during a typical heating season, only 155,000 Btu struck while the pump was operating. Of this amount, only 67,000 Btu was actually collected and brought into the house for heating. Only *thirty percent* of the solar radiation which struck the collector was eventually used for heat—a percentage that is difficult to better with indirect systems. The accompanying chart indicates the performance of MIT Solar House III for three heating seasons. The percentage of heating load supplied by solar radiation includes energy which entered the house via the windows. The improved performance in the 1951-1952 season was the result of changes in piping that equalized the circulation of water over the absorber surface.

In the article, "Heating by Sun Power: A Progress Report," August L. Hesselschwerdt noted that the house was located where heavy heating loads and poor atmospheric conditions do not favor solar heating. Despite these limitations, he concluded that "solar energy can be successfully used for space heating," but warned that an economically competitive system would require much more research and development. Three critical problems in the design of the heat transport system would always require expert attention:

- uneven distribution of water over the absorber surface;
- incomplete drainage or filling of the collectors when the pump stops or starts; and
- high electrical costs for pumping.

The close collaboration required between engineer, architect, and contractor to build such a solar house is often prohibitively expensive. Simpler designs requiring less collaboration are one alternative to these high costs.

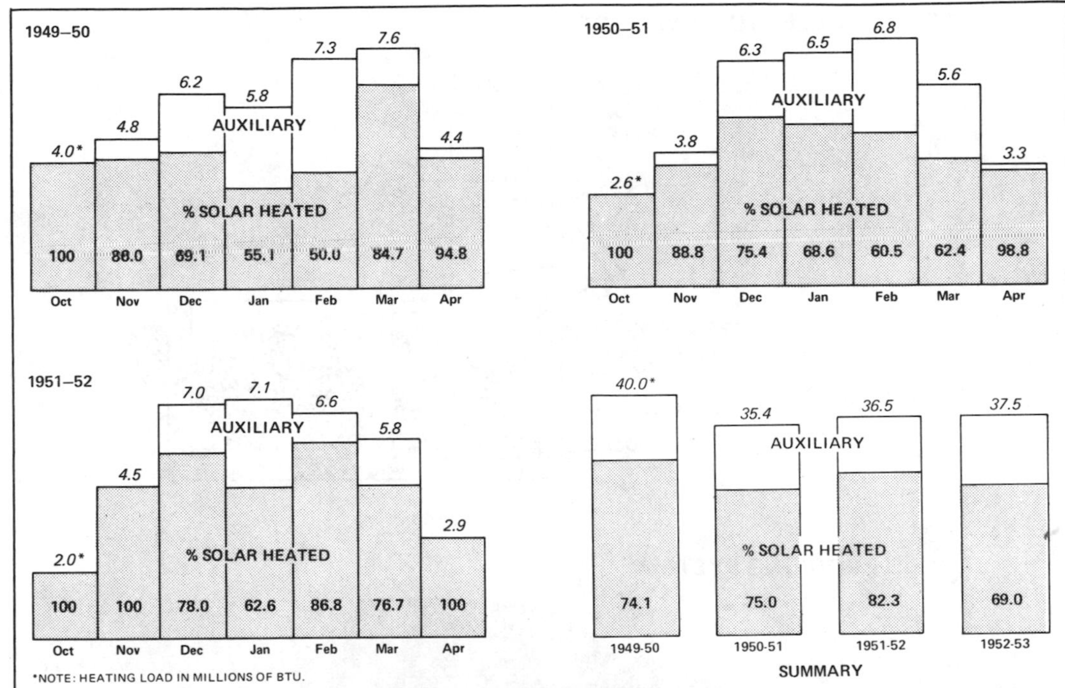

Performance of MIT Solar House III for three heating seasons. Charts indicate the percentage of heating load carried by solar energy and auxiliary heating.

Desert Grassland Station

In 1954, Raymond Bliss and Mary Donovan completed a 100 percent solar heating system attached to a 25 year old house owned by the U.S. Forest Service. This "small, rather dilapidated dwelling" was located about 30 miles south of Tucson, Arizona—an area noted for its mild winters and hot summers. The primary purpose of the project was to develop a "stepping stone towards design of a complete solar air-conditioning system, capable of high-quality performance the year round." Of secondary importance was the desire to show that a house could be heated entirely with solar energy.

SOLAR COLLECTOR AND STORAGE

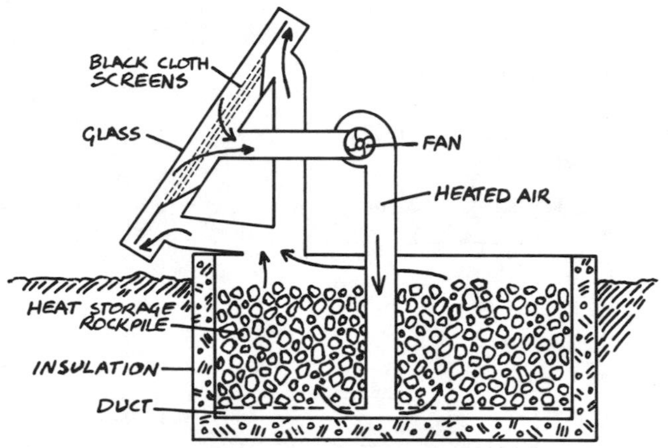

NIGHT COOLING SYSTEM

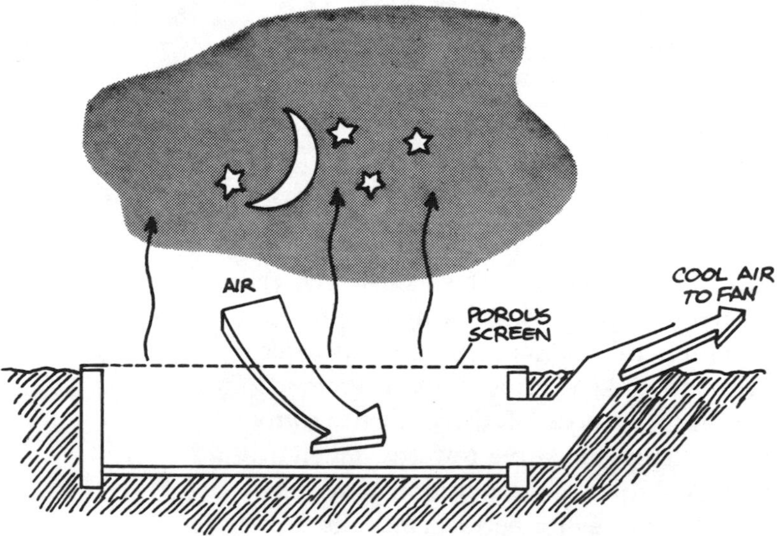

Solar collection and night cooling systems in Desert Grassland Station.

For financial and architectural reasons, the collector-storage system was built on its own structure, separate from the house. The single-glazed collector, which was tilted to face the midday sun squarely on January 15, had four layers of black cotton screening to absorb the sunlight. The screens easily transferred their collected heat to circulating air, which subsequently passed through a rockpile and deposited this heat in the rocks. Air flowed upward through the rockpile when it was being heated by the collector, but downward when house heating was needed (not shown). This arrangement insured that the house always received air at the higher temperatures available at the bottom of the rockpile.

Primarily because of the large surface area of screen filaments, such hot air systems are efficient and economical, as long as collector temperatures are less than 100°F above the outdoor temperatures. On a clear winter day, the system collected about 1000 Btu per square foot for a total of 315,000 Btu, or twice the average daily heat requirement of the house. At noon on a typical winter day, air entered the collector at 100°F and returned to the rockpile at temperatures ranging from 130°F to 140°F. The average rockpile temperature in mid-winter was 120°F, providing sufficient heat for about four cold sunless days. Heat loss from the storage pile was estimated to be about 25 percent of the heat collected. Auxiliary electric heat was available if the storage temperature fell below 85°F, but this never happened.

For summer cooling, the night air was drawn through a 280 square foot porous screen exposed horizontally so that it lost heat by radiation to the night sky. This air was then drawn through the rockpile to cool the rocks. Performance of this cooling system was good, but the comfort level attained was not comparable to that of a high quality air conditioner. Operating costs of the cooler were comparable to those of an evaporative cooler. However, the heating system realized a savings of about $70 per year over a conventional butane heating system (1954 prices). The fans and controls required a large amount of power, about 250 kilowatt-hours more than used in a regular furnace. Ray Bliss estimated that the realistic cost of a similar system in a moderately sized house was about $1500 more than the cost of a conventional heating and cooling system. He also estimated that such a solar heating and nocturnal cooling system could save about $100 of fuel per year.

The Desert Grassland Station solar collector.

Desert Grassland Station
Amado, Arizona (32°N)
Donovan and Bliss, 1954-55

Heated floor area:		$672\ ft^2$
Collector -	type:	air
	area:	$315\ ft^2$
	tilt:	53° S
Storage -	type:	rocks
	volume:	$1300\ ft^2$
Heat distribution:		warm air
Auxiliary heating:		electric
Percent solar heated:		100%
Winter heating needs:		14.2 MBtu
Solar heat supplied:		14.2 MBtu

Bridgers and Paxton Office Building

In 1956, the engineering firm of Bridgers and Paxton completed the world's first solar-tempered commercial building in Albuquerque, New Mexico. The heating system used a flat-plate collector which slopes at 60° and forms one wall of the building. Water was used as the heat transport fluid. Solar heat was stored in a 6000 gallon tank and distributed to the rooms by radiant panels in the floor.

The Bridgers and Paxton office building.

Bridgers and Paxton Office Building
Albuquerque, New Mexico (35°N)
Bridgers and Paxton, 1956-62

Heated floor area:		$4300\ ft^2$
Collector -	type:	water
	area:	$750\ ft^2$
	tilt:	60° S
Storage -	type:	water
	volume:	6000 gal
Heat distribution:		radiant panels
Auxiliary heating:		heat pump
Percent solar heated:		92%

In the heat pump system pictured below, a compressor circulates a heat transfer fluid from a chiller to a condenser. This fluid evaporates in the chiller—absorbing heat from the water in it. This heat is deposited in the condenser, where the transfer fluid liquefies under pressure. In this way low grade solar heat from the storage tank can still be used to heat the building—although indirectly.

The system included a water chiller that functioned as a *heat pump* to boost the temperature of the water being sent to the floor panels. A heat pump is a device that moves heat against the grain—from cold areas to warm. A household refrigerator is a heat pump that removes heat from the food compartment and deposits it in the kitchen air. More information about the operation of heat pumps is available in Chapter 6. Together with an evaporative cooler and a ceiling panel distribution system, the heat pump also helped provide cooling in the summer. During the spring and fall, the system could supply both heating and cooling within the same day.

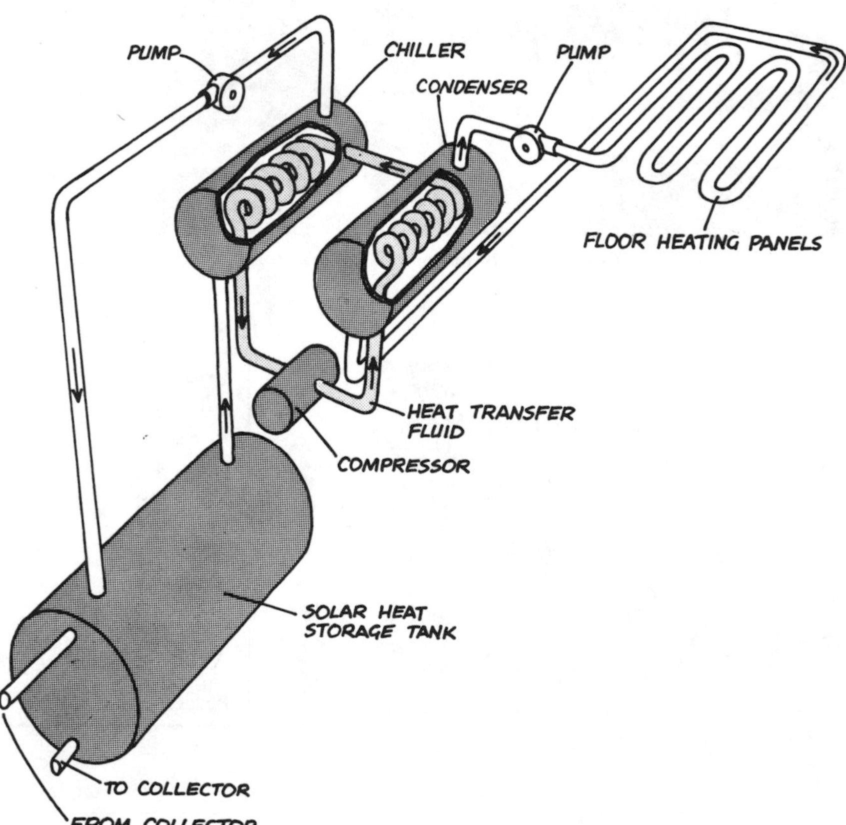

Solar heat distribution aided by a heat pump—Bridgers and Paxton office building. This portion of the system was used when temperatures in the storage tank were too low for direct use.

There were some preliminary difficulties with the collector. Pinhole leaks developed in the original aluminum collector plates having integral water passages *(tube-in-plate)*. Water froze in some of the panels due to improper drainage. Air was trapped in the panels when they were refilled, causing uneven circulation of water through the collector. But all of these problems were eventually corrected. The final version of the collector was made of blackened aluminum sheets with ½-inch copper tubes soldered to the back and covered with one sheet of ordinary double strength glass.

During the 1956-57 heating season, 63 percent of the total heating requirements was supplied without help from the heat pump. The remaining 37 percent was supplied by the heat pump, which extracted low grade heat from the solar-heated water in the storage tank. The electrical energy required to run the heat pump was only 8 percent of the total heating requirements. Unfortunately, the cost and complexity of the system led to its early termination in 1962. The economics of the system have improved with rising fuel prices, however, and the National Science Foundation (NSF) has awarded a grant to Pennsylvania State

University to rejuvenate this solar heating and cooling system and gauge its performance.

Odeillo Houses

Located high in the French Pyrenees is the Solar Energy Laboratory of the *Centre Nationale de la Recherche Scientifique* (CNRS). Under the leadership of Felix Trombe, this laboratory has become world famous for its 1000 kilowatt solar furnace, which can generate temperatures in excess of 6000°F. But their less glamorous work on the applications of solar energy to housing may prove to be much more valuable to society.

Together with architect Jacques Michel, Professor Trombe has built a series of *maisons solaires* with solar collection directly incorporated into the fabric of the buildings themselves. The first prototype, built in 1956, was a squat, single story house with a concrete south wall approximately one foot thick. Covered with a single pane of glass and painted black on its roughcast outer surface, the wall doubled as a solar collector and heat storage container. Unlike the storage walls in the second MIT solar house, this wall is also a structural member of the house.

Solar heat was collected and distributed to the rooms without using any pumps, fans, or blowers. Sunlight penetrated the glass and was absorbed in the blackened wall, heating both the concrete and the air in the gap between glass and concrete. This heated air would expand, becoming lighter, and rise. Ducts along the top of the wall allowed this hot air to flow into the rooms, while ducts at the bottom allowed cool room air to replace it. Such a process of *gravity convection,* or *thermosiphoning,* carried most of the absorbed solar energy into the house during the day The rest of the heat migrated through the concrete, with a time delay of 6 to 8 hours. At night, the interior was warmed by thermal radiation from the inner surface.

This uninhabited house was used to experiment with various duct sizes and air gaps, and very little performance data are available. Later versions used other storage media, such as sand and water, for the south wall. The

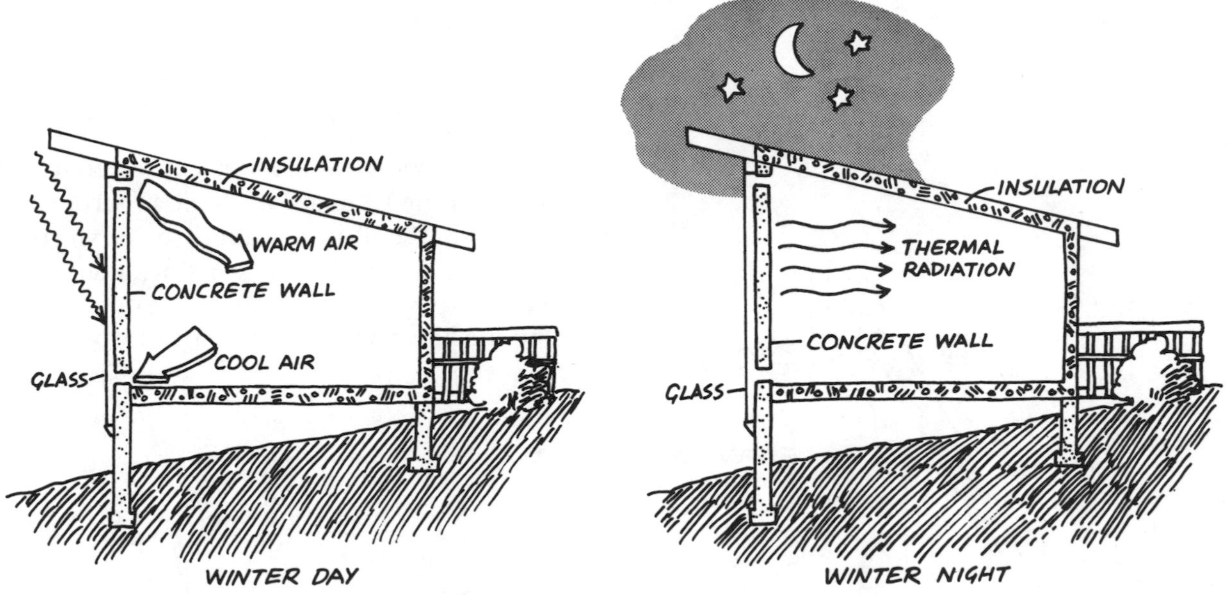

Concrete wall collector in the first Odeillo House—day and night operation. Gravity convection and thermal radiation carried the absorbed solar heat into the rooms.

Odeillo Houses	
Odeillo, France (43°N)	
Trombe and Michel, 1956-	
Heated floor area:	960 ft^2
Collector - type:	concrete
area:	660 ft^2
tilt:	90° S
Storage - type:	concrete
volume:	600 ft^3
Heat distribution:	radiation and gravity convection
Auxiliary heating:	electric
Percent solar heated:	about 75%

series of experiments has culminated in a reliable and effective solar house design.

MIT Solar House IV

After almost twenty years of research, the solar heating team at MIT built a fourth solar house in the Boston suburb of Lexington. The first floor of this two story house was partially below grade (below ground level) and the house was well insulated throughout.

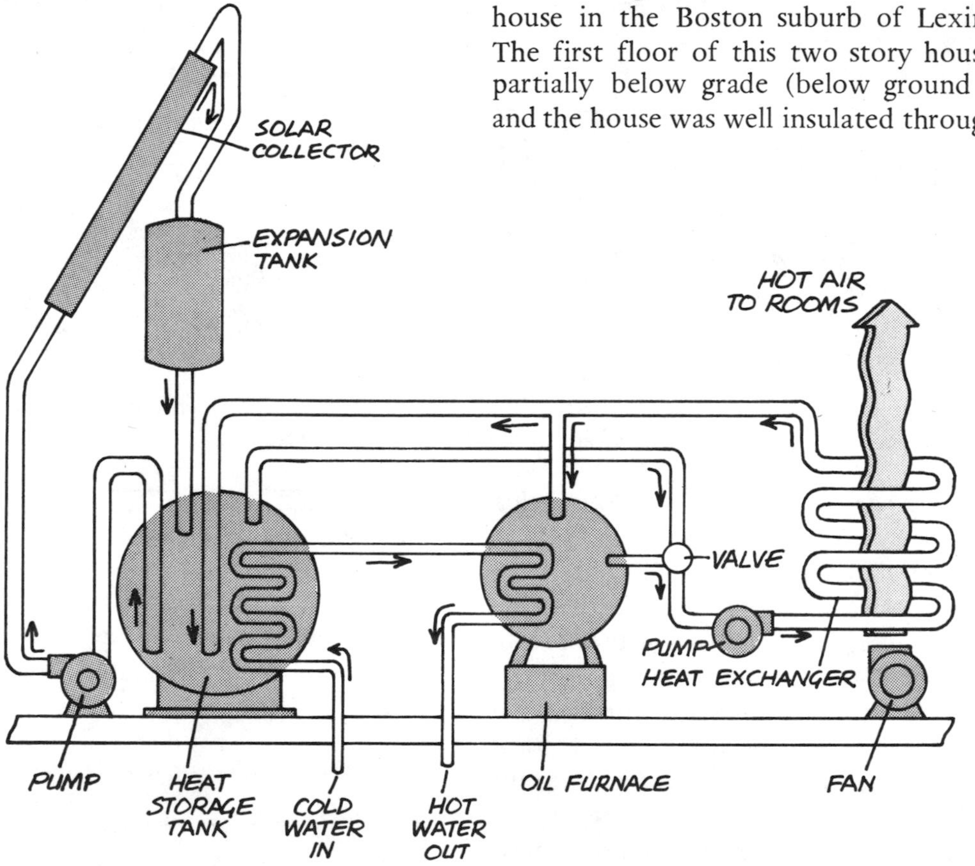

Solar heating system in MIT Solar House IV. About half of household heating needs, including domestic hot water, were provided by this system during two successive winters.

The solar heating system was designed to supply 75 to 80 percent of the house heat and a large part of the domestic hot water. The collector consisted of two layers of glass covering a thin aluminum sheet painted black. Water from the 1500 gallon storage tank was warmed as it circulated through 3/8-inch copper tubes clamped to the aluminum plate at intervals of 5 inches. Aluminum was used as the collector plate primarily because of its low cost rather than its high conductivity; copper tubes were used because of their corrosion resistance. A 200 gallon expansion tank allowed draining of the collector when it was not in use.

The solar heated water passed from the storage tank through a heat exchanger to warm the passing air which then warmed the house. An oil furnace provided auxiliary heat when the tank temperature fell below 95°F. Domestic hot water was provided by passing city water through coils that were submerged in the storage and auxiliary tanks. In the summer, a small refrigeration compressor chilled the water in the storage tank to 45°F in order to provide house cooling.

The MIT researchers knew that economical solar heating might be difficult, but they intended to achieve a measure of success upon which to build yet another house. A series of setbacks forced them to abandon this plan. Costs of the system were greater than anticipated. It provided less than half of the heat required by the house in its first winter, and many minor breakdowns kept the MIT people busy as repairmen.

Although the system itself worked well, routine components such as valves and gauges failed to function properly.

The first complete heating season, from October of 1959 to March of 1960, was colder than predicted with less sunshine than usual. Of the total solar energy hitting the collector, 32.4 million Btu was too diffuse to justify collection. Of the remaining 90 million Btu, only 40.9 million Btu was actually extracted from the collector and brought to the storage tank. This heat provided 34.4 million Btu, or 46 percent, of the total heating needs. In the following heating season, the weather was not quite so severe, and solar energy provided more than half of the total heating load. The house was converted to standard heating in 1962, bringing to an end MIT's program of solar house construction.

MIT Solar House IV in Lexington, Massachusetts.

Denver House

After his experience with the Boulder house, Dr. George Löf designed and built himself a single story contemporary home in Denver with a solar heating system "added to it as an appliance rather than being made an integral part of the house." The house had a flat roof on which were placed two banks of sloping solar collectors. Other than influencing the choice of a flat roof, the solar heating objective was not as important in the planning of the house as were the living needs of his family.

Basic characteristics of this system are the same as those used in the Boulder house. An overlapping plate solar air heater serves as the energy collector, and the heat is stored in gravel beds. But specific details of the system are quite different. The tilted solar collector troughs are shorter and mounted on a flat roof rather than a sloping one. Air circulates through two adjacent 6-foot troughs before returning to storage. A second innovation is the reversal of hot and cold areas within the gravel beds. Solar heated air enters the lower end of each vertical storage column so that heat can be readily withdrawn there and supplied to the rooms at floor level. Because the natural circulation of air inside a closely packed bed is limited, the stratification of heat can be maintained for reasonably long

MIT Solar House IV
Lexington, Massachusetts (42°N)
MIT, 1959-61

Heated floor area:		1450 ft^2
Collector -	type:	water
	area:	640 ft^2
	tilt:	60° S
Storage -	type:	water
	volume:	1700 gal
Heat distribution:		warm air
Auxiliary heating:		oil furnace
Percent solar heated:		46%
Winter heating needs:		74.5 MBtu
Solar heat supplied:		34.4 MBtu

Denver House
Denver, Colorado (40°N)
Löf, 1958-

Heated floor area:	2100 ft²
Collector - type:	air
area:	600 ft²
tilt:	45° S
Storage - type:	gravel
volume:	250 ft³
Heat distribution:	warm air
Auxiliary heating:	gas heater
Percent solar heated:	26%
Winter heating needs:	194 MBtu
Solar heat supplied:	51.7 MBtu

periods of time. A third innovation was the use of two dense, laminated fiberboard cylinders to contain the gravel. These tubes, 3 feet in diameter and 16 feet high, are also used as forms for large concrete columns in heavy building construction. They were lined with aluminum and left uninsulated, as they are entirely within the heated area of the house. The total cost of this simple storage scheme was about one hundred dollars.

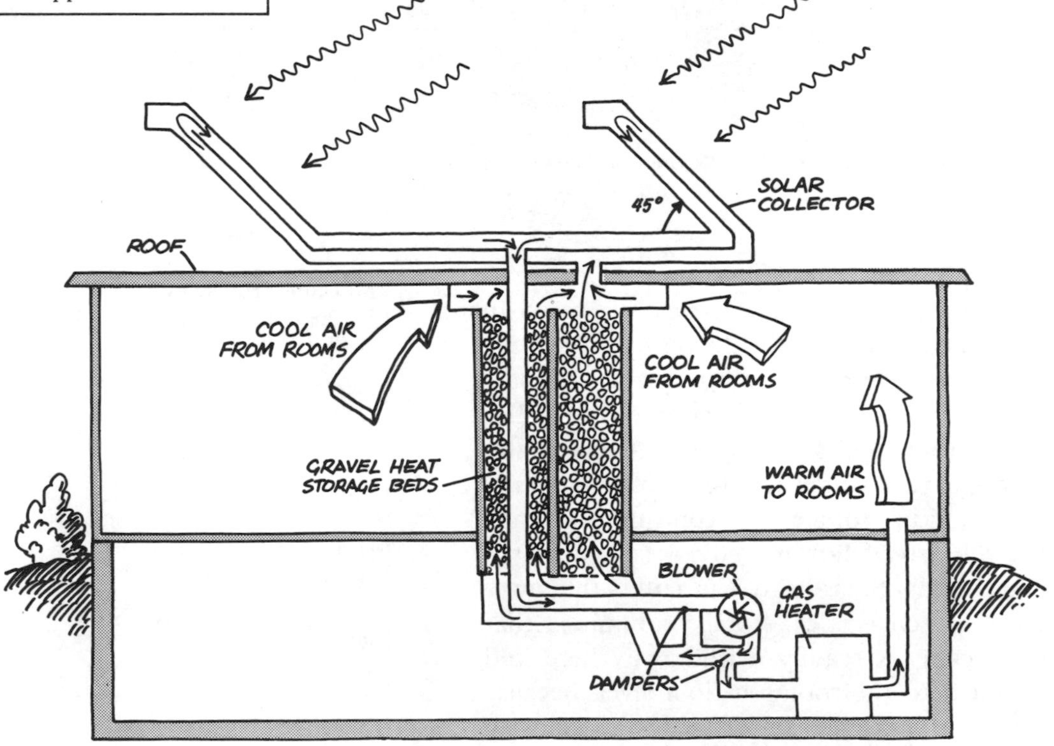

Solar heating system in Dr. George Löf's Denver house.

Auxiliary equipment, including a blower, control panel, and a duct heater, is located in the basement at the foot of the storage beds. Solar heated air passes down through a hot air duct inside one of the gravel cylinders to the bottom of the beds before flowing up through the gravel and returning to the collector. Automatic dampers divert this flow to the rooms when heat is needed. Withdrawal of heat from storage is accomplished by another damper adjustment that permits house air to flow down through the heated gravel to the blower and back to the rooms. If air entering the rooms is not warm enough, the auxiliary duct heater provides extra heat. During the winter of 1959-60, this system provided only 26 percent of the heating needs, well below the 60 to 70 percent predicted, plus a portion of the heat needed for domestic hot water. The system is still in operation.

The Tucson Laboratory

Mary Donovan and Raymond Bliss directed a solar laboratory for several years in Tucson, Arizona. Beginning in 1959, the building had a copper tube-in-plate collector covering its entire roof. This unglazed panel faced south at a 7° tilt. Water was heated by circulation through the panel during sunny winter periods, and cooled during the summer by nighttime circulation. The roof panel was painted a dark green, which is perhaps more attractive than black and absorbs almost as much solar energy. The heat was transferred to the building through a second flat-plate heat exchanger made of a similar tube-in-plate material that formed the ceiling of the interior. Warm water circulated through the ceiling panel to heat the building, and cold water circulated to cool it.

A vertical 4500-gallon water storage tank was divided into two compartments by a horizontal baffle across the mid-portion of the tank. The division between the two compartments was purposely not watertight, allowing convective mixing to equalize temperatures between the two compartments whenever the temperature of the bottom exceeded that of the top. When hot water was stored in the upper tank and cold water in the lower one, no convective mixing occurred. Both heating and cooling temperatures were simultaneously available with this system. A heat pump was used to move additional heat from the lower tank to the upper tank as needed.

The controls had a manual selector switch which could be set at any one of three positions: heating only, cooling only, and heating-cooling. With the switch set for heating only, the lower tank received heat from the collectors and hot water from the upper tank circulated through ceiling panels to heat the offices. With the switch set for cooling only, the upper tank was cooled whenever possible through the roof collectors, and cool water from the lower tank circulated through the ceiling panels. With the switch set for both heating and cooling, solar heated water was stored in the upper tank during the day and cool water was delivered to the lower tank at night. The lab was heated or cooled as required, all within the same 24 hour period. In all three modes of operation, the heat pump operated whenever higher temperatures were needed in the upper tank or lower temperatures were needed in the lower tank.

During the 1959-60 heating season, the

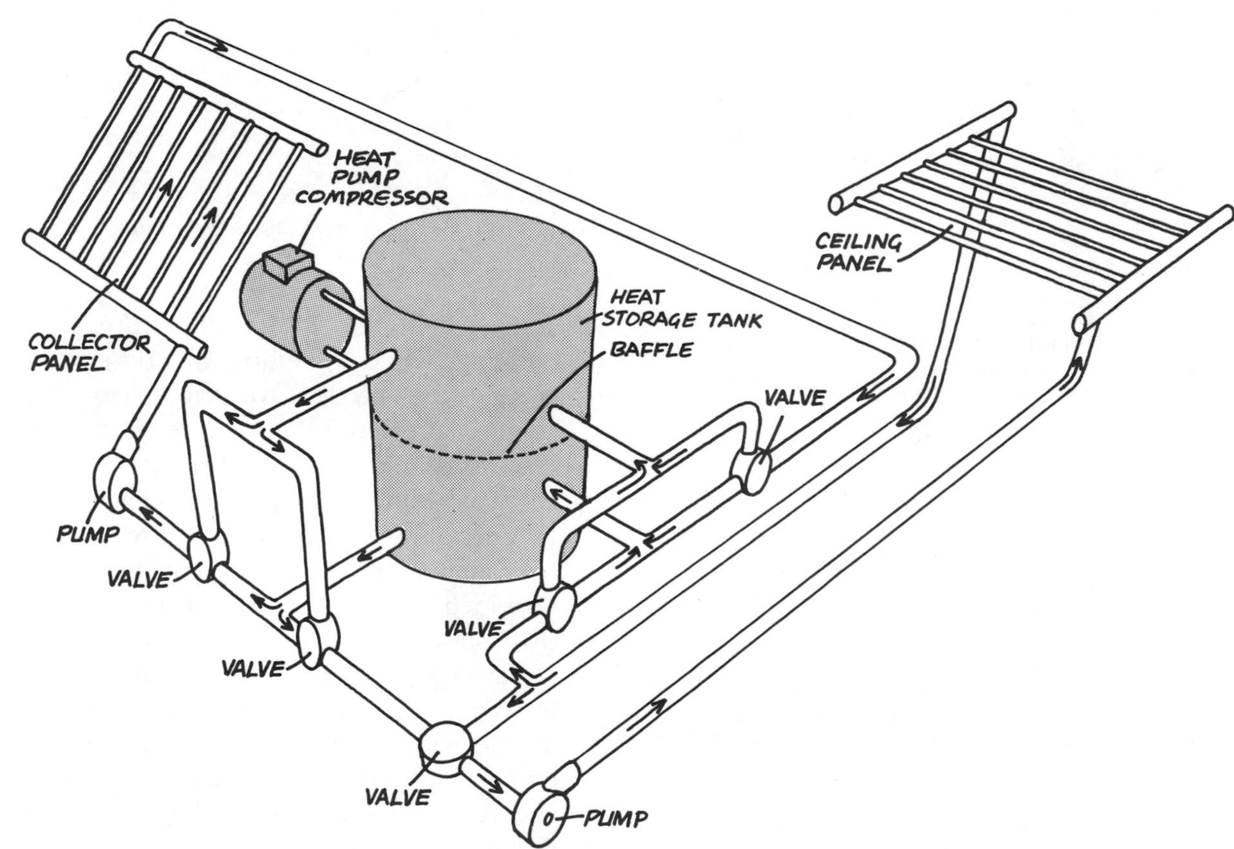

Solar heating and cooling system in the Tucson Laboratory.

The Tucson Laboratory with storage tank outside the building.

Tucson Laboratory	
Tucson, Arizona (32°N)	
Donovan and Bliss, 1959-69	
Heated floor area:	1440 ft^2
Collector - type:	water
area:	1623 ft^2
tilt:	7° S
Storage - type:	water
volume:	4500 gal
Heat distribution:	radiant panels
Auxiliary heating:	heat pump
Percent solar heated:	86%
Winter heating needs:	35.3 MBtu
Solar heat supplied:	30.3 MBtu

system delivered 35.3 million Btu to the building, while 5 million Btu of electrical energy was needed to drive the pumps, compressor, controls, and ventilating fan. About 85 percent of the solar energy collected during December, January, and February was at a high enough temperature for immediate use, while the heat pump upgraded the remaining 15 percent. During the following summer, the cooling system removed a total of 6.6 million Btu from the building, while consuming 2.5 million Btu of electrical energy. Almost all the heat removed from the building during June, July, and August required upgrading by the heat pump prior to nighttime rejection.

Difficulties encountered with this system included condensation on the cool ceiling panels during summer months, overheating of offices on the south side due to the large south windows, and troubles in refilling the collectors after draining them to prevent freezing. Estimates showed that the system probably cost about $4500 more than a combination gas furnace and air conditioner (1960 prices). The operating cost of the system, with electricity at 2.5 cents per kilowatt-hour, was about $215 per year, as compared with the conventional system operating cost of about $400 per year. From their experiences with this building, Mary Donovan and Ray Bliss advocated systems which would require no electricity, use auxiliary fuel moderately, allow fairly wide variations of interior temperatures, and have a low initial cost. Their most significant conclusion was that "conversion of solar energy to thermal energy external to the building is an obviously roundabout and awkward solution. It seems most likely that an economically practicable solution will come via systems using conversion inside the building."

Thomason Solar House I

One of the mavericks in the field of solar energy is Dr. Harry Thomason, a Washington patent attorney who claims more patents in this field than anyone else in the world. During a sudden rainstorm, he took shelter under the rusty corrugated roof of a shed and noticed that the runoff was hot because the sun had warmed the roof. He realized that the same effect could be used to heat a house, and in 1959 he built a three bedroom house in District Heights, Maryland. The simple designs of his collectors and heating system, and the lack of detailed cost information and performance data made available to the public, have left his work open to skepticism. However, he seems to have succeeded in constructing an inexpensive, easy-to-build collector and in keeping the entire heating system free from complexity.

Thomason built four solar houses prior to 1963, and his corporation is now building houses to sell to the public. His patented system, called the "Thomason Solaris (TM) System," uses blackened corrugated aluminum absorber panels on the south slope of the roof. A fraction of an inch above the aluminum is a layer of glass. At the peak of the roof is a pipe perforated with holes which are aligned so that low pressure water runs out of the holes, down each corrugation, and into a gutter or trough at the bottom. From the trough, the warm water flows to a 1600 gallon tank surrounded by 50 tons of egg-size stones in the basement. Air from the house heating system is blown through the stones, which have been heated by the hot tank. An oil burner provides auxiliary heat for long cold periods of no sunshine.

A 1965 article by Thomason evaluating the first house reported "no major flaw in design or construction." There were a few leaks that resulted in deterioration of wood supports. A plastic film over the corrugated aluminum collector plate had disintegrated and the collector was rebuilt without it, leaving a single layer of glass over the aluminum. Thomason reported that the collector efficiency was only slightly lower without the plastic film—44 percent rather than 47 percent. Two cover plates are recommended for cold climates, but most "trickle-type" collectors have used only one. In the original design, water flowed through the collector at a rate of seven to ten

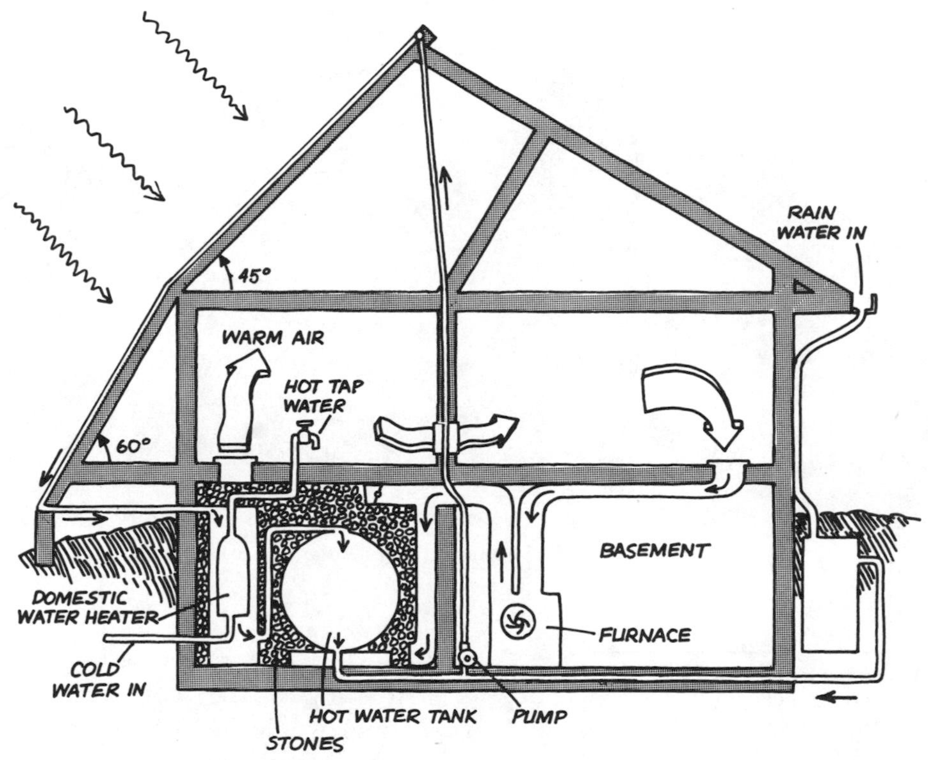

Solar heating system in Thomason Solar House I.

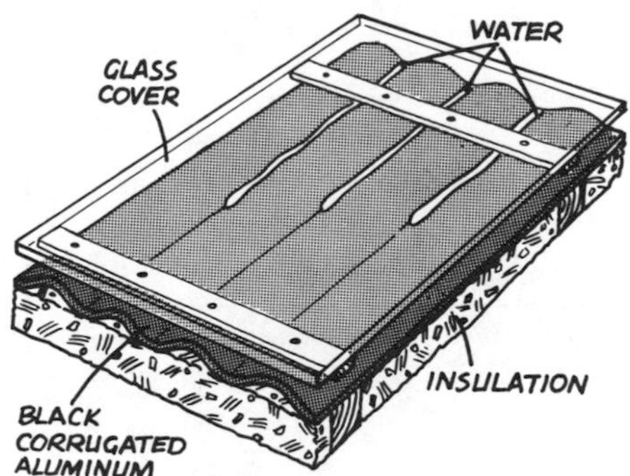

Trickle-type collector.

Thomason's first solar house, District Heights, Maryland.

Thomason House I
District Heights, Maryland (39°N)
Thomason, 1959-

Heated floor area:	900 ft^2
Collector - type:	water
area:	840 ft^2
tilt:	45°, 60° SSW
Storage - type:	water and rocks
volume:	1600 gal, 50 tons
Heat distribution:	warm air
Auxiliary heating:	oil furnace
Percent solar heated:	about 95%

gallons per minute. According to Thomason, "On a bright, fairly cold winter day, typical readings would be: flow rate, seven gallons per minute; temperature of water to the collector, 80 degrees; temperature of water from the collector, 105 to 110 degrees (Fahrenheit)."

During the first winter, Thomason's first house required only 31 gallons of fuel oil for auxiliary heating at a total cost of $4.65. Without the use of the sun, the house would have needed $100 to $125 worth of oil (1960 prices). The reported cost of $2500 includes the furnace, the solar water heater, a small electric water heater, the controls, and the 840 square foot collector. The five day storage tank and basement rock storage space may be included in this figure since the total house cost was only $13,000. The published maintenance costs are less than $600 in 15 years, and the electricity used to operate the system is about equal to that saved on water heating.

Harry Thomason has put together an impressive record for domestic solar heating. His relatively simple solutions, whether or not they perform to the standards of scientists and engineers, satisfy his needs and those of many other people who can adapt his system to their own purposes. A license and plans for the "Thomason Solaris System" may be obtained from the Edmund Scientific Company (150 Edscorp Building, Barrington, New Jersey 08006). The Federal Housing Administration has approved the basic design for government-insured mortgage loans.

RECENT SOLAR HOMES

The solar heating of homes has only recently begun to emerge as a very real alternative to the use of gas, oil, or electricity. As home heating costs continue to rise, more and more people are turning to the sun for their warmth. Since the Arab oil embargo of 1973-74, the number of solar homes in this country has risen from a mere handful to several hundred, with thousands more on the drawing boards. All across the land, people of radically different persuasions are rushing to harness the sun's warm rays.

What surprises many is the fact that solar

heating is often most applicable in *cold* climates. The longer and colder the heating season, the greater the demand for solar heat—and the greater the savings to be gained from solar heating systems. But areas with plenty of sunshine will usually fare better than cloudy ones. Quite naturally, Colorado and New Mexico are in the vanguard of solar home construction. But Arizona, California, New England, and the Mid-Atlantic states aren't exactly hanging back.

In some of the most successful solar homes the solar apparatus has been used for purposes other than space heating. Domestic water heating, needed the year round, can usually be incorporated into the solar heating system. Where the solar heater can also *cool* the home, as in the Desert Grassland Station, fuel savings are possible in winter *and* summer. Shading, ventilation, and even the production of food (in a greenhouse attached to the house) are a few of the other possible functions of a comprehensive system. Often the success or failure of a solar home depends upon the degree to which the solar apparatus can be integrated into the entire domestic scene. The problem is one of engineering, and also one of architecture.

The solar homes described in the remainder of this chapter illustrate the architectural implications of designing with the sun. They demonstrate the intimate relationship between climate and design that is necessary for good solar architecture. Some illustrate how solar energy is being used today in real living situations, while others indicate promising new directions. All of these houses are either under construction or already functioning.

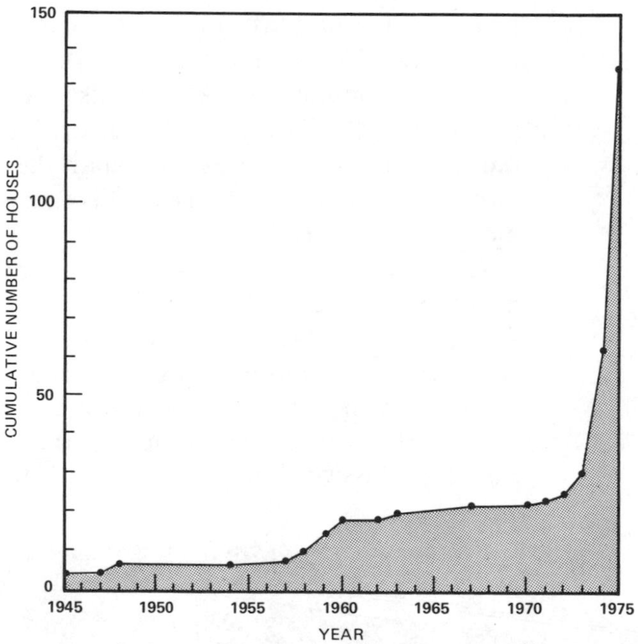

SOURCE: W.A. Shurcliff, *Solar Heated Buildings: A Brief Survey*, 1976.

The growth of solar homes in the United States.

Saunders House

A remarkable solar house that is still working well after fifteen years was built in 1960 by Norman Saunders in Weston, Massachusetts. This two story home with 2700 square feet of floor space primarily uses south windows for its solar collector.

It also has massive walls and floors, with banked earth against the north, east, and west walls. The exterior walls are made of foot-thick pumice block that also acts as insulation. The heavy insulation drastically reduces the overall heat loss, and the sun directly supplies more than half this establishment's heating needs.

On sunny winter days, solar energy penetrates directly through the 325 square feet of double-glazed south windows. These windows are surrounded by overhangs and baffles which block the summer sun and reduce the winter winds across the glass surface. Massive interior walls and floors absorb the solar energy and act as the heat storage. The floor of the first story is a 6-inch concrete slab resting on masonry blocks. According to Saunders' estimate, the house can absorb 75,000 Btu with a temperature rise of only 1°F.

Norman Saunders' house in Weston, Massachusetts. This house gets more than half its winter heat from sunlight streaming through the south windows.

The open interior design of the house encourages natural circulation and even distribution of heat. When needed, fan and duct systems can aid this circulation. Another fan sends excess hot air to the channels in the masonry blocks below the concrete slab. This fan can also withdraw cool air from these channels in summer. Electric heaters in the ceiling provide auxiliary heat when needed.

This house was 40 percent solar heated during the winter of 1972-73, and 58 percent the following winter. This figure improved to 63 percent in 1974-75, after installation of a single layer of glass enclosing the area below the second floor balcony. During the 1960's, Saunders tried several rooftop collectors of his own design, but they have since been removed. The fact that the house still achieves better than 60 percent solar heating is a testament to his foresight in the initial design.

Winters House

Another house relying on passive solar heating was built in 1974 in Winters, California. Designed by Jonathan Hammond, the house is very well adapted to the mild (but often foggy) winters and hot summers of the Central Valley. Sunlight is collected in 300 square feet of "roof ponds," which are oriented east-west along the peak of the house. The solar heat is stored in these ponds and transferred to the rooms by thermal radiation. Because this unique house can hold most of the heat it gains from the sun, it needs only a small wood stove for backup.

The roof ponds are contained in galvanized sheet steel pans supported on timbers that run the length of the house. A transparent vinyl plastic sheet covers the 10 inches of water sitting in each pan. On sunny winter days, insulated covering lids are opened to reflect extra sunlight onto the ponds. The sunlight penetrates the vinyl and water and is absorbed in an asphalt coating on the bottom of the pans. Water temperatures as high as 90°F have been recorded in mid-winter, and enough heat can be stored in the ponds for two average winter days. Thermal radiation from the inside surface of the pans supplies a very even and comfortable warmth to the house. Solar energy also enters the house through 100 square feet of skylights set between the roof ponds and through 118 square feet of south windows.

Once inside the building skin, the solar heat is retained by a number of innovative features. A bare concrete slab floor and a number of recycled oil drums filled with water help store this heat. Styrofoam shutters are provided on the windows and the insulating covers are closed on the ponds to "shut the heat in" at night and on cloudy winter days.

Despite summer daytime temperatures as high as 110°F, the house is fully cooled by the pond system operated in reverse. The insulating covers are kept closed by day, and heat from the room is absorbed in the ponds. With the lids open at night, the ponds radiate this heat to the night sky. Natural ventilation through doors and windows aids cooling during the day. Deciduous trees on

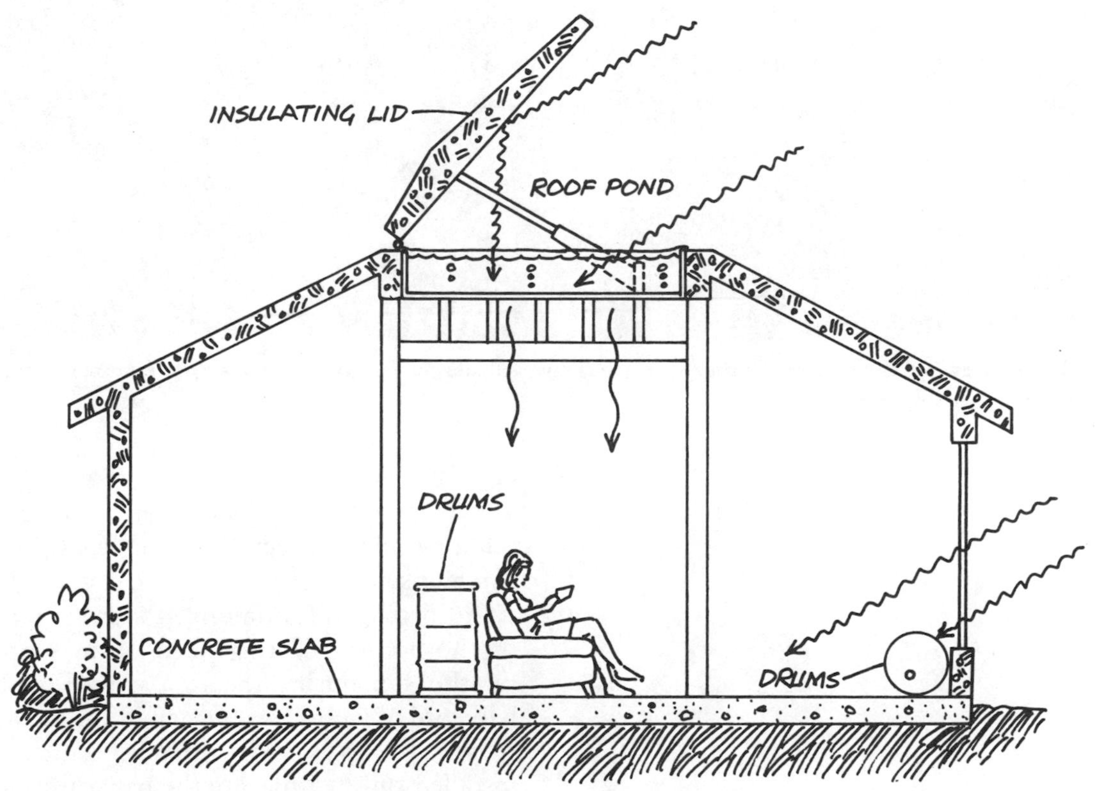

Roof pond heating and cooling system in the Winters House. Solar energy is absorbed in the ponds and stored there at night, when the insulating lids are closed.

Southwest view of the Winters House. Trees and vines shade the east, west and south windows from intense summer sunlight.

east and west sides and a deciduous vine on a trellis on the south side shade the building and protect the windows from the summer sun.

According to Hammond, about 90 percent of the house's heating needs were supplied by the sun during the first winter. Only a half cord of wood was burned in the Franklin stove. The gas wall furnace, which was installed only to placate the bank, was never used. Indoor temperatures stayed between 65°F and 78°F throughout that first year. The total cost for this 1200 square foot house was $27,000, excluding the cost of land. Of this amount, $4000 went for the roof pond system and another $1000 was spent for shutters, drums, and other energy-conserving features.

Mathew House

One of the most difficult areas of the country for solar heating is the Oregon Coast, which is much better known for lumber than for sunshine. Although temperatures of 40°F to 50°F are common in December and January, only two or three hours of direct sunlight can be expected on an average day. Nevertheless, inventor Henry Mathew has designed and built himself a very successful solar home in Coos Bay.

The single story house, which was built in 1967, has a flat-plate collector on the roof and a smaller collector in the yard. The main collector measures 5 feet high and runs 80 feet east-to-west along the roof ridge. It slopes back at an angle of 8° from the vertical to catch both direct sunlight and rays reflected from the south portion of the roof, which is covered with aluminum foil. Mathew built the collector from plate glass, corrugated aluminum panels, and galvanized iron pipe. The pipe is wired snugly to the corrugations, which run horizontally along the ridge. The secondary collector has an area of 325 square feet, for a total collection surface of 725 square feet.

A small pump delivers water to the

collectors from the bottom of an enormous 8000 gallon tank located beneath the house. When conditions are unfavorable for solar collection, the pump stops and the water drains back to the tank. Around the outside of the tank is a large air space surrounded by a 10-inch layer of fiberglass insulation. The air in this space warms and rises into the house by gravity convection controlled by dampers. Mathew estimates the cost of materials for the collector and storage at only $1000.

The unique feature of this house is the large reflecting surface, which was made from household aluminum foil held in place by an asphalt roofing compound. The extra sunlight reflected onto the collector permits the pump to operate even during the periods of diffuse sunshine that abound in the Oregon winter. During the summer, the aluminized roof keeps the home cool by reflecting away much unneeded sunlight.

A study team from the University of Oregon and Oregon State University found that because of the reflector about 50 to 70 percent more sunlight hits the collector, and almost twice as much solar heat is collected. Electric baseboard heaters provide supplementary heat, but Mathew estimates that 80 percent of the heating needs of this 1650 square foot house are supplied by the sun.

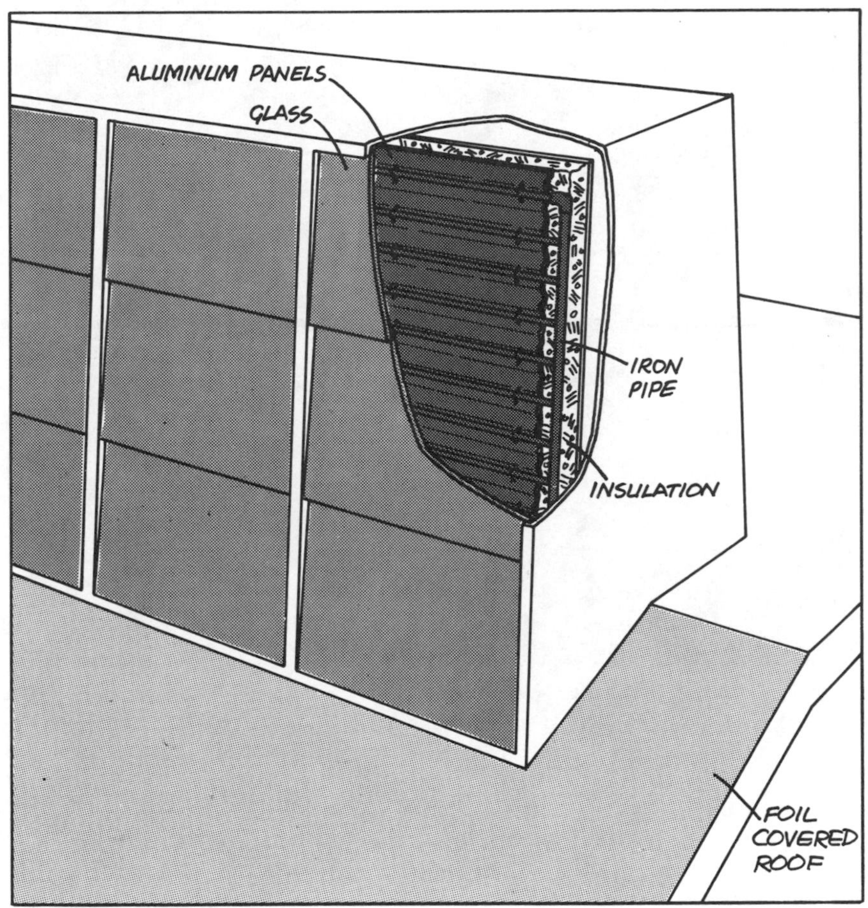

Rooftop collector in the Mathew House, Coos Bay, Oregon. The foil-covered south roof reflects about 50 percent more sunlight onto this collector.

Van Dresser House

High fuel costs and the burdensome costs of new houses have led many homeowners to apply solar heating to their present homes. Although the design limitations are severe, a number of successful solar heating systems have been added to existing structures. One of the earliest examples of such a *retrofit*

Peter van Dresser's house, Santa Fe, New Mexico. This converted adobe is one of the oldest operating solar houses in the U.S.A.

solar house was built in Santa Fe, New Mexico, by Peter van Dresser. In 1957, he extensively renovated an old adobe house and converted it to solar heating.

The small single-story house with a floor area of 500 square feet was oriented from north to south—hardly ideal for solar collection. Van Dresser surmounted this problem by adding two banks of solar collectors with a total area of 230 square feet to the flat roof of the structure. Both collectors, one tilted at 60° and the other at 45°, are flat-plate collectors with air as the transfer fluid. In each one a galvanized sheet steel absorber plate sits one inch behind a sheet of double strength glass. Sunlight heats the black coating of this plate, warming the air in the gap behind the plate.

Three small blowers suck this warm air from the solar collectors and deliver it to a gravel bed located beneath the south portion of the house. The 4-foot deep bed is surrounded by 8-inch thick pumice blocks and covered by the brick floor of the house. Solar heat is also stored in a bed of sand under the north portion of the house. The stored heat percolates up through the brick floor and warms the rooms by thermal radiation. The massive adobe walls of the house, which are at least a foot thick, help to hold this warmth and prevent rapid temperature fluctuations. Supplementary heat comes from a small gas furnace and a fireplace.

This system has worked well since 1958, providing solar heat for a succession of tenants. Van Dresser estimates that 65 percent of the house's heat comes from the sun in a typical year. No cooling is needed in summer because of the massive walls and floors and the lack of any large windows facing the sun. With its adobe walls and solar collectors, this house is a happy marriage of ancient building arts with modern technologies.

Barber House

In addition to heating their homes with solar energy, some people are turning to wind-generated electricity to escape the clutches of the power companies. An example of this trend is the two story house built by Yale professor Everett Barber in Guilford, Connecticut. First occupied in the fall of 1975, the house will eventually get 70 percent of its heat from the sun and 80 percent of its electricity from the wind.

The 1300 square foot house was designed by Charles W. Moore Associates and the environmental systems provided by Barber's firm, Sunworks, Inc. The roof faces south-southwest at a tilt of 57°, holding about 400 square feet of Sunworks collectors. Their copper absorber plates are coated with a black *selective surface*—one of a number of compounds that absorb virtually all incident sunlight while emitting very little thermal radiation. Barber claims that these single-glazed collectors have an efficiency of 60 percent when the absorber temperature is 100°F above the outside temperature.

A solution of ethylene glycol in water (to prevent freezing) circulates through copper tubes soldered to the absorber surface. This solution then passes through a coiled heat exchanger in the 2000 gallon water storage tank sitting vertically behind the living room fireplace. Over a temperature range of 60°F, this tank can store about a million Btu. Extra heat is provided by the fireplace and by a large oil-fired water heater. Eventually, Barber plans to use excess electricity from two windmills to heat the tank water.

Another million Btu of solar heat can be stored in a bed of rocks beneath the concrete floor slab and in the masonry block walls. The outside of these walls are insulated by 3 inches of sprayed-on polyurethane foam to keep the heat inside. Every window has double-thick insulating glass, and most have south exposure to make the best use of the winter sun. Insulating shutters on the insides

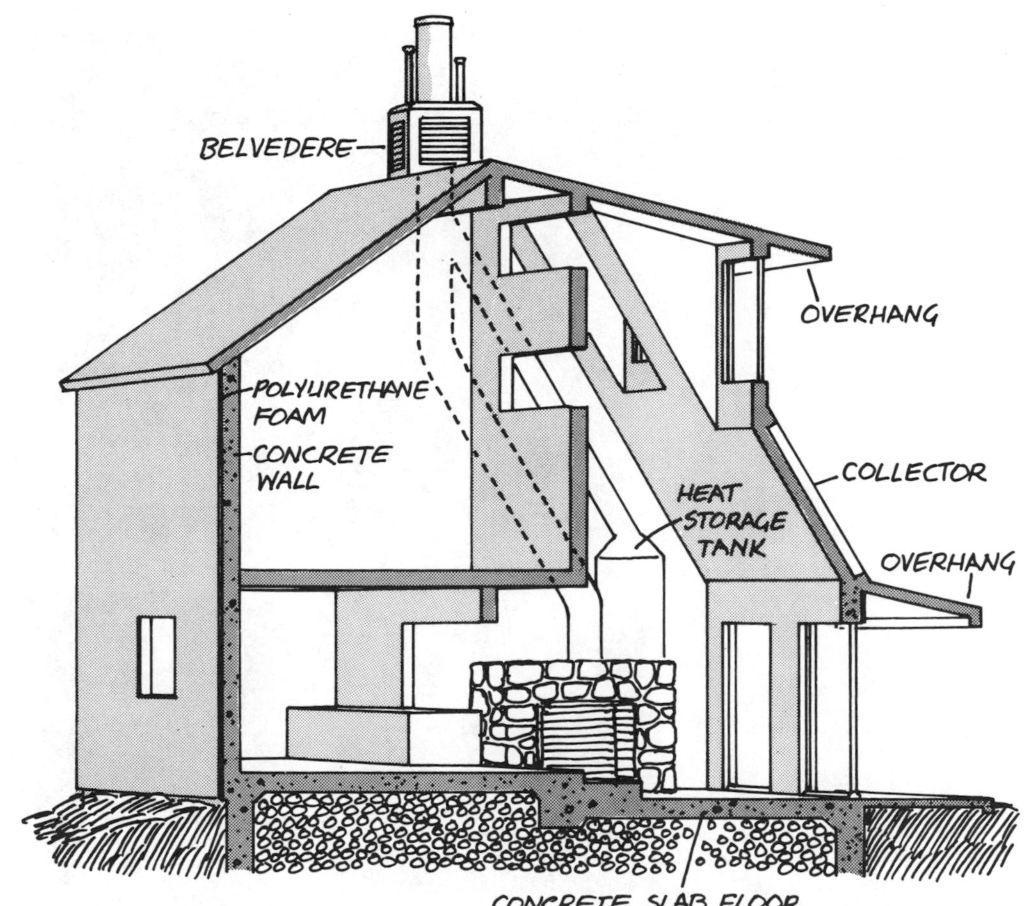

Rough cut-away view of the Barber House in Guilford, Connecticut.

Winter and summer at Project Ouroboros in Rosemont, Minnesota.

of these windows are used instead of drapes. Overhangs above the south-facing windows shield them from the summer sun. A belvedere at the peak of the house vents warm air during the summer, allowing cooler air to be drawn in at the windows.

Ouroboros

Autonomous living is the theme of an experimental house built in Rosemount near the University of Minnesota. Called Ouroboros after a mythical dragon that survived by eating its tail and regenerating itself, the house began as a design project under architecture professor Dennis Holloway. A total of 160 students contributed designs and labor toward making the house a reality. It has been occupied since June of 1975 by a student and his family. An evolving laboratory for energy conservation and self-sufficiency, Ouroboros already has such "novel" features as a sod roof, a windmill, and a composting toilet.

The shape and orientation of the house are adapted closely to the Minnesota climate, with its cold, blustery winters. At its base, Ouroboros is a trapezoid with its longest side facing due south. Earth is piled against the north, east, and west walls. The sod roof slopes backward almost to the level of the ground, protecting the house against fierce north and west winds. The walls and roof have at least 9 inches of fiberglass insulation throughout. In winter, snow drifts collect on the roof and around these walls to provide extra insulation where it is most needed.

The entire south side of the 1500 square foot house is devoted to collecting solar energy. Its upper part is tilted at 60°, the optimum for winter collection at this latitude (45°N). Vertical south windows and a greenhouse, both double-glazed, occupy the lower part. The 590 square feet of trickle-type collectors originally installed in the upper wall were covered with two panes of glass because of the extremely cold weather. Solar heated water drained to a 1000 gallon basement tank surrounded by 35 tons of crushed rock. Air blown through the rock carried heat to the house.

But John Ilse, a graduate student in mechanical engineering, replaced half of the trickle-type collectors with a *sandwich-type* collector of his own design. Water flows between two sheets of cold-rolled steel that have been dimpled and spotwelded into a single unit. Water is pumped *upwards* through the cavity between these plates and drains from the roof ridge to the basement tank. The two types of collectors began working side-by-side in March of 1975, and Ilse's collectors performed consistently better than the trickle-type collectors. In February of 1976, the earlier collectors were removed and completely replaced with Ilse's collectors.

Although not as crucial as winter heating in the Minnesota climate, summer cooling is part of the comprehensive design. The upper part of the south wall forms an overhang that keeps almost all summer sun off the south windows. Natural ventilation through

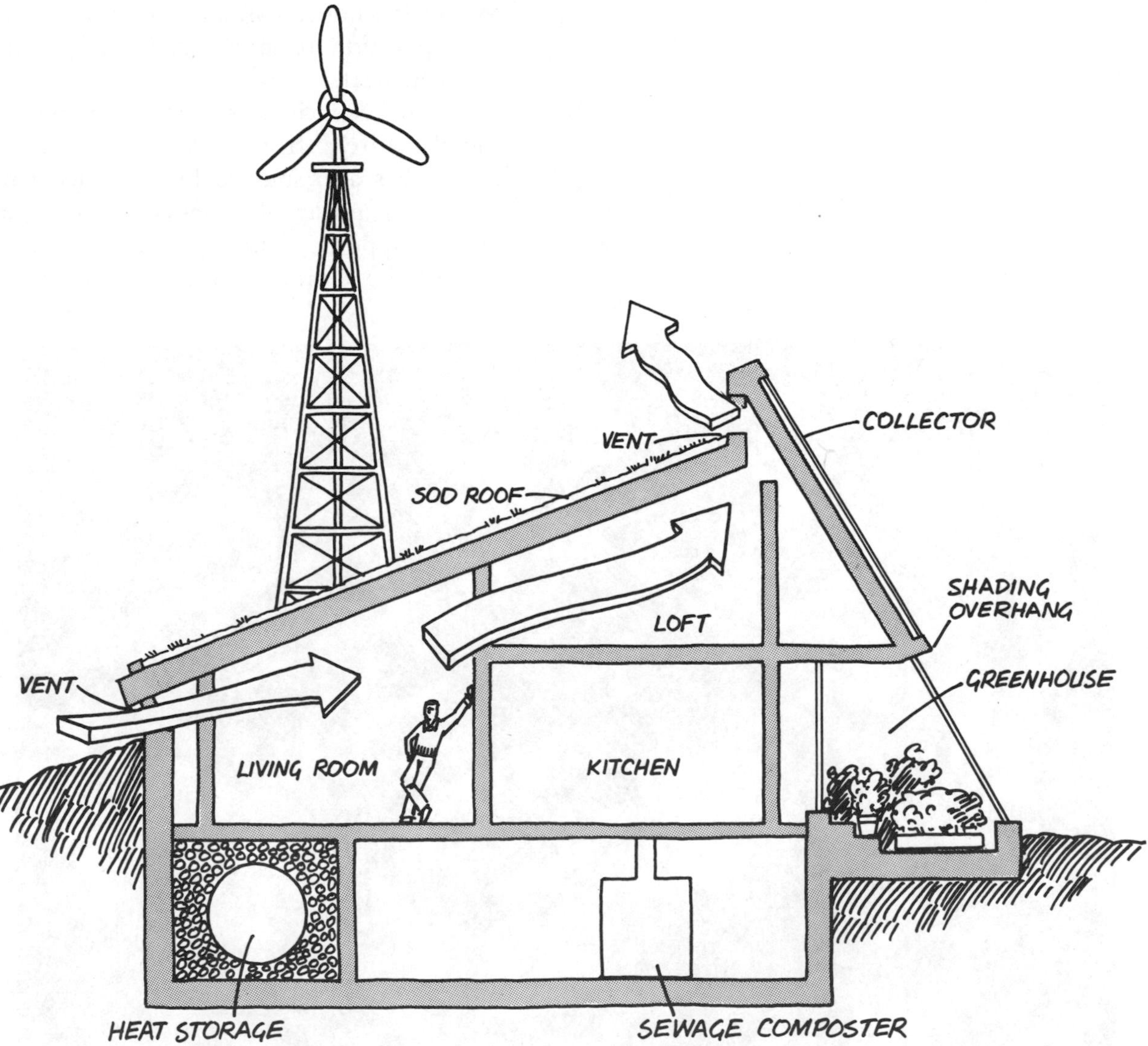

Ouroboros—a model house for energy conservation and self-sufficiency.

vents in the north wall and roof peak, aided by evaporation from the sod roof, carries off excess heat.

The entire project cost $95,000 and was funded entirely from local sources. Local funding has also spurred local involvement. Holloway is proud of its independence from federal support, claiming that "the house wouldn't be built by now" if he'd waited for such money. Together with his students and co-workers, he sees many more improvements in Ouroboros' future.

Experimental solar house at the University of Florida is both heated and cooled by the sun.

University of Florida House

One of the oldest operating solar houses is a building located just off the campus of the University of Florida in Gainesville. The 1400 square foot house was built in 1955 in order to study home heating and air conditioning. In 1963, it was converted to solar heating by a group from the University's Solar Energy and Energy Conversion Laboratory, headed by Dr. Erich Farber. Solar energy was collected in 300 square feet of flat-plate collectors and used only for heating.

This first solar heating system was replaced in 1973 by a system that could provide both heating and cooling. Mounted on the roof are 12 flat-plate collectors, each 4 feet by 10.5 feet, for a total collection area of 500 square feet. Their tilt angle of 30° is more suitable for collection in summer than in winter. Heated water from these collectors is stored in a 3000 gallon tank of water sitting in the yard. The heat is distributed to the house by baseboard radiators or by forced-air convection. A 4 foot by 12 foot collector supplies the domestic hot water.

Both systems have worked well for more than a decade. Over the years they have supplied the house with almost 100 percent of its heat and hot water. Solar energy is also used to heat a swimming pool in the winter. In the summer it powers an ammonia-water air conditioner which can supply the residence with all of its needed cooling. Dr. Farber estimates that this system would cost about $5000 if it was installed in an average single-family home.

Colorado State University Houses

Under a National Science Foundation grant, Colorado State University is building and evaluating three 3000 square foot laboratory-residential structures in Fort Collins, Colorado. The three houses are nearly identical in construction but use markedly different collector and storage systems. This project is a comprehensive study and comparison of several indirect methods of solar heating and cooling.

The first structure, completed in July of 1974, has about 75 percent of its heating and cooling needs supplied by solar energy. Rather than being designed for energy conservation, the house is conventionally framed and insulated. Under the severest winter conditions, it has a heating load of 55,000 Btu per hour—typical for a house this size.

The 750 square feet of aluminum tube-in-plate collector surface is double glazed and tilted at an angle of 45°. The collector materials cost about $4.00 per square foot, with labor an added $2.65 per square foot (no overhead or profit). A water-antifreeze solution circulates through the collector and transfers solar heat to an 1100 gallon tank insulated with 6 inches of fiberglass.

The first Colorado State University solar house, completed in July, 1974.

Heat is supplied to the living space through a fan-coil heat exchanger. The heating system design allows storage temperatures as low as 80°F. Auxiliary heat, up to the design capacity of 55,000 Btu per hour, is provided by a gas-fired boiler.

Summer cooling is provided by an Arkla-Servel unit powered by 175-200°F water from either the collector, storage, or boiler. In this *absorption cooling* unit, solar heat

Evacuated tube collector used in the third Colorado State University house.

evaporates water from a lithium bromide solution. After condensing, this water passes through a heat exchanger exposed to the room air. There the water evaporates once again, chilling the room air and carrying off the heat.

The other two structures have two very different types of collectors. The collector in the second house has the same area and orientation as that in the first house, but it is an air-type collector. Sunlight passes through two sheets of glass and is absorbed by blackened galvanized steel plates. Air passing through a 1/2-inch gap on the underside of this absorber carries heat to a storage bin containing 25 tons of stones. Air is blown through the bin to heat the rooms.

In the third house, the collectors are parallel arrays of concentric glass tubes. The outer tube is evacuated and the two inner tubes carry the heat transfer fluid, which can be either a liquid or a gas. Because the outer tube is evacuated, there is little conduction or convection heat loss from the inner tubes. High temperatures (200° F) and collection efficiencies (70 percent) are possible with this collector, and it is particularly well suited for powering an absorption cooling unit. The heat storage, heat distribution, and cooling systems are similar to those in the first house.

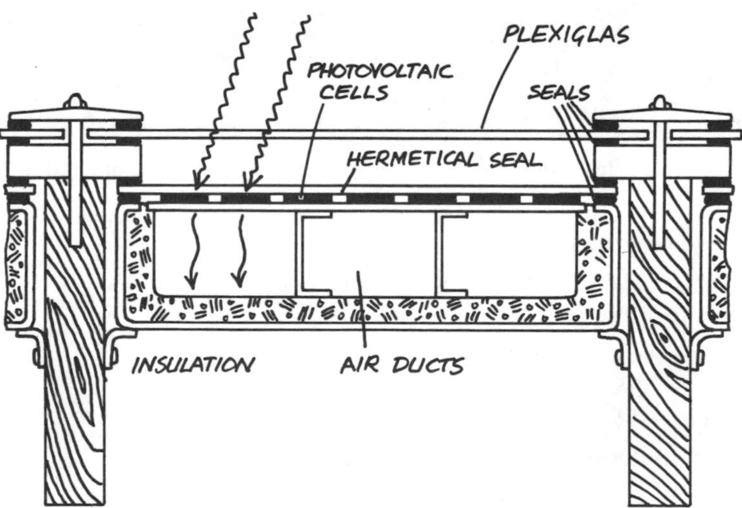

One of the solar collectors used in the University of Delaware House. Both heat and electricity are generated in this collector.

University of Delaware House

Probably the most complex solar house now in existence is a research project at the University of Delaware. This is the first house in which there has been a substantial effort to convert sunlight into both heat and electricity. Solar heat is collected in 850 square feet of rooftop and wall-mounted panels. Solar energy is also converted into electrical energy by cadmium sulfide photovoltaic cells located in three of the rooftop panels. Between 5 and 10 percent of the incident solar energy is converted to electricity in these cells, with the remaining energy being converted to heat as in any normal collector. An airstream moving behind the cells picks up this heat, transferring it to storage and keeping the cells cool.

The heat storage system was developed by Dr. Maria Telkes, who engineered the Dover house in the late 1940's. Heat is stored in three *eutectic* (low melting point) salts with a total weight of 6 tons. These salts absorb large amounts of heat when they melt and release this heat when they solidify. Sodium thiosulfate pentahydrate ($Na_2S_2O_3 \cdot 5H_2O$), which melts at 120°F, is used to store heat during the winter heating season. Sodium sulfate dodecahydrate or Glaubers salt ($Na_2SO_4 \cdot 10H_2O$), which melts at 75°F, is used to store heat when the collector temperature is not high enough to melt the 120° salts. A heat pump then extracts the heat from the 75° salts and transfers it to the main 120° storage. An electric heater provides auxiliary heat to the main storage when needed.

A mixture of sodium chloride, sodium sulfate, ammonium chloride, and borax is used during the summer. These salts, which melt at 55°F, are chilled by a heat pump operating in an air-conditioning mode and dumping heat into the cool night air. During the day, room air is cooled by passage through the bin of 55° salts, which absorb heat from the air as they melt.

The performance of these systems has been gauged since 1973. In the winter of 1973-74, the system provided 67 percent of the house heating needs. The photovoltaic cells originally achieved a conversion efficiency of only 3 to 5 percent, but this figure has been improved to as much as 7 percent. The other collectors also performed below expectation, but continued development has increased their efficiencies. After 30 years of effort, Dr. Telkes seems to have succeeded in making eutectic salts perform properly. But a solar powered and heated home such as this is impractical today because of the high cost of solar cells. However, the cadmium sulfide cells may eventually be cheap enough for widespread use.

Advanced methods of using solar energy are being tested at the University of Delaware House.

Grassy Brook Village

One of the most farsighted community projects for using solar energy is Grassy

Brook Village in Brookline, Vermont. This project was designed by People/Space, Co., and engineered by Dubin-Mindell-Bloome Associates. In the first phase of the project,

Grassy Brook Village in Brookline, Vermont—a community approach to solar energy.

10 housing units are being built, with an additional 10 planned. According to developer Richard Blazej, they are designed to "create a minimum of impact on the natural environment and provide a pattern of utilities and life-support systems that derive their energy from natural and non-polluting sources." When complete, the development will feature the use of solar energy for house heating and hot water, the generation of electricity from wind power, and the on-site handling of wastes.

The three-bedroom houses, each with about 1200 square feet of floor space, are clustered in two groups of 10. Each cluster has its own solar heating and waste handling systems, as well as walkways, decks, a central green, and a pond for recreation and fire protection.

The solar heating system is unique in that a single collector serves the entire cluster of 10 houses. The collector sits on three separate support structures, separated from the houses to allow freedom of architectural design. The three structures are arranged in a low sawtooth configuration in order to reduce the effect of wind loads and to permit easier servicing. The total collector area of 4500 square feet faces south at a tilt angle of 57°.

A water-glycol solution carries the solar heat to the heat storage tanks (with a total capacity of about 15,000 gallons) located under some of the housing units. A heat pump, supplemented by a oil-fired boiler, provides extra heat during extended periods of cloudy or extremely cold weather. A wood stove in each house permits the residents to use the abundant backup fuel supply in the surrounding woods.

Some of the best examples of solar heated and cooled homes were described in this section in order to introduce you to the possible residential uses of solar energy. In subsequent chapters, many more homes and buildings will be used to demonstrate specific design principles.

There are many more solar tempered homes and buildings than we could cover here. Two admirable efforts to chronicle this building surge make worthwhile reading: *Solar Heated Buildings—A Brief Survey,* by William Shurcliff; and *Solar Oriented Architecture,* a report to the American Institute of Architects by a research team under John Yellott. The former is a catalog of the solar heating methods and systems being used in over 200 buildings throughout the world. The latter work examines the architectural modifications required to integrate solar heating into 70 dwellings in the United States. Both are invaluable references for anyone planning to build a solar home.

FURTHER READING

Association for Applied Solar Energy. *Proceedings of the World Symposium on Applied Solar Energy, Phoenix, Arizona 1955.* Menlo Park, California: Stanford Research Institute, 1956.

Hamilton, R. *Space Heating with Solar Energy.* Cambridge, Massachusetts: MIT Press, 1954. (proceedings of the Space Heating Symposium, MIT-1950)

Hesselschwerdt, A. L. Jr. "Heating by Sun Power: A Progress Report." in *Heating and Air Conditioning Contractor,* October 1956.

Shurcliff, William A. *Solar Heated Buildings: A Brief Survey.* Cambridge, Massachusetts, 1975. (available for $8 from W. A. Shurcliff, 19 Appleton St., Cambridge, Mass. 02138)

Steadman, Philip. *Energy, Environment, and Building.* Cambridge, England: Cambridge University Press, 1975.

Telkes, Maria. "Space Heating with Solar Energy." in *Science Monthly,* Vol. 69, December 1949.

Thomason, Harry. "Experience with Solar Houses," in *Solar Energy,* Vol. 10, No. 1, January-March 1966.

United Nations Department of Economic and Social Affairs. *New Sources of Energy and Energy Development.* New York: United Nations, 1964. (proceedings of the UN Conference on New Sources of Energy, Rome-1961)

Yellott, John I., et al. *Solar Oriented Architecture.* Tempe, Arizona: Arizona State College of Architecture, 1975. (report to AIA Research Corporation and the National Bureau of Standards)

3
Fundamentals

Understanding and Calculating the Energy Flows around a House

Sunrise at Stonehenge on the summer solstice.

Is it not by the vibrations given to it by the sun that light appears to us; and may it not be that every one of the infinitely small vibrations, striking common matter with a certain force, enters its substance, is held there by attraction and augmented by successive vibrations, till the matter has received as much as their force can drive into it?

Is it not thus that the surface of this globe is heated by such repeated vibrations in the day, and cooled by the escape of the heat when those vibrations are discontinued in the night?

Benjamin Franklin,
Loose Thoughts on a Universal Fluid

Before you design and build a solar tempered home, you need to acquire some familiarity with the natural energy flows of your surroundings. You need to know the position of the sun in order to orient a house and collector to receive its warm rays. To gauge the solar heat flows into a house, you must calculate the solar radiation hitting the walls, windows, roofs, and collector surfaces. You also need to calculate the heat escaping from a house in order to select the best methods to stem this leakage. Only when you have grasped a few of these fundamentals can you begin to take advantage of these natural energy flows.

First you need to understand some of the language others use to describe and measure these flows. You must also become familiar with climatic data and the properties of common building materials. The aim of this chapter is to acquaint you with these and other essentials that will help you use the abundance of solar energy falling all around you. Some of this chapter may seem tedious, but it is all very important to good solar home design.

SOLAR PHENOMENA

After centuries of observation, ancient astronomers could accurately predict the sun's motion across the sky. Stonehenge was probably a gigantic "computer" that recorded in stone the movements of the sun and moon. From their earthbound viewpoint, early peoples reckoned that the sun gave them night and day by moving in a path around the earth. But today, thanks to the work of the sixteenth-century Polish astronomer Copernicus, we know that the earth travels in an orbit around the sun and that the rotation of the earth, not the motion of the sun, gives us the cycles of night and day.

The earth actually follows an elliptical (egg-shaped) path around the sun. As it travels this orbit, its distance from the sun changes slightly—it is closest in winter and most distant in summer. The amount of solar radiation striking the earth's atmosphere is consequently most intense in winter. Then why are winters so dreadfully cold?

This seeming paradox is readily explainable. The earth's axis is tilted relative to the

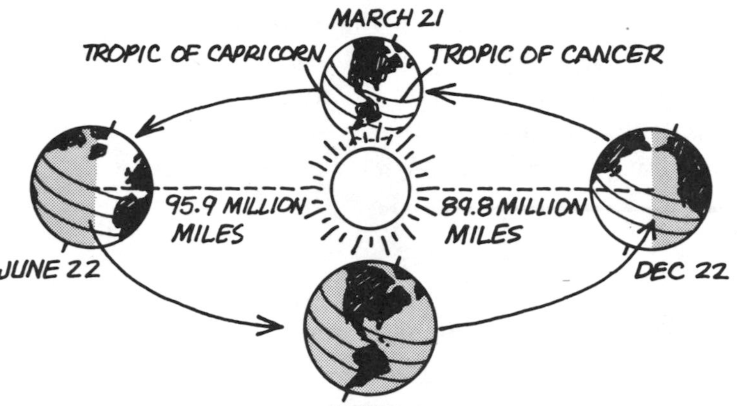

The earth's elliptical path around the sun. The tilt of the earth's axis results in the seasons of the year.

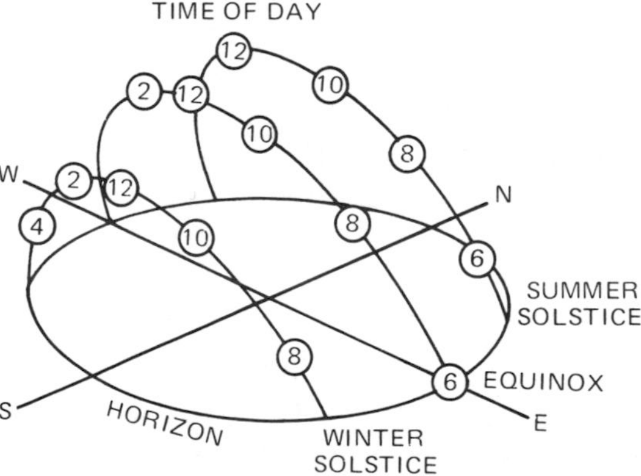

The sun's daily path across the sky. The sun is higher in the sky in summer than in winter due to the tilt of the earth's axis.

plane of its orbit, as shown in the first diagram. The north pole is tilted *toward* the sun in summer and *away from* the sun in winter. From our viewpoint here on earth, this tilt means that the sun is higher in the sky in summer, and lower in winter. Consequently, the sun's rays have a greater distance to travel through the atmosphere in winter, and they strike the earth's surface at a more glancing angle. The amount of solar radiation eventually striking a horizontal surface is less during the winter, and the weather is colder.

This tilt of the earth's axis results in the seasons of the year. If the axis were perpendicular to the orbital plane, there would be no noticeable change of seasons. Each day the sun would follow the same path across the sky, and the weather would be uniformly dull. Likewise, if the earth did not rotate on its axis, the sun would creep slowly across the sky, and a single day would last a whole year. The diurnal (daily) and seasonal cycles that we take for granted are a direct result of this rotation of the earth about a tilted axis.

Solar Position

Most people have probably noticed that the sun is higher in the sky in summer than in winter. Some also realize that it rises south of due east in winter and north of due east in summer. Each day the sun travels in a circular path across the sky, reaching its highest point at noon. As winter proceeds into spring and summer, this circular path moves higher in the sky. The sun rises earlier in the day and sets later.

The actual position of the sun in the sky depends upon the latitude of the observer. At noon on March 21 and September 23, the vernal and autumnal *equinoxes,* the sun is directly overhead at the equator. At 40°N latitude, however, its angle above the horizon is 50° (= 90° − 40°). By noon on June 22, the summer *solstice* in the Northern Hemisphere, the sun is directly overhead at the Tropic of Cancer, 23½°N latitude. Its angle above the horizon at 40°N is 73½° (= 90° + 23½° − 40°), the highest it gets at this latitude. At noon on December 22, the sun is directly overhead at the Tropic of Capricorn, and its angle above the horizon at 40°N latitude is only 26½°.

A more exact description of the sun's position is needed for most applications. In the language of trigonometry, this position is expressed by the values of two angles—the solar *altitude* and the solar *azimuth.* The solar altitude θ is measured up from the horizon to the sun, while the solar azimuth ϕ is the angular deviation from true south.

These angles need not be excessively mysterious—you can make a rough measurement of them with your own body. Stand facing the sun with one hand pointing toward it and the other pointing due south. Now drop the first hand so that it points to the horizon directly below the sun. The angle that your arm drops is the solar altitude θ and the angle between your arms in the final position is the solar azimuth ϕ. Much

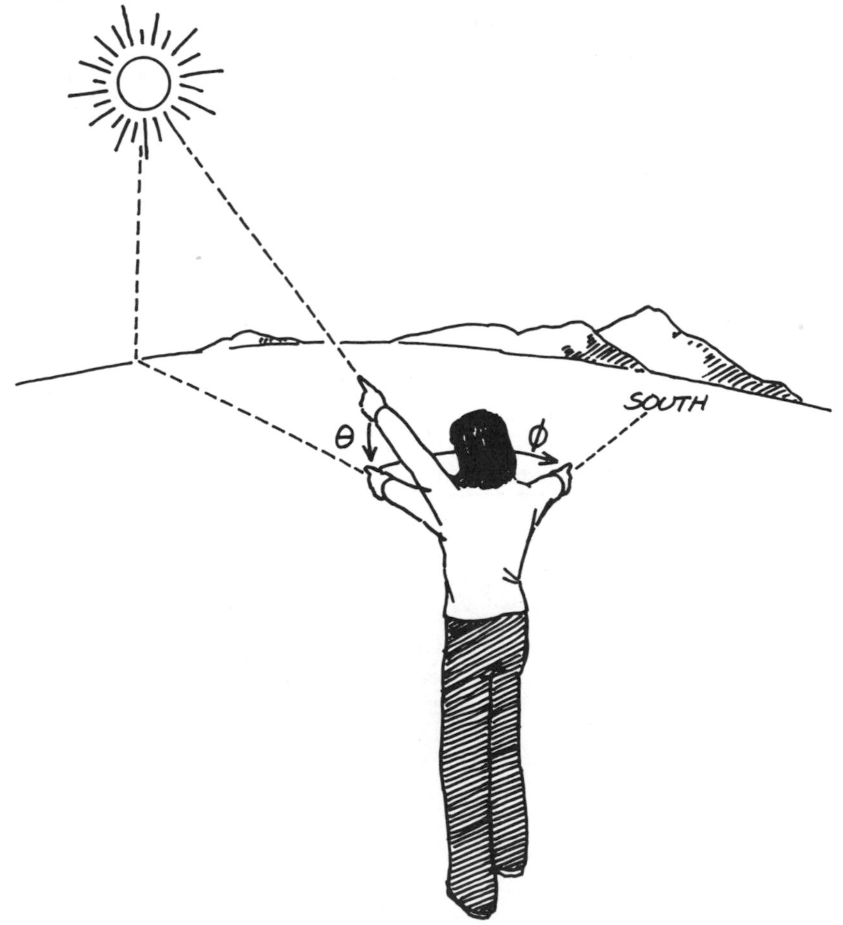

Measuring the sun's position. The solar altitude θ is the angle between the sun and the horizon, and the azimuth ϕ is measured from true south.

better accuracy can be obtained with better instruments, but the measurement process is essentially the same.

The solar altitude and azimuth can be calculated for any day, time, and latitude. For 40°N latitude (Philadelphia, for example),

SOLAR POSITIONS FOR 40°N LATITUDE														
AM	PM	ANGLE	Jan 21	Feb 21	Mar 21	Apr 21	May 21	Jun 21	Jul 21	Aug 21	Sep 21	Oct 21	Nov 21	Dec 21
5	7	ALT θ					1.9	4.2	2.3					
		AZI ϕ					114.7	117.3	115.2					
6	6	ALT θ				7.4	12.7	14.8	13.1	7.9				
		AZI ϕ				98.9	105.6	108.4	106.1	99.5				
7	5	ALT θ		4.3	11.4	18.9	24.0	26.0	24.3	19.3	11.4	4.5		
		AZI ϕ		72.1	80.2	89.5	96.6	99.7	97.2	90.0	80.2	72.3		
8	4	ALT θ	8.1	14.8	22.5	30.3	35.4	37.4	35.8	30.7	22.5	15.0	8.2	5.5
		AZI ϕ	55.3	61.6	69.6	79.3	87.2	90.7	87.8	79.9	69.6	61.9	55.4	53.0
9	3	ALT θ	16.8	24.3	32.8	41.3	46.8	48.8	47.2	41.8	32.8	24.5	17.0	14.0
		AZI ϕ	44.0	49.7	57.3	67.2	76.0	80.2	76.7	67.9	57.3	49.8	44.1	41.9
10	2	ALT θ	23.8	32.1	41.6	51.2	57.5	59.8	57.9	51.7	41.6	32.4	24.0	20.7
		AZI ϕ	30.9	35.4	41.9	51.4	60.9	65.8	61.7	52.1	41.9	35.6	31.0	29.4
11	1	ALT θ	28.4	37.3	47.7	58.7	66.2	69.2	66.7	59.3	47.7	37.6	28.6	25.0
		AZI ϕ	16.0	18.6	22.6	29.2	37.1	41.9	37.9	29.7	22.6	18.7	16.1	15.2
12 noon		ALT θ	30.0	39.2	50.0	61.6	70.0	73.5	70.6	62.3	50.0	39.5	30.2	26.6
		AZI ϕ	0.0	0.0	0.0	0.0	0.0	0.0	0.0	0.0	0.0	0.0	0.0	0.0

NOTES: Altitudes θ are measured from the horizon, and azimuths ϕ are measured from true south. Angles are given in degrees, and solar times are used.

SOURCE: Koolshade Corporation.

the values of θ and ϕ are given at each hour for the 21st day of each month in the accompanying table. Note that ϕ is always zero at solar noon and that θ varies from 26.6° at noon on December 21 to 73.5° at noon on June 21. You can find similar data for latitudes 24°N, 32°N, 48°N, 56°N, and 64°N in the tables titled "Clear Day Insolation Data" in Appendix 1. This appendix also shows you how to calculate these angles directly for any day, time, and latitude.

Exactly why you need to know these solar positions may still seem vague, but it should become obvious by the next chapter. A knowledge of the sun's position helps you to determine the orientation of a house and placement of windows to collect the most winter sunlight. This knowledge is also helpful in positioning shading devices and vegetation to block the summer sun. Often the available solar radiation data only applies to horizontal or south-facing surfaces, and exact solar positions are needed to convert these data into values that are valid for other surfaces.

Insolation

Arriving at a quantitative description of the solar radiation striking a surface, or the *insolation* (not to be confused with ins*u*lation), is a more difficult task. Most of this difficulty arises from the many variables that affect the amount of solar radiation striking a particular spot. Length of day, cloudiness, humidity, elevation above sea level, and surrounding obstacles all affect the insolation. Compounding this difficulty is the fact that the total solar radiation striking a surface is the sum of three contributions: the *direct* radiation from the sun, the *diffuse* radiation from the entire sky, and the *reflected* radiation from surrounding terrain and vegetation. Fortunately, however, we do not need exact insolation data for most low-temperature applications of solar energy.

Although insolation data has been recorded at about 80 weather stations across the country, much of it is inaccurate and incomplete. The information is usually provided in units of *langleys* striking a horizontal surface over a period of time, usually a day. A langley is one calorie of radiant energy per square centimeter, and one langley is equivalent to 3.69 Btu per square foot, the more familiar English measure. An example of the information available is the map of "Mean Daily Solar Radiation, Annual" presented here. You can find monthly maps of the mean daily solar radiation in Appendix 1. These data apply only to horizontal surfaces, and can be misleading. Complicated trigonometric conversions, which involve assumptions about the ratio of direct to diffuse radiation, are necessary to apply these data to vertical or tilted surfaces. These trigonometric conversions are also discussed in that appendix.

The weather bureau also provides information about the percentage of possible sunshine, defined as the percentage of time the sun "casts a shadow." An example of these data is the map titled "Mean Annual Percentage of Possible Sunshine." In Appendix 1, you will find monthly maps that are probably more useful for calculations of insolation. By themselves, these maps tell us little about the amount of solar radiation falling on a surface, but when coupled with the "Clear Day Insolation Data," they make a powerful design tool.

Clear Day Insolation tables, prepared by the American Society of Heating, Refrigerating, and Air-Conditioning Engineers (ASHRAE), provide hourly and daily insolation (and solar positions) for a variety of latitudes. Tables for 24°N, 32°N, 40°N,

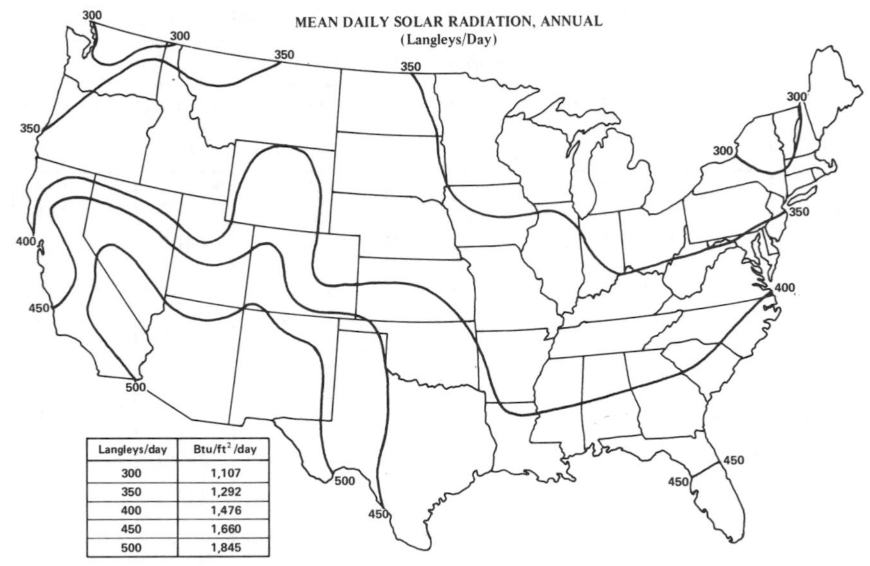

Langleys/day	Btu/ft²/day
300	1,107
350	1,292
400	1,476
450	1,660
500	1,845

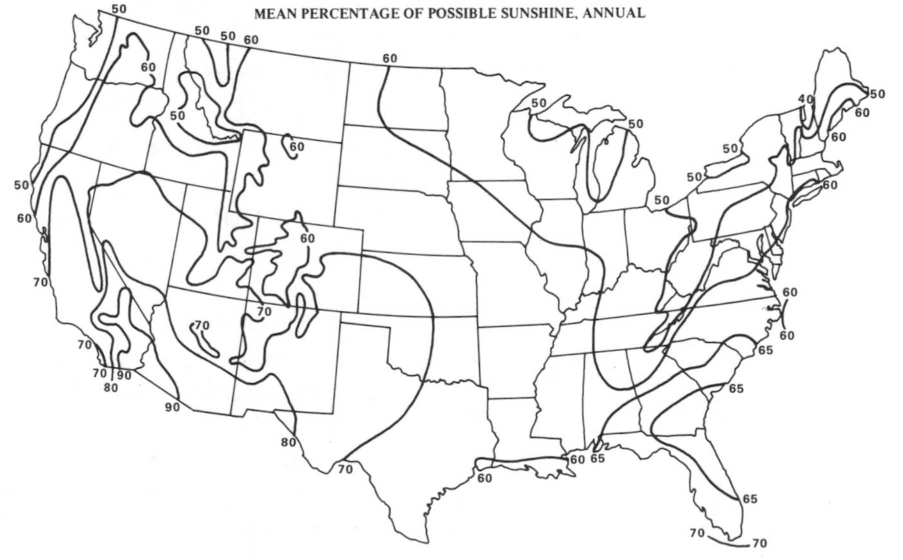

CLEAR DAY INSOLATION FOR 40°N LATITUDE							
TOTAL INSOLATION, Btu/ft²							
21st of:	Normal Surface	Horiz. Surface	South Facing Surface at Tilt Angle of:				
			30°	40°	50°	60°	90°
January	2182	948	1660	1810	1906	1944	1726
February	2640	1414	2060	2162	2202	2176	1730
March	2916	1852	2308	2330	2284	2174	1484
April	3092	2274	2412	2320	2168	1956	1022
May	3160	2552	2442	2264	2040	1760	724
June	3180	2648	2434	2224	1974	1670	610
July	3062	2534	2409	2230	2006	1728	702
August	2916	2244	2354	2258	2104	1894	978
September	2708	1788	2210	2228	2182	2074	1416
October	2454	1348	1962	2060	2098	2074	1654
November	2128	942	1636	1778	1870	1908	1686
December	1978	782	1480	1634	1740	1796	1646

SOURCE: ASHRAE, *Handbook and Product Directory: 1974 Applications.*

48°N and 56°N latitude are reprinted in Appendix 1. The values of the daily insolation from the 40°N latitude table are included here for your convenience. These tables list the *average clear day insolation* on horizontal and normal (perpendicular to the sun) surfaces, and on five south-facing surfaces tilted at a variety of angles (including vertical). The insolation figures quoted include a diffuse contribution for an "average" clear sky, but do not include any contribution for reflections from the surrounding terrain.

Hourly and daily insolation data are given in Appendix 1 for the 21st day of each month. You can readily interpolate between these numbers to get values of the insolation for other days, times, latitudes, and south-facing orientations. Trigonometric conversions of these data to other surface orientations are explained there.

When multiplied by the appropriate "percentage of possible sunshine," these data provide a good estimate of the hourly and daily insolation on a variety of surface orientations. You will note immediately, for example, that the total clear day insolation on a vertical south-facing wall in Philadelphia (40°N) is 610 Btu/ft² on June 21 and 1726 Btu/ft² on January 21, or almost *three* times greater! Multiplied by the percentage of possible sunshine for this locale (about 65% in June and 49% in January), the total insolation becomes 396 Btu/ft² in June and 846 Btu/ft² in January, or still a factor of *two* greater. On the other hand, the clear day insolation on a horizontal roof is 2648 Btu/ft² in June and only 948 Btu/ft² in January. For average days, the insolation is 1721 Btu/ft² in June and 464 Btu/ft² in January, or almost a factor of *four* smaller. Clearly, the roof is taking the heat in summer and the south walls are getting it in winter.

Limitations of the Insolation Data

You must be careful to note the limitations of the Clear Day Insolation tables. These data are based upon "average" clear day conditions, but "average" can vary with locale. Many locations are 10 percent clearer, such as deserts and mountains, and others, such as industrial and humid areas, are not as clear as the "average". Reflected sunlight from vegetation and ground cover is *not* included in the values given in the tables. Another 10 to 20 percent more sunlight may be reflected onto a surface than the amount listed. In the winter, even more radiation will be reflected onto south-facing walls because the sun is lower in the sky and snow may be covering the ground.

Other difficulties arise from the subjective evaluations of "percentage of possible sunshine." In the method of calculating average insolation described above, an assumption was made that the sun is shining full blast during the "sunshine" period and not at all during other times. In reality, up to 20 percent of the clear day insolation may still be hitting the surface during periods of total cloudiness. During hazy periods when the sun still casts a shadow, as little as 50 percent of the clear day insolation may be striking the surface. More accurate calculations, in which the diffuse and direct components of solar radiation are treated separately, are provided in Appendix 1.

Another problem is the variability of weather conditions with location and time of day. The weather maps provide only area-wide averages of the percent of possible sunshine. The actual value in your exact building location could be fairly different from your county average. On the other hand, the cloudiness in some areas, particularly

Diffuse and Reflected Radiation

The total solar radiation striking a plane surface is the sum of three components: the direct solar radiation I_D, the diffuse sky radiation I_d, and the radiation reflected from surroundings I_r. The direct component consists of rays coming straight from the sun—casting shadows on a clear day. If all our days were clear, we could simply use the Clear Day Insolation Data, add a small percentage for ground reflection, and have a very good estimate of the total insolation on our walls, roofs, and collectors. But all of us can't live in Phoenix or Albuquerque, so we must learn to deal with cloudy weather.

As it passes through the atmosphere, sunlight is scattered by air molecules, dust, clouds, ozone, and water vapor. Coming uniformly from the entire sky, this scattered radiation makes it blue on clear days and grey on hazy days. Although this diffuse radiation amounts to between 10 and 100 percent of the radiation reaching the earth's surface, little is known about its strength and variability.

The Clear Day Insolation Data aren't much help on a cloudy day. But frequently we only need to know the average daily insolation over a period of a month. In such a case we can use the monthly maps of the percent of possible sunshine to help us estimate this average. If P is the percentage of possible sunshine for the month and location in question, then we compute a factor F according to:

$$F = 0.30 + 0.65 \times (\frac{P}{100}) .$$

The numbers 0.30 and 0.65 are coefficients that actually vary with climate, location, and surface orientation. But their variation is not severe, and we can use these average values without too much error. If I_o is the Clear Day Insolation (whole day total) on a plane surface, then we compute the average daily insolation I_a according to:

$$I_a = F \times I_o .$$

These formulas are an attempt to account for the diffuse radiation that still strikes the surface on cloudy and partly cloudy days. Even in a completely cloudy month (P = 0), we would still be receiving 30% (F = 0.30) of the clear day insolation, according to these equations. This is perhaps a bit high, but the coefficients have been selected to produce accurate results under normal conditions, not blackouts.

For example, calculate the average daily insolation striking a horizontal roof in Philadelphia during the months of June and January. Using the first equation and P = 65 (June) and 49 (January) from before, we get:

$$\text{June:}\quad F = 0.30 + 0.65 \times (\frac{65}{100}) = 0.72 ;$$

$$\text{Jan:}\quad F = 0.30 + 0.65 \times (\frac{49}{100}) = 0.62 .$$

Therefore, the average daily insolation in each month is:

$$\text{June:}\quad I_a = 0.72 \times 2648 = 1907\ Btu/ft^2 ;$$

$$\text{Jan:}\quad I_a = 0.62 \times 948 = 588\ Btu/ft^2 .$$

These numbers may be compared with 1721 Btu/ft^2 and 464 Btu/ft^2 calculated earlier in our naive analysis. If we include diffuse radiation during cloudy weather, our results are 10 to 20% higher than before.

The diffuse and reflected radiation striking a surface also depend upon the orientation of the surface. Under the same sky conditions, a horizontal roof (which faces the entire sky) receives about twice the diffuse radiation hitting a vertical wall (which faces only ½ of the sky). Tilted surfaces receive some average of these two. Ground reflection depends a lot upon the shape and texture of the surroundings and the altitude of the sun. Snow reflects much more sunlight than green grass, and more reflection occurs when the sun is lower in the sky. During the winter, as much as 30% of the horizontal clear day insolation may be reflected up onto the surface of a south facing wall. But a roof receives no reflected radiation in any season, because it faces the sky—not the ground.

coastal areas, can occur at specific times of the day, rather than being distributed at random over the entire day. There may be a morning fog when the sun is low on the horizon, and a clear sky from mid-morning on, but this would be recorded as 75 percent of possible sunshine, while 90 percent of the total clear day insolation was actually recorded that day.

You may need more detailed information than is available from national weather maps. Occasionally, friendlier than usual personnel will assist you at the local weather station, but you will almost always be referred to the National Weather Records Center in Asheville, North Carolina. This center collects, stores, and distributes weather data from the entire country, and makes it available in many forms. You should first obtain their "Selective Guide to Climatic Data Sources," Key to Meteorological Records, Documentation No. 4.11 to give you an overview of the types of data available. You may obtain a copy for $1.00 from the Superintendent of Documents there.

But, in the last analysis, you should search out local information and opinion or rely on your own observations and intuition to temper the weather data for your design purposes. Becoming more involved with the environment you inhabit is an integral part of working with the sun.

HEAT FLOW

Heat energy is simply the motion of the atoms and molecules in a substance—their twirling, vibrating, and banging against each other. It is this motion that brings different atoms and molecules together in our bodily fluids, allowing the chemical reactions that sustain our lives. This is why our bodies need warmth. Seventeenth-century natural philosophers thought heat was a fluid—"phlogiston" they called it—that was released by fire and flowed from hot bodies to cold. They were correct about this last observation, for heat always flows from warm areas to colder ones, warming them in turn.

The rate of heat flow is proportional to the temperature difference between the source of the heat and the object or space to which it is flowing. Heat flows out of a house at a faster rate on a cold day than on a mild day. If there is no internal source of heat, such as a furnace or wood stove, the temperature inside the house approaches that of the outdoor air. Heat always flows in a direction that will equalize temperatures.

While the rate of heat flow is proportional to the temperature difference, the quantity of heat actually flowing depends on how much resistance there is to the flow. Since we can do little about the temperature difference between inside and outside, most of our effort goes into increasing a house's resistance to heat flow. The actual mechanisms of heat flow are numerous, and so are the methods of resisting them. Therefore, we will review briefly the three basic methods of heat flow—conduction, convection and radiation.

In our youth, we all learned about heat conduction intuitively by touching a hot stove or skillet. When an iron skillet sits on a hot stove for a while, the handle heats up. Heat from the burner flows through the metal to the handle. But the rate of flow to the handle of an *iron* skillet is much slower than if the skillet were made of *copper.* The heat flow through copper is quicker because it has greater conductance (less resistance to heat flow) than cast iron. It also takes less heat to warm copper than iron, and therefore less time to heat all the metal between the burner and the handle. These principles are basic to the concept of conduction heat flow.

Convection is heat flow through the movement of fluids—liquids or gases. When a fluid is heated, it can move or be moved to a cooler area where it transfers this heat to warm that area. In a kettle of water on a stove, the heated water at the bottom rises and mixes with the cooler water above, spreading the heat and warming the entire volume of water far more quickly than could have been done by heat conduction alone. A house with a warm air furnace is heated in much the same way. Air is heated in the firebox and rises up to the living spaces. Since the house air is cooler than the hot air, the heat is transferred from the air to the rooms.

Heated fluids move by natural convection or forced convection. As a fluid is warmed, it expands and becomes less dense, making it buoyant in the surrounding cooler fluid. It rises and the cooler fluid that flows in to

Radiation: sun, skillet handle, stove. Conduction: frying pan to handle. Convection: air passing around stove; steam.

How many paths of heat flow can you find? (Answer above.)

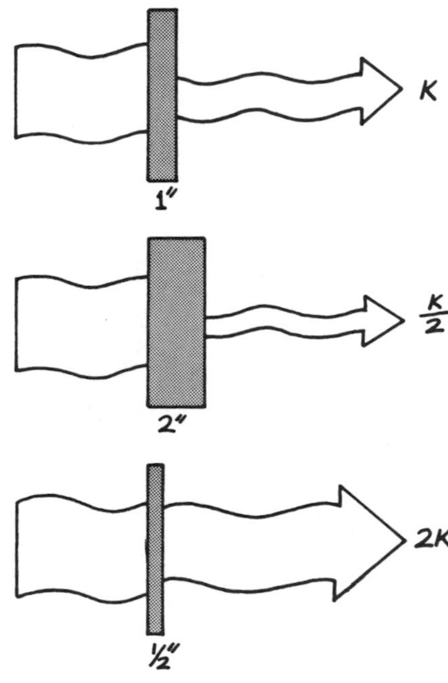

The thicker a slab, the less heat it conducts.

replace it is heated in turn. The warmed fluid moves to a cooler place where its heat is absorbed. Thus the fluid cools down, becomes heavier and sinks. This movement is known as *natural convection* or *thermosiphoning.* When we want more control over the heat flow, we use a pump or a blower to move the heated fluid. This is called *forced convection.*

Note that convection works hand-in-hand with conduction. Heat from a warm surface is conducted to the adjacent fluid before it is carried away by convection, and heat is also conducted from a warm fluid to a cool surface nearby. The greater the temperature difference between the warm and cool surfaces, the greater the heat flow between them.

Thermal radiation is the flow of heat energy through an open space by electromagnetic waves. This flow occurs even in the absence of any material in that space—just as sunlight can leap across interplanetary voids. Most objects which stop the flow of light also stop thermal radiation, which is primarily invisible longwave radiation. Warmer objects constantly radiate their thermal energy to cooler objects (as long as they can "see" each other) at a rate proportional to their temperature difference. We experience radiation heat flow to our bodies when we stand in front of a fireplace or hot stove. The same transfer mechanism, although more subtle and difficult to perceive, is what makes us feel cold while sitting next to a window on a winter night. Our warm bodies are radiating energy to the cold window surface—and we are chilled.

Of the three basic kinds of heat loss, radiation is the most difficult to calculate at the scale of a house. Calculation of convection heat loss through open doors or cracks and around windowframes is educated guesswork. Conduction heat loss through the exterior skin of the house (roofs, walls, and floors) is perhaps the easiest to estimate. Fortunately, this is the villain, culprit, rake, and thief who usually pilfers the most heat from our homes.

Conduction Heat Loss

The ability of a material to *permit* the flow of heat is called its thermal conductivity or conductance. The *conductance* of a slab of material is the quantity of heat that will pass through one square foot of that slab per hour with a 1°F temperature difference maintained between its two surfaces. Conductance is measured in units of Btu per hour per square foot per degree Fahrenheit, or Btu/hr/ft^2/°F. The conductance of a slab of material decreases as its thickness increases. While 10 Btu per hour may flow through a 1-inch slab of polystyrene, only 5 Btu per hour will flow through a 2-inch slab under the same conditions.

The opposite of conductance is *resistance*, the tendency of a material to *retard* the flow of heat. All materials have some resistance to heat flow – those with high resistance we call insulation. The resistance R of a slab of material is the inverse of its conductance, or

$$R = \frac{1}{C}$$

The higher the R-value of a material, the better its insulating properties. In the table you can find R-values for a few common building materials (and air spaces!). More detailed lists are provided in Appendix 2 under "Insulating Value of Materials."

A related quantity, the overall *coefficient of heat transmission* U, is a measure of how well a wall, roof, or floor conducts heat. The lower the U-value of a wall, the higher its insulating ability. Numerically, U is the rate of heat loss in Btu per hour through a square foot of surface with a 1° F temperature difference between the inside and outside air. Similar to conductance, U is expressed in units of Btu/hr/ft²/° F. To find the conduction heat loss, ΔH_{con}, through an entire wall, we multiply its U value by the number of hours, h, the wall area, A, and the temperature difference, ΔT, between the inside and outside air

$$\Delta H_{con} = U \times h \times A \times \Delta T$$

To find the heat lost through a 50-square foot wall with a U-value of 0.12, over an 8 hour time span when the inside temperature is 65° F and the outside temperature is 40° F, we multiply

$$\Delta H_{con} = 0.12 \times 8 \times 50 \times (65\text{-}40) = 1200 \text{ Btu}$$

RESISTANCES OF COMMON BUILDING MATERIALS		
Material	**Thickness (inches)**	**R-value (ft^2-°F-hr/Btu)**
Hardwood siding	1.0	0.91
Softwood siding	1.0	1.25
Gypsumboard	0.5	0.45
Wood shingles	lapped	0.87
Wood bevel siding	lapped	0.81
Brick, common	4.0	0.80
Concrete (sand and gravel)	8.0	0.88
Concrete blocks (filled cores)	8.0	1.93
Gypsum fiber concrete	8.0	4.80
Mineral Wool (batt)	3.5	10.9
Mineral Wool (batt)	6.0	18.8
Fiberglass board	1.0	4.35
Corkboard	1.0	3.85
Expanded polyurethane	1.0	5.88
Expanded polystyrene	1.0	4.17
Molded polystyrene beads	1.0	3.85
Loose fill insulation:		
Cellulose fiber	1.0	3.70
Mineral wool	1.0	4.00
Sawdust	1.0	2.22
Flat glass	0.125	0.89
Insulating glass (¼" space)	—	1.54
Vertical air space	0.75	0.87
Vertical air space	4.0	1.01

SOURCE: ASHRAE, *Handbook of Fundamentals*, 1970.

If the inside temperature is 70° F instead of 65° F, then the heat loss is 1440 Btu over the same time span.

The U-value includes the thermal effects of all the materials in a wall, roof, or floor—including air gaps inside and air films on the

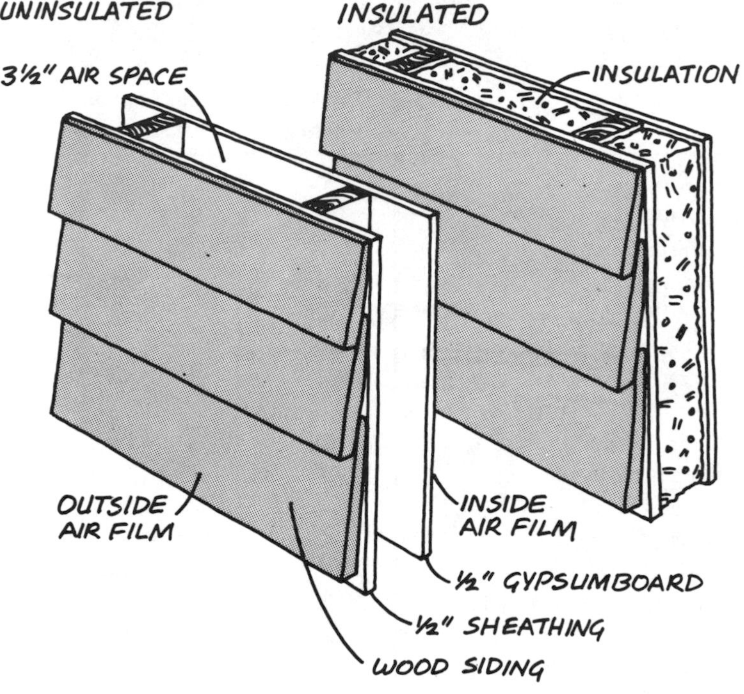

SAMPLE CALCULATIONS OF U-VALUES		
Wall Construction Components	R-values	
	Uninsulated	Insulated
Outside air film, 15 mph wind	0.17	0.17
Wood bevel siding, lapped	0.81	0.81
½" Sheathing, regular density	1.32	1.32
3½" Air space	1.01	–
3½" Mineral wool batt	–	10.90
½" Gypsumboard	0.45	0.45
Inside air film	0.68	0.68
TOTALS (R_t)	4.44	14.33
U-values ($U = 1/R_t$)	0.23	0.069

Calculation of U-values for two walls—uninsulated and insulated.

inner and outer surfaces. It can be computed from the conductances or resistances of all these separate components. The total resistance R_t is the sum of the individual resistances of these components. As U is the conductance of the entire building section, it is the inverse of R_t, or

$$U = \frac{1}{R_t} = \frac{1}{R_1 + R_2 + R_3 + \ldots + R_n}$$

Thus, computation of U involves adding up all the R-values, including R-values of inside and outside air films, any air gap greater than ¾-inch, and all building materials.

As an example, the U-values of two typical walls—one insulated and the other uninsulated—are calculated here. Note that the uninsulated wall conducts heat about three times more rapidly than the insulated wall. For a 50-square-foot wall, this is the difference between a loss of 3700 Btu and 1100 Btu.

Once you have calculated the U-values of all exterior surfaces (windows, walls, roofs, and floors) in a house, you can begin calculating the total conduction heat losses. One important quantity is the hourly heat loss of the house at outside temperatures close to the lowest expected—these extreme temperatures are called *design temperatures.* A listing of the recommended design temperatures for a number of U.S. cities is provided here; those for many other locations in the United States are provided in Appendix 1 under "Degree Days and Design Temperatures." The following approach is used to find the number of Btu per hour that your heating system will have to supply in order to keep your house warm under all but the most extreme conditions. Subtract the design temperature from the normal inside temperature to find the temperature difference, ΔT. Next, determine the total area, A, of each type of exterior building surface and multiply it by the temperature difference and the appropriate U-value, U_s, to get the total conduction heat loss, ΔH_s, of that surface per hour:

$$\Delta H_s = U_s \times A_s \times \Delta T$$

The total conduction heat loss of the house is merely the sum of the conduction heat losses through all these building surfaces. For example, the conduction heat loss of the 50-square-foot insulated wall with a U-value of 0.069 under design temperature conditions (0° F) in Denver, Colorado, is $\Delta H_s = 0.069 \times 50 \times (70\text{-}0) = 242$ Btu per hour.

To compute the total conduction heat loss for a single heating season, you must first grasp the concept of degree days. They are somewhat analogous to man-days of work. If a man works one day, the amount of work he does is often called a man-day. Similarly, if the outdoor temperature is one degree below the indoor temperature of a building for one day, we say one *degree day* has accumulated. Standard practice uses an indoor temperature of 65° F as the base from which to calculate degree days, because most buildings do not require heat until the outdoor air temperature falls between 60° F and 65° F. If the outdoor temperature is 40° F for one day then 65 − 40 = 25 degree days result. If the outdoor temperature is 60° F for 5 days, then 5 × (65 − 60) = 25 degree days again result.

The Weather Service publishes degree day information in special maps and tables. Maps showing the monthly and yearly total degree days are available in the *Climatic Atlas*. Tables of degree days, both annual and monthly, are provided for many cities in Appendix 1 under "Degree Days and Design Temperatures." Your local oil dealer or propane distributor should also know the number of degree days for your town.

To compute the total conduction heat loss during the heating season, you first multiply the total degree days for your locality by 24 hours per day to get the total *degree hours* during that time span. Now your calculation proceeds as in the earlier example: you multiply the area of each section, A_s, by its U-value, U_s, and the number of degree hours, 24D, and get the seasonal heat loss through that section:

$$\text{Seasonal } \Delta H_s = 24 \times U_s \times A_s \times D$$

The seasonal conduction heat loss from the entire house is the sum of seasonal heat losses through all the building surfaces. But to get the *total* seasonal heat loss, you must include the convection heat losses described in the next section.

The total cost of heating a house (assuming

DEGREE DAYS AND DESIGN TEMPERATURES (HEATING SEASON)

State	City	Design Temperature (°F)	Degree Days (°F-days)
Alabama	Birmingham	19	2,600
Alaska	Anchorage	-25	10,800
Arizona	Phoenix	31	1,800
Arkansas	Little Rock	19	3,200
California	Los Angeles	41	2,000
California	San Francisco	35	3,000
Colorado	Denver	-2	6,200
Connecticut	Hartford	1	6,200
Florida	Tampa	36	600
Georgia	Atlanta	18	3,000
Idaho	Boise	4	5,800
Illinois	Chicago	-3	6,600
Indiana	Indianapolis	0	5,600
Iowa	Des Moines	-7	6,600
Kansas	Wichita	5	4,600
Kentucky	Louisville	8	4,600
Louisiana	New Orleans	32	1,400
Maryland	Baltimore	12	4,600
Massachusetts	Boston	6	5,600
Michigan	Detroit	4	6,200
Minnesota	Minneapolis	-14	8,400
Mississippi	Jackson	21	2,200
Missouri	St. Louis	4	5,000
Montana	Helena	-17	8,200
Nebraska	Lincoln	-4	5,800
Nevada	Reno	2	6,400
New Hampshire	Concord	-11	7,400
New Mexico	Albuquerque	14	4,400
New York	Buffalo	3	7,000
New York	New York	12	5,000
North Carolina	Raleigh	16	3,400
North Dakota	Bismark	-24	8,800
Ohio	Columbus	2	5,600
Oklahoma	Tulsa	12	3,800
Oregon	Portland	21	4,600
Pennsylvania	Philadelphia	11	4,400
Pennsylvania	Pittsburg	5	6,000
Rhode Island	Providence	6	6,000
South Carolina	Charleston	23	2,000
South Dakota	Sioux Falls	-14	7,800
Tennessee	Chattanooga	15	3,200
Texas	Dallas	19	2,400
Texas	San Antonio	25	1,600
Utah	Salt Lake City	5	6,000
Vermont	Burlington	-12	8,200
Virginia	Richmond	14	3,800
Washington	Seattle	28	5,200
West Virginia	Charleston	9	4,400
Wisconsin	Madison	-9	7,800
Wyoming	Cheyenne	-6	7,400

no free heat from the sun, people, or other sources such as lights and appliances) is the cost of providing the total number of Btu lost by the house over the course of the heating season. We commonly express the cost of heat in dollars per million Btu (MBtu). The actual cost of delivering heat includes the price of the fuel, the efficiency of heat delivery, and the number of Btu provided by that fuel. The chart "Cost per Million Btu of Energy" will aid you in comparing the actual costs of various heating alternatives. The "Heat Conduction Cost Chart" provided in Appendix 2 is an invaluable aid in calculating the seasonal heating costs of various building surfaces. This simplified procedure facilitates the study of alternative construction and the savings you can realize from added insulation.

This chart can also help you compare the energy costs of two different methods of insulation. In the previous example of the insulated and uninsulated stud walls, for example, the difference in U-value is 0.23 − 0.069 = 0.16. This *difference* can be run through the chart in the same way as done for a single U-value. Assuming 5,000 degree days, 100 square feet of wall surface, and $9 per million Btu, the savings in heat costs for *one* year is about $21, or more than the cost of the insulation.

The foregoing analysis includes only the effects of conduction heat loss. Radiation and convection heat losses are also quite significant and must be included in the calculations of building heat loss. A complete

Cost per Million Btu of Energy

This chart will help you to calculate the actual cost of providing the heat lost by a house. It converts the unit price of the energy source—whether gas, oil, or electricity—into the cost per million Btu produced inside the house.

The efficiency of heat delivery determines the percentage of available heat that actually is used inside the house. Typically, only 65 to 80 percent of the heat content of gas or oil ever gets into the house. The rest goes up the chimney. A poorly adjusted furnace loses another 5 to 10 percent more heat. By contrast, electrical resistance heating is 100 percent efficient—every kilowatt-hour you pay for is used to heat the house. But it takes about 11,000 Btu (from coal, oil or gas) at the power plant to generate 1 kilowatt-hour of electricity (the equivalent of 3412 Btu) for your home. This is why electric heat is so expensive—in reality, it's only about 30 percent efficient!

An example will guide you in the use of this chart:

1) *Find the point on the appropriate vertical scale that corresponds to the retail price of the fuel you are using—for example, oil at $1.00/gallon (or, equivalently, electricity at $0.025/kwh and gas at $0.74/therm);*
2) *Move right to find the retail cost for 1 million Btu of that fuel, or $7.40;*
3) *Continue right to intersect the oblique line representing the efficiency of heat delivery, or 60 percent in this case;*
4) *Drop down to find the actual cost per million Btu of heat produced, or $12.35.*

While this price may seem a little exaggerated (heating oil costs about 50 cents/gallon at this writing), it may be very realistic in the near future. And in some areas of the country, electric heat already costs this much per million Btu.

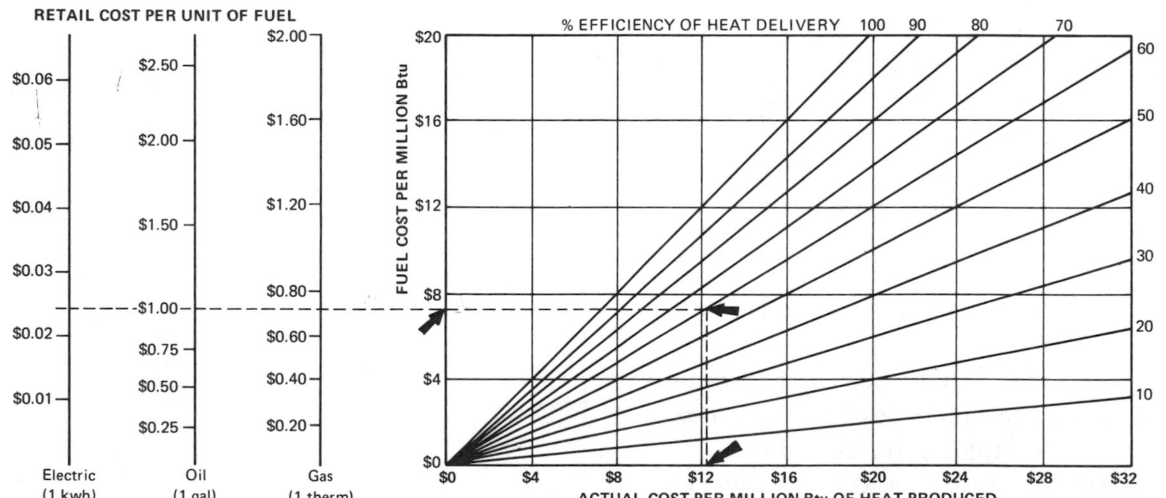

economic analysis, which is beyond the scope of this book, would also include long-term costs,predictions of future fuel supplies, and moral and social considerations about the use of non-renewable resources.

Convection Heat Loss

There are three modes of convection which influence the heat loss from a building. The first two have already been included in the calculation of conduction heat losses through the building skin. They are the convection heat flow across air gaps in the wall and heat flow to or from the walls through the surrounding air. These two effects have been included in the calculation of U-values by assigning insulating values to air gaps or air films. The third mode of convection heat flow is *air infiltration* through openings in walls (such as doors and windows) and through cracks around doors and windows. In a typical house, heat loss by air infiltration is often comparable to heat loss by conduction.

The first mode of convection heat loss occurs within the walls and between the layers of glass in the skin of the building. Wherever there is an air gap, and whenever there is a temperature difference between the opposing surfaces of that gap, natural air convection results in a heat flow across that gap. This process is not very efficient, and air gaps are considered to have a fair insulating value—though not a very large value compared to normal insulation materials. For the insulating value to be significant, the width of the air gap must be greater than ¾ inch. However, a quick glance at the insulating values of air gaps in Appendix 2 reveals that further increases in the width don't produce significant increases in insulation. Wider air gaps allow freer circulation of the air in the space, offsetting the potentially greater insulating value of the thicker air blanket. Most common forms of insulation do their job simply by trapping air in tiny spaces and preventing air circulation in the space they occupy. Fiberglass blanket insulation, polystyrene boards, cotton, feathers,

Natural convection in an air gap adds to the heat loss through a wall.

WINDOW

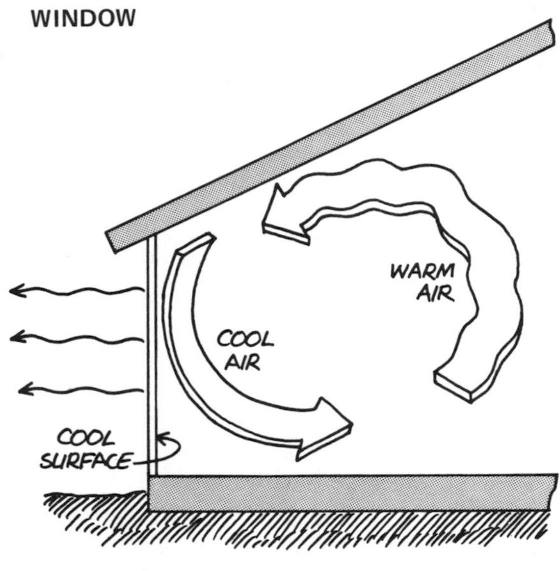

WELL-INSULATED WALL

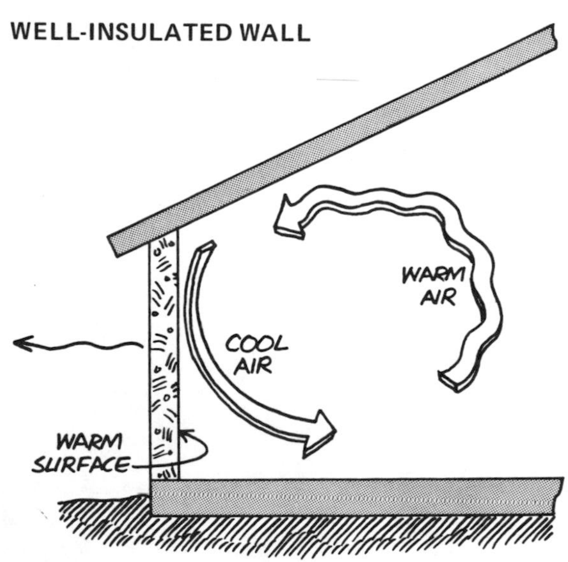

Air movements within a house.

crumpled newspaper, and even popcorn make good insulators because they create tiny air pockets to slow down the convection flow of heat.

Conduction heat flow through the exterior skin of a house works together with air movements within the rooms and winds across the exterior surface to siphon off even more heat. Since the interior surfaces of perimeter walls are usually cooler than the room air, they cool the air film right next to the wall. This cooled air sinks down and runs across the floor, while warmer air at the top of the room flows in to take its place, thereby accelerating the cooling of the entire room. The inside surface of a well-insulated wall will have about the same temperature as the room air. But the inside surface of a window will be much colder, and the air movement and cooling effects are severe. Heating units or warm air registers are commonly placed beneath windows in an effort to eliminate the cold draft coming down from the glass surfaces. While this practice improves the comfort of the living areas, it substantially increases heat losses to the outdoors.

Though not very large, the insulating value of the air films on either side of a wall or roof do make a contribution to the overall U-value. The air films on horizontal surfaces provide more insulation than those on vertical surfaces. Convection air flow, which reduces the effective thickness of the still air insulating film, is greater down a vertical wall than across a horizontal surface. Similarly, the air film on the outside surface is reduced by wind blowing across the surface. The heat that leaks through the wall is quickly transmitted to the moving air and carried away. The outer surface is cooled—drawing more heat through the wall. These heat losses can be reduced by wind screens or plantings that prevent fast-moving air from hitting the building skin.

Air infiltration heat losses through openings in buildings and through cracks around doors and windows are the primary convection losses that are not included in the calculations of conduction heat losses. Infiltration losses are not easy to calculate because they vary greatly with tightness of building construction and the weatherstripping detail of the windows, doors, and other openings. In the following calculations, we assume that the general wall construction is air-tight, that only the infiltration through windows and doors needs to be considered. You must take great care to insure that this assumption is valid. Small openings such as holes around outside electrical outlets or hose faucets can channel large amounts of cold air into the heated rooms. This cold air must be heated to room temperature, as must the air infiltrating around windows and doors.

The magnitude of air infiltration through cracks around doors and windows is somewhat predictable. It depends upon wind speeds and upon the linear footage of cracks around each window or door—usually the perimeter of the opening. If the seal between a window frame and the wall is not air-tight, you must also consider the length of this crack. From the table "Air Infiltration Through Windows," you can approximate the. volume of air leakage, Q, per hour per foot of crack. Next, determine the temperature difference, ΔT, between inside and outside. You can then determine the amount of heat required to warm this air to room temperature, ΔH_{inf}, according to

$$\Delta H_{inf} = c \times Q \times L \times h \times \Delta T$$

where $c = 0.018$ Btu/ft³/°F is the heat capacity of air, L is the total crack length in feet, and h is the time span in hours.

With 10 mph winds beating against an average double-hung, non-weatherstripped, wood-sash window, the air leakage is 21 cubic feet per hour for each foot of crack. Assuming the total crack length is 16 feet and the temperature is 65° F inside and 40° F outside, the total infiltration heat loss during an 8-hour time span is

$$\Delta H_{inf} = 0.018 \times 21 \times 16 \times 8 \times (65\text{-}40) = 1210 \text{ Btu}$$

If the same window is weatherstripped (Q = 13 instead of 21), then the infiltration heat loss is 749 Btu over the same time span. You can make a multitude of other comparisons using the Q-values given in the table.

Apply the above formula to the total crack length for each different type of crack leakage. The total crack length varies with room layout: for rooms with one exposure, use the entire measured crack length; for rooms with two or more exposures, use the length of crack in the wall having most of the cracks; but in no case use less than one-half of the total crack length.

You can also use this formula to calculate the heat loss through infiltration under the worst, or "design", conditions your house will undergo. For these conditions, use the outdoor design temperatures and average

AIR INFILTRATION THROUGH WINDOWS

Window Type	Remarks	Air Leakage (Q)[1] at Wind Velocity (mph) 5	10	15	20	25
Double-hung wood sash	Average fitted,[2] non-weather stripped	7	21	39	59	80
	Average fitted,[2] weatherstripped	4	13	24	36	49
	Poorly fitted,[3] non-weatherstripped	27	69	111	154	199
	Poorly fitted,[3] weatherstripped	6	19	34	51	71
Double-hung metal sash	Non-weatherstripped	20	47	74	104	137
	Weatherstripped	6	19	32	46	60
Rolled-section steel sash	Industrial pivoted[2]	52	108	176	244	304
	Residential casement[4]	14	32	52	76	100

[1] Air leakage, Q, is measured in cubic feet of air per foot of crack per hour.
[2] Crack = 1/16 inch.
[3] Crack = 3/32 inch.
[4] Crack = 1/32 inch.

SOURCE: ASHRAE, *Handbook of Fundamentals*, 1970.

wind speed for your area. Fortunately, the design temperature does not usually accompany the maximum wind speed. Average winter wind velocities are given for a number of localities in the *Climatic Atlas of the Unites States.*

The total seasonal heat loss through air infiltration is calculated by replacing $h \times \Delta T$ with the total number of degree hours, or 24 times the number of degree days:

$$\text{Seasonal } \Delta H_{inf} = 24 \times c \times Q \times L \times D$$

The "Air Infiltration Cost Chart" provided in Appendix 2 is a powerful design tool similar to the chart for heat conduction. You can use this chart to make quick evaluations of the yearly savings resulting from changes in window construction. For example, if a wooden double-hung window is weather-stripped, the air infiltration rate will change from 39 to 24 ft^3/hr/ft under 15 mph winds. By moving through the chart from a starting point of 15 ft^3/hr/ft, you will arrive at the savings resulting from the weatherstripping. Assuming 5000 degree days, 16 feet of crack length, and $9 per million Btu costs, we arrive at an immediate savings of about $6 per season. Since weatherstripping costs a few cents per foot, it can pay for itself in fuel savings in a few weeks.

Radiation Heat Flow

Radiation works together with conduction to accelerate the heat flow through walls, windows, and roofs. If surrounding terrain and vegetation are colder than the outside surfaces of your house, there will be a net flow of thermal radiation to these surroundings. Your roof will also radiate substantial amounts of energy to the cold night sky in winter. If the relative humidity is low, as much as 30 Btu per hour can be radiated to the sky per square foot of roof. This radiation can rapidly cool your roof surface to temperatures lower than the outside air temperature, thereby increasing the temperature difference across the roof section and the overall heat flow through the roof.

In summer, this radiative heat flow provides desirable nocturnal cooling–particularly in arid areas. Harold Hay's house in Atascadero, California, which will be discussed in Chapter 5, is the most significant solar house yet to make use of this phenomenon. Ray Bliss and Jonathan Hammond also used nocturnal radiation systems in solar buildings described in Chapter 2. In the winter, however, this nocturnal cooling is an *undesirable* effect. Well-insulated roofs are necessary to prevent excessive losses of heat.

If the interior surfaces of walls and windows are colder than the objects (and people!) inside a room, there will be a net flow of thermal radiation to these surfaces. A substantial flow of heat radiates to the inside surfaces of windows, which are much colder during winter than adjacent walls. This flow warms the inside surface of the glass, and more heat is pumped to the outside air because of the greater temperature difference across the glass. Insulated window shutters, discussed in the next chapter, can reduce this flow drastically.

In both examples above, radiation heat flow enhances the transfer of heat from warmer to cooler regions. Its effects are included in the calculation of conduction heat loss through surfaces of the house. But don't ignore radiation heat flow when taking preventive measures.

Seasonal and Design Heat Loads

The total heat escaping from a house is the sum of the conduction heat loss and the convection heat loss through air infiltration, because the effects of radiation heat flow have already been included in these two contributions. The total conduction heat loss is itself the sum of conduction losses through all the exterior surfaces, including walls, windows, floors, roofs, skylights, and doors. The total conduction heat loss is generally one to four times the total convection heat loss through air infiltration, which includes all convection heat losses through cracks in walls and around windows and doors.

For example, the total conduction heat losses from a typical poorly-insulated 1200-square-foot house in Oakland, California, may be 1000 Btu per hour per °F temperature difference between the inside and outside air, while the convection heat loss is only 250 Btu per hour per °F. If the temperature drops to

UNINSULATED ATTIC

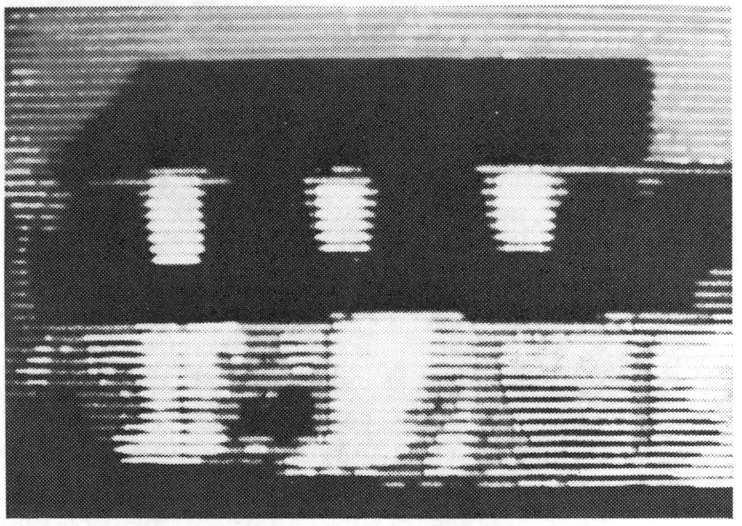

INSULATED ATTIC

Infrared photographs showing thermal radiation from a conventional house. Note that more heat escapes from an uninsulated attic than from an insulated one.

Heat Load Calculations

So far, you have learned to calculate the heat losses through the individual surfaces and cracks of a house. To calculate the overall heat loss (or heat load) of a house, you merely sum the losses through all surfaces and cracks. The heat load of a house depends on its construction and insulation and varies with the outside temperature and wind velocity.

To indicate just how bad things can get, let's use a drafty, uninsulated, wood-frame house as an example. Assume it's 40 feet long and 30 feet wide as shown in the plan diagram. It has uninsulated stud walls and a hardwood floor above a ventilated crawl space. The ceiling has acoustical tile but is otherwise uninsulated, and it sits below a low, pitched roof of plywood and asphalt shingles. The house has 8 single-pane, double-hung, wood-sash windows (each 4 feet high by 2.5 feet wide) and 2 solid oak doors (each 7 feet by 3 feet).

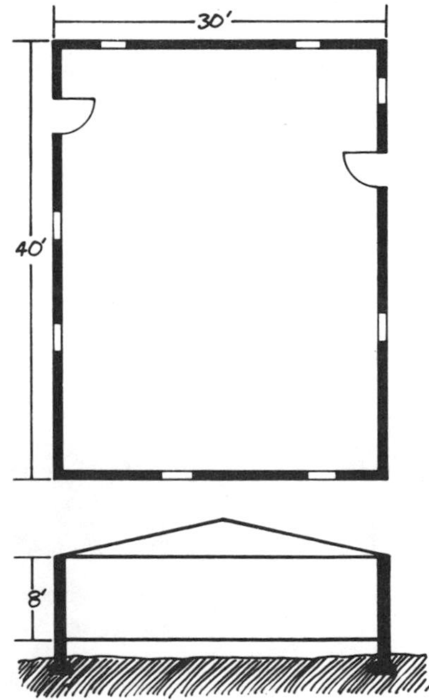

45°F on a typical winter night, the house loses a total of 1250 X (65-45) = 25,000 Btu per hour, assuming the indoor temperature is 65°F.

The design temperatures introduced earlier allow us to estimate the maximum likely heat loss from a house. The design temperature for a locality is the lowest outdoor temperature likely to occur during winter. Houses are often rated in their thermal performance by the number of Btu per hour that the heating system must produce to keep the building warm during these conditions. The design temperature for Oakland is 35°, so that 1250 X (65-35) = 37,500 Btu per hour is the *design heat load* that the heating system must be able to produce in the above house. The same house would have design heat loads of 62,500 Btu/hr in Chattanooga, Tennessee, where the design temperature is 15°F, and 100,000 Btu/hr in Sioux Falls, South Dakota, where the design temperature is −15°F. At $9 per million Btu, that's 90 cents an hour to heat this house in Sioux Falls, and one might be persuaded to add some insulation!

Degree day information allows us to calculate the amount of heat a house loses in a single heating season. The greater the number of degree days for a particular location, the greater the total heat lost from a house. Typical houses lose from 15,000 to 40,000 Btu per degree day, but energy conservation measures could cut these in half. Our house in Oakland loses 24 X 1250 = 30,000 Btu per degree day, for example. As there are 2800 degree days for that city, the total heat loss over an entire heating season is 84 million Btu (30,000 X 2800) or about 1200 therms of gas burned at 70 percent efficiency. But a therm of gas costs less than $0.20, and the seasonal heat cost is under $240 (1976) for this uninsulated California house. In most other regions of the country, where seasonal heat loads are much greater and energy costs higher, insulation is mandatory.

For even more detailed explanations of heat flows in a house, the reader is encouraged to consult the recent Sierra Club book, *Other Homes and Garbage.* In the chapter on solar energy of this very readable work, the conduction and air infiltration losses from a house are examined very closely. This book guides you step-by-step through the heat loss calculations for a number of sample California homes. Not so readable, but much more comprehensive, is the *Handbook of Fundamentals* published by the American Society of Heating, Refrigerating and Air-Conditioning Engineers (ASHRAE). Anyone interested in solar home design should get a copy—a good library should have one.

FURTHER READING

ASHRAE. *Handbook of Fundamentals.* New York: American Society of Heating, Refrigerating and Air Conditioning Engineers, 1972. (see also the 1967 edition)

Baldwin, J. L. *Climates of the United States,*

Washington: National Oceanic and Atmospheric Administration, 1973. (available for $5.25 from National Technical Information Service, U.S. Department of Commerce, Springfield, VA 22161)

Bennett, Iven. "Monthly Maps of Mean Daily Insolation in the United States." *Solar Energy,* Vol. 9, 1965.

Cramer, R. D. and L. W. Neubauer. "Diurnal Radiant Exchange with the Sky Dome." *Solar Energy,* Vol. 9, 1965.

Environmental Science Services Administration. *Climatic Atlas of the United States.* Washington: U.S. Department of Commerce, 1968. (available for $4.50 from Superintendent of Documents, U.S. government Printing Office, Washington, DC 20402)

Leckie, J. et al. *Other Homes and Garbage.* San Francisco: Sierra Club Books, 1975.

Liu, B.Y.H. and R. C. Jordan. "Availability of Solar Energy for Flat-Plate Solar Heat Collectors," in *Low Temperature Engineering Applications of Solar Energy.* New York: American Society of Heating, Refrigerating and Air Conditioning Engineers, 1967.

Moorcraft, Colin. "Solar Energy in Housing." *Architectural Design,* Vol 42, October 1973.

Rogers, Benjamin T. "Using Nature to Heat and Cool." *Building Systems Design,* Vol. 70, October-November 1973.

Severns, W. H. and J. R. Fellows. *Air Conditioning and Refrigeration.* New York: John Wiley and Sons, Inc., 1966.

Heat Load Calculations (continued)

First we need the U-values of each surface. From the "Sample Calculations of U-values" earlier in this chapter, we know that an uninsulated stud wall has a U-value of 0.23. From Appendix 2.1, we get U = 1.13 for single-pane windows, and R = 0.61 for a 1" oak door. Adding the resistance of the inside and outside air films, we get R_t = 1.69 or U = 1/1.69 = 0.59 for the doors.

The calculation of the U-values of the floor and ceiling is a bit more involved. The hardwood floor has 3 layers—interior hardwood finish, felt, and wood subfloor—and still air films above and below. The resistances of all 5 layers are added to give R_t = 2.94, or U = 1/2.94 = 0.34. However, about half the floor area is covered by carpets (R = 1.23 including the rubber pad), and this half has a U-value of 0.24. Finally, the total resistance of the ceiling and roof is the sum of the resistances of 8 different layers, including the acoustical tile, gypsumboard, attic space, plywood, building paper, asphalt shingles, and the inside and outside air films. These add to R_t = 4.12, and the U-value of the ceiling is U = 1/4.12 = 0.24.

For a 1°F temperature difference between indoor and outdoor air, the conduction heat loss through each surface is the product of the (area of the surface) times the (U-value of the surface). If the design temperature is 35°F, for example, we multiply by 30 (= 65 – 35) to get the design heat loss through that surface. The conduction heat losses through all surfaces are summarized in the table.

Infiltration heat losses are calculated using Q-values from the table "Air Infiltration Through Windows" on page 71. Double-hung poorly fitted, wood-sash windows have a Q-value of 111 in a 15 mph wind. Around poorly fitted doors, the infiltration rate is twice that: 220 ft³/hr for each crack foot. And there is still some infiltration through cracks around window and door frames as well—with a Q-value of 11.

These Q-values are then multiplied by the heat capacity of a cubic foot of air (0.018 Btu/ft³/°F) and the total length of each type of crack to get the infiltration heat loss. Only windows and doors on 2 sides of the house (that is, 4 windows and 1 door) are used to get total crack lengths. The infiltration heat losses through all cracks are also summarized in the table.

In a 15 mph wind, the conduction heat loss of this house is 981 Btu/hr for a 1°F temperature difference between indoor and outdoor air. Under the same conditions, the infiltration loss is 244 Btu/hr—or a total heat load of 1225 Btu/hr/°F. Over an entire day, the house loses 24 (hours) times 1225 (Btu per hour) for each 1°F temperature difference, or 29,400 Btu per degree day. Under design conditions of 35°F and a 15 mph wind, the heat load of this house is 36,732 Btu/hr (= 29,421 + 7311). The furnace has to crank out almost 37,000 Btu/hr to keep this house comfy during such times.

Surface:	Area (ft²)	U-value (Btu/hr/ft²/°F)	Conduction Heat Losses:* 1°F temp diff (Btu/hr/°F)	35°F outside (Btu/hr)
Walls	998	0.23	230	6,886
Windows	80	1.13	90	2,712
Doors	42	0.59	25	743
Bare Floor	600	0.34	204	6,120
Carpeted Floors	600	0.24	144	4,320
Ceiling	1200	0.24	288	8,640
Total Conduction Heat Losses			981	29,421

Crack Around:	Length (ft)	Q-value (ft³/hr/ft)	Infiltration Heat Losses:* 1°F temp diff (Btu/hr/°F)	35°F outside (Btu/hr)
Window sash	64	111	128	3,836
Door	20	220	79	2,376
Window & door frames	84	11	17	499
Other	---	------	20	600
Total Infiltration Heat Losses			244	7,311

*NOTE: All calculations assume 15 mph wind.

4
Direct Solar Heating

Designing a House to Collect and Store the Sun's Heat

Mimbres spirits guard the entrance to an Anasazi kiva.

As the position of the heavens with regard to a given tract on the earth leads naturally to different characteristics, owing to the inclination of the circle of the zodiac and the course of the sun, it is obvious that designs for houses ought similarly to conform to the nature of the country and the diversities of climate.

Vitruvius,
Ten Books on Architecture

Energy conservation is essential to good shelter design. Only the house that loses heat begrudgingly can use sunlight to make up most of these losses. Some people might think it rather dull to let sunlight in through the windows and keep it there, but others of us delight in the simplicity of this approach. In fact, conserving the sun's energy can often be more challenging than inventing elaborate systems to capture it.

Nature uses simple designs to compensate for the irregularities of solar radiation and temperature. Many flowers open and close with the rising and setting sun. Many animals find shelters to shield themselves from intense summer heat, and bury themselves in the earth to stay warm during the winter.

Primitive peoples took a hint or two from nature in order to design shelters and clothing. But as we have learned to protect ourselves from the elements, we have lost much of this intuitive understanding and appreciation of natural phenomena. We rely more on technology than nature and the two are often in direct conflict.

The earth's heat storage capacity and an atmospheric greenhouse effect help to moderate temperatures at ground level. These temperatures fluctuate somewhat, but the large heat storage capacity prevents the earth's cooling off too greatly at night and its heating up too much during the day. The atmosphere retards the escape of heat by thermal radiation, slowing the cooling process even further. Because of these phenomena, afternoon temperatures are warmer than morning, and summer temperatures reach their peak in July and August.

A shelter design should reflect similar principles. Weather variations from one hour to the next or from cold night hours to warm daytime hours should not affect a shelter's internal climate. Ideally, not even the wide extremes of summer and winter would affect it. There are countless examples of indigenous architecture based on these criteria. Perhaps the most familiar of these is the heavy adobe home of the Pueblo Indian. The thick walls of hardened clay absorb the sun's heat during the day and prevent it from penetrating the interior of the home. At night, the stored heat continues its migration into the interior, warming it as the temperatures of the desert night plummet. The coolness of the night air is then stored in the walls and keeps the home cool during the hot day. In many climates houses made of stone, concrete, or similar heavy materials will perform in a like fashion.

A house should moderate extremes of temperature. In winter it absorbs solar heat by day and retains it at night; in summer, the house excludes heat by day and releases it at night.

A shelter should moderate extremes of temperature that occur both daily *and* seasonally. Caves, for example, have relatively constant temperatures and humidities year round. Likewise, you can protect a house from seasonal temperature variations by piling earth against the outside walls or molding the structure of the house to the side of a hill.

On sunny winter days, you should be able to open a house up to the sun's heat. At night, you should be able to close out the cold and keep this heat in. In the summer, you should be able to do just the opposite: during the day close it off to the sun, but at night open it up in order to release heat into the cool night air.

A house is a solar collector, or at least should be designed on the principle of one. A solar collector gathers heat when the sun is shining and stores it for later use. It also ceases to operate when the sun is not shining or when enough heat has already been stored. During the summer, when the house does not need heat, the collector might operate in reverse, cooling the house instead of heating it.

The best way to use the sun for heating is to have the house collect the sun's energy itself, without burdening it with a solar collector. To achieve this, a house must meet three basic requirements:

- *The house must be a solar collector*

  It must let the sunlight in when it needs heat and keep the sunlight out when it doesn't; it must also let coolness in as needed. These feats may be accomplished by orienting and designing the house so as to let the sun penetrate through the walls and windows during the winter and by using shading to keep it out during the summer.

- *The house must be a heat storehouse*
  It must store the heat for times when the sun isn't shining. Houses built of heavy materials such as stone and concrete do this best.
- *The house must be a heat trap*
  It must make good use of the heat (or coolness) and let it escape only very slowly. This is done by using insulation, weatherstripping, shutters, and storm windows.

Each of these requirements will be discussed further in the next three sections.

In addition to heating and cooling, many other processes and factors determine the energy needed in a building: the ventilation and movement of air; the purification of that air; the control of humidity; and the use of electricity for lighting, pumps, fans, controls, and other mechanical equipment. Some of the physiological factors affecting human comfort are:

- production and regulation of heat in the body
- moisture losses from the body
- radiant heat exchange with cold or hot surfaces
- conductive and convective heat exchange with the room air

As designers and builders try to insure human comfort by influencing these factors, they must be careful to use the earth's resources as wisely as possible. The remainder of this chapter examines direct methods of solar heating that can decrease your dependence upon complex technologies which require continual expenditures of resources. Some discussion of cooling methods is also included.

THE HOUSE AS A SOLAR COLLECTOR

The best way of using the sun's energy to heat a house is to let it penetrate directly through the roof, walls, and windows. You should attempt to maximize your heat gain from insolation during cold periods, and minimize it during hot weather. You can do this by altering the color of your house, its orientation and shape, the placement of windows, and the use of shading, both natural and artificial.

Customarily, solar heat gains have not entered into the computation of seasonal heating supply or demand. Unfortunately, most research done on solar gain applies to hot weather conditions and to reducing the energy required for cooling. The data which apply to heating are difficult to understand and difficult to use in building design. This chapter is an attempt to translate these data into useful design tools, but more extensive work in this area would help us to lower our energy needs.

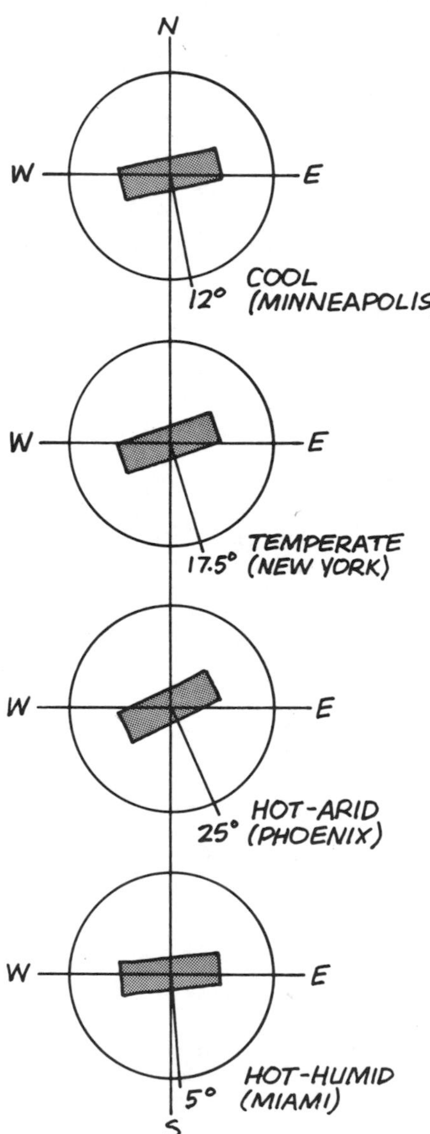

Optimum house orientations for four different U.S. climates.

Orientation and Shape

Since solar radiation strikes differently oriented surfaces with varying intensity, a house will benefit if its walls and roofs are oriented to receive this heat in the winter and shed it in the summer. After much detailed study of this matter, a number of researchers have reached the same conclusion that primitive peoples have always known: The principal facade of a house should face within 30° of due south (between south-southeast and south-southwest), with due south being preferred. With this orientation of the house, the south-facing walls can absorb the most radiation from the low winter sun, while the roofs, which can reject excess heat most easily, catch the brunt of the intense summer sun.

In his *Design With Climate,* however, Victor Olgyay cautions against generalizing to all building locations. He promotes the use of "sol-air temperatures" to determine the optimal orientation. These temperatures recognize that solar radiation and outdoor air temperatures act *together* to influence the overall heat gain through the surfaces of a building. Because the outdoor air temperatures are lower in the morning and peak in the mid-afternoon, he suggests that a house be oriented somewhat east of due south in order to take advantage of the early morning sun when heat is most needed. In the summer, the principal heat gain comes in the afternoon, from the west and southwest, so the house should face *away* from this direction to minimize the solar heat gain in that season. Depending upon the relative needs for heating and cooling, as well as upon other factors such as the winds, the optimum orientation will vary for different regions and building sites. The accompanying diagram gives the best orientations for four typical U.S. climate zones, as determined by Olgyay's sol-air approach.

A house also benefits in solar heat gain because of different ratios of its length to its width to its height. The ideal shape loses the minimum amount of heat and gains the maximum amount of insolation in the winter, and does just the reverse in the summer. Olgyay has noted that:

- In the upper latitudes (greater than 40°N), south sides of houses receive nearly twice as much solar radiation in winter as in summer. East and west sides receive 2½ times more in summer than they do in winter.
- At lower latitudes (less than 35°N) houses gain even more on their south sides in the winter than in the summer. East and west walls can gain two to three times more heat in summer than the south walls.
- The square house is not the optimum form in any location.
- All shapes elongated on the north-south axis work with less efficiency than the square house in both winter and summer.
- The optimum shape in every case is a form elongated along the east-

west direction.

Of course, other factors must influence the shape of a house, including the demands of the site and the needs of the inhabitants. But energy conservation can often be successfully integrated with these factors.

The relative insolation for houses with various shapes, sizes, and orientations can be a very useful aid at the design stage, particularly for placement of the windows. The first chart lists the relative insolation for different combinations of house shape, orientation, and floor and wall area. Values in this chart are for January 21, and are based on the next chart—Solar Heat Gain Factors for 40°N latitude. The ASHRAE *Handbook of Fundamentals* and the Koolshade Corporation provide similar information for many other latitudes. These factors represent the clear day solar heat gain through a single layer of clear, double-strength glass. But they can be used to estimate the insolation on the walls of a house.

From the relative solar insolation data, you may note that a house with its long axis oriented east-west has the greatest potential for total solar heat gain, significantly greater than that for a house oriented north-south. The poorest shape is the square oriented NNE-SSW or ENE-WSW. In doubling the ground floor area, the optimal east-west gain increases by about 40 percent because the perimeter increases by 40 percent. If you doubled the floor area of a house by adding a second floor, the wall area would double as would the total solar insolation.

FACADE ORIENTATIONS		INSOLATION ON WALL (Btu/day)				
		a	b	c	d	Total
N (0°)	A	118	508	1630	508	2764
	B	84	722	1160	722	2668
	C	168	361	2320	361	3210
	DOUBLE B	118	1016	1630	1016	3780
	DOUBLE C	236	508	3260	508	4612
22½°	A	123	828	1490	265	2406
	B	87	1180	1060	376	2703
	C	174	590	2120	188	3072
	DOUBLE B	123	1656	1490	530	3799
	DOUBLE C	246	828	2980	265	4319
45°	A	127	1174	1174	127	2602
	B	90	1670	835	180	2775
	C	180	835	1670	90	2775
	DOUBLE B	127	2348	1174	254	3903
	DOUBLE C	254	1174	2348	127	3903
67½°	A	265	1490	828	123	2406
	B	188	2120	590	174	3072
	C	376	1060	1180	87	2703
	DOUBLE B	265	2980	828	246	4319
	DOUBLE C	530	1490	1656	123	3799

Relative insolation on houses of different shape and orientation – January 21, 40°N Latitude. Listed values represent the insolation on a hypothetical house with w = 1 square foot. To get the daily insolation on a house of similar shape with w = 100 square feet, multiply these numbers by 100.

This study does not account for the color of the walls, the solar impact on the roof, the variations in window location and sizes,

SOLAR HEAT GAIN FACTORS FOR 40°N LATITUDE, WHOLE DAY TOTALS Btu/ft^2/day (Values for 21st of each month)												
	Jan	Feb	Mar	Apr	May	Jun	Jul	Aug	Sep	Oct	Nov	Dec
N	118	162	224	306	406	**484***	422	322	232	166	122	98
NNE	123	200	300	400	550	**700***	550	400	300	200	123	100
NE	127	225	422	654	813	**894***	821	656	416	226	132	103
ENE	265	439	691	911	1043	**1108***	1041	903	666	431	260	205
E	508	715	961	1115	1173	**1200*†**	1163	1090	920	694	504	430
ESE	828	1011	1182	**1218*†**	**1191†**	1179	**1175†**	**1188†**	1131	971	815	748
SE	1174	1285	**1318***	1199	1068	1007	1047	1163	1266	1234	1151	1104
SSE	1490	**1509***	1376	1081	848	761	831	1049	1326	1454	1462	1430
S	**1630*†**	**1626†**	**1384†**	978	712	622	694	942	**1344†**	**1566†**	**1596†**	**1482†**
SSW	1490	**1509***	1370	1081	848	761	831	1049	1326	1454	1462	1430
SW	1174	1285	**1318***	1199	1068	1007	1047	1163	1266	1234	1151	1104
WSW	828	1011	1182	**1218*†**	**1191†**	1179	**1175†**	**1188†**	1131	971	815	748
W	508	715	961	1115	1173	**1200*†**	1163	1090	920	694	504	430
WNW	265	439	691	911	1043	**1108***	1041	903	666	431	260	205
NW	127	225	422	658	813	**894***	821	656	416	226	132	103
NNW	123	200	300	400	550	**700***	550	400	300	200	123	100
HOR	706	1092	1528	1924	2166	**2242***	2148	1890	1476	1070	706	564

*month of highest gain for given orientation(s)

†orientation(s) of highest gain in given month

SOURCE: ASHRAE, *Handbook of Fundamentals,* 1970; Koolshade Corporation.

or the effects of heat loss. A detailed analysis would also include the actual weather conditions. However, this study does produce relative values to enable you to make preliminary choices.

Color

The color of the roofs and walls strongly affects the amount of heat which penetrates the house, since dark colors absorb much more sunlight than light colors do. Color is particularly important when little or no insulation is used, but it has less effect as the insulation is increased. Ideally, you should paint your house with a substance that turns black in winter and white in summer. But no such substance yet exists and you will have to be content with simpler methods. In warm and hot climates, the exterior surfaces on which the sun shines during the summer should be light in color. In cool and cold climates, use dark surfaces facing the sun to increase the solar heat gain.

Two properties of surface materials, their *absorptance* α and *emittance* ϵ, can help you to estimate their radiative heat transfer qualities. The absorptance of a surface is a measure of its tendency to absorb sunlight. The emittance gauges its propensity to emit thermal radiation. These properties are explained further in the accompanying material and sample values of α and ϵ are available in the table.

Substances with large values of α are good absorbers of sunlight; those with large values of ϵ are good emitters of thermal radiation. Substances with a small value of α, particularly those with a small value of α/ϵ like white paint, are good for surfaces that will be exposed to the hot summer sun (your roof and east and west walls, for example). Those that have a large value of α, particularly those with large α/ϵ like black paint, are good for south-facing surfaces, which you want to absorb as much winter sunlight as possible.

Windows

Although the color, orientation, and shape of the house are important, the most significant factors in capturing the sun's energy are size and placement of windows. Openings in shelters are the origin of present day windows; they were used for the passage of people and possessions, and for natural ventilation and lighting. These openings also allowed the people to escape from indoor drudgeries by gazing off into sylvan surroundings. But the openings also had their discomforts and inconveniences. Animals and insects had free access, the inside temperature was difficult to regulate, and humidity and air cleanliness could not be controlled.

Although glass has been dated as early as 2300 BC, its use in windows did not occur until about the time of Christ. And only in the present century have the production and use of glass panes larger than eight or twelve inches on a side become possible. As the technology and economics improve, glass is replacing the traditional masonry or wood exterior wall. But the design problems accompanying this substitution have often been ignored or underrated.

Besides reducing the amount of electricity needed for lighting, glass exposed to sunlight captures heat through the greenhouse effect explained earlier. Glass readily transmits the short-wave visible radiation, but does not transmit the long-wave thermal radiation

Absorptance, Reflectance, and Emittance

Sunlight striking a surface is either absorbed or reflected. The absorptance α of the surface is the ratio of the solar energy absorbed to the solar energy striking that surface:

$$\alpha = \frac{I_a}{I} = \frac{\text{absorbed solar energy}}{\text{incident solar energy}}$$

A hypothetical "blackbody" has an absorptance of 1—it absorbs all the radiation hitting it, and would be totally black to our eyes.

But all real substances reflect some portion of the sunlight hitting them—even if only a few percent. The reflectance ρ of a surface is the ratio of solar energy reflected to that striking it:

$$\rho = \frac{I_r}{I} = \frac{\text{reflected solar energy}}{\text{incident solar energy}}$$

A hypothetical blackbody has a reflectance of 0. The sum of α and ρ is always 1.

All warm bodies emit thermal radiation—some better than others. The emittance ϵ of a material is the ratio of thermal energy being radiated by that material to the thermal energy radiated by a blackbody at that same temperature:

$$\epsilon = \frac{R}{R_b} = \frac{\text{radiation from material}}{\text{radiation from blackbody}}$$

Therefore, a blackbody has an emittance of 1.

The possible values of α, ρ, and ϵ lie in a range from 0 to 1. Values for a few common surface materials are listed in the accompanying table. More extensive listings can be found in Appendix 2, under "Absorptances and Emittances of Materials."

	α	ρ	ϵ	α/ϵ
White Plaster	0.07	0.93	0.91	0.08
Fresh Snow	0.13	0.87	0.82	0.16
White Paint	0.20	0.80	0.91	0.22
White Enamel	0.35	0.65	0.90	0.39
Green Paint	0.50	0.50	0.90	0.56
Red Brick	0.55	0.45	0.92	0.60
Concrete	0.60	0.40	0.88	0.68
Grey Paint	0.75	0.25	0.95	0.79
Red Paint	0.74	0.26	0.90	0.82
Dry Sand	0.82	0.18	0.90	0.91
Green Roll Roofing	0.88	0.12	0.94	0.94
Water	0.94	0.06	0.96	0.98
Black Tar Paper	0.93	0.07	0.93	1.00
Flat Black Paint	0.96	0.04	0.88	1.09
Granite	0.55	0.45	0.44	1.25
Graphite	0.78	0.22	0.41	1.90
Aluminum Foil	0.15	0.85	0.05	3.00
Galvanized Steel	0.65	0.35	0.13	5.00

The values listed in this table (and those given in Appendix 2) will help you compare the response of various materials and surfaces to solar and thermal radiation. For example, flat black paint (with α = 0.96) will absorb 96% of the incoming sunlight. But green paint (with α = 0.50) will absorb only 50%. Both paints (with emittances of 0.88 and 0.90) emit thermal radiation at about the same rate if they are at the same temperature. Thus, black paint (with a higher value of α/ϵ) is a better absorber of sunlight and will become hotter when exposed to the sun.

SOLAR BENEFIT VALUES				
City	Percent possible sunshine	Average temperature during heating season—°F	Net heat gain, Btu/hr/ft² through: Single glass	Double glass
Albany, N.Y.	46	35.2	-12.8	5.6
Albuquerque, N.M.	77	47.0	18.0	30.2
Atlanta, Ga.	52	51.5	9.0	18.8
Baltimore, Md.	55	43.8	2.0	15.9
Birmingham, Ala.	51	53.8	10.9	19.5
Bismarck, N.D.	55	24.6	-20.1	4.0
Boise, Id.	54	45.2	22.9	16.0
Boston, Mass.	54	38.1	5.2	11.7
Burlington, Va.	42	31.5	-19.5	.9
Chattanooga, Tenn.	50	49.8	5.9	16.7
Cheyenne, Wyo.	67	41.3	5.7	20.9
Cleveland, Ohio	41	37.2	-13.7	3.7
Columbia, S.C.	51	54.0	11.2	19.6
Concord, N.H.	52	33.3	-12.0	7.4
Dallas, Texas	47	52.5	7.1	16.4
Davenport, Iowa	54	40.0	-3.1	12.8
Denver, Colo.	70	38.9	5.2	21.7
Detroit, Mich.	43	35.8	14.1	44.0
Eugene, Ore.	44	50.2	2.7	13.2
Harrisburg, Pa.	50	43.6	-1.5	12.5
Hartford, Conn.	53	42.8	-.3	14.1
Helena, Mont.	52	40.7	-3.3	12.2
Huron, S.D.	58	28.2	-14.1	8.0
Indianapolis, Ind.	51	40.3	-4.6	11.2
Jacksonville, Fla.	40	62.0	13.9	18.1
Joliet, Ill.	53	40.8	2.9	12.8
Lincoln, Neb.	61	37.0	-2.2	15.3
Little Rock, Ark.	51	51.6	8.5	18.3
Louisville, Ky.	51	45.3	1.5	14.6
Madison, Wis.	50	37.8	-7.6	9.5
Minneapolis, Minn.	53	29.4	-15.7	5.8
Newark, N.J.	55	43.4	1.4	15.5
New Orleans, La.	37	61.6	11.7	16.1
Phoenix, Ariz.	59	59.5	21.9	27.5
Portland, Me.	52	33.8	-7.2	12.0
Providence, R.I.	54	37.2	-6.1	11.3
Raleigh, N.C.	57	50.0	-10.0	20.6
Reno, Nev.	64	45.4	8.6	21.7
Richmond, Va.	59	47.0	8.0	20.2
St. Louis, Mo.	57	43.6	2.6	16.6
Salt Lake City, Utah	59	40.0	0.0	15.9
San Francisco, Cal.	62	54.2	17.3	25.7
Seattle, Wash.	34	46.3	-7.3	5.2
Topeka, Kan.	61	42.3	3.8	18.4
Tulsa, Okla.	56	48.2	7.4	19.0
Vicksburg, Miss.	45	56.8	-10.7	17.7
Wheeling, W.Va.	41	46.1	3.7	9.0
Wilmington, Del.	56	45.0	3.7	16.9

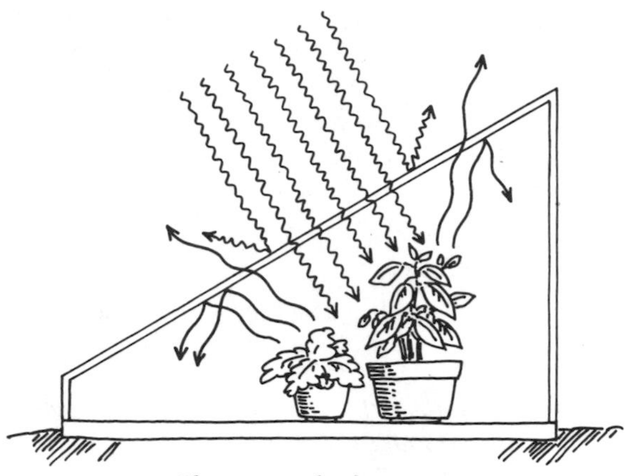

Glass as a solar heat trap.

emitted after the light changes to heat when it hits an interior surface. Almost all this thermal radiation is absorbed in the glass and a substantial part of it is returned to the interior space.

Experimental houses were built in the 1930's and 40's with the major parts of south-facing walls made entirely of glass. The most extensive work with these "solar houses" was done by F. W. Hutchinson at Purdue University. In 1945, under a grant from Libbey-Owens-Ford Glass Company, he built the two nearly identical houses described in Chapter 2. Based on the performance of these two houses, Hutchinson reported that "the available solar gain for double windows in south walls in most cities in the U.S. is more than sufficient to offset the excess transmission loss through the glass."

Hutchinson also concluded that more than twice as much solar energy is transmitted through south-facing windows in winter than in summer. If the windows are shaded in summer, the difference is even greater. For a fixed latitude, the solar intensity does not vary strongly with the outside air temperature, but heat loss does. Consequently, the use of glass has greater potential for reducing winter heating demand in mild climates than in cold climates.

The table of "Solar Benefit Values" gives us plenty of evidence for this potential. Many of the cities studied showed net energy gains through single glass (a negative number represents a net loss), and all 48 cities studied showed net gains through double glass. The losses through single glass in some cities should be compared to the heat loss through a typical wall that the glass replaces.

There are a number of reasons that the quantity of solar energy which gets through a south window on a sunny day in the winter is *more* than that which is received through that same window on a sunny day in summer. Among them are:

- There are more hours when the sun shines directly on a south window in winter than in summer. At 40°N latitude, for example, there are 14 hours of possible sunshine on July 21, but the sun remains north of east until 8:00 a.m. and goes to north of west at 4:00 p.m., so that direct sunshine occurs for only eight hours on the south wall. But on January 21, the sun is shining on the south wall for the full ten hours that it is above the horizon.

- The intensity of sunlight hitting a surface perpendicular to the sun's rays is about the same in summer and winter. The extra distance that the rays must travel through the atmosphere in the winter is offset by the sun's being closer to the earth in that season.
- Since the sun is closer to the southern horizon during the winter, the rays strike the windows more perpendicularly than they do in the summer when the sun is higher in the sky. At 40°N latitude, two hundred Btu strikes a square foot of window surface during an average hour on a sunny winter day, whereas 100 Btu is typical for an average summer hour.
- The more nearly the sun's rays hit the window at right angles, the more energy transmitted. Such is the case in winter, and most of the sunlight striking a south window makes it inside.

In addition to these effects, the diffuse radiation from the winter sky is double that from the summer sky.

The *type* of glass you use can have a significant effect on energy gains and losses. All types of glass—clear, heat-absorbing, or reflecting—lose about the same amount of heat by conduction. On the other hand, there is a great difference in the amount of solar heat which is transmitted through these three types of glass. The accompanying diagrams will give you an idea of the net

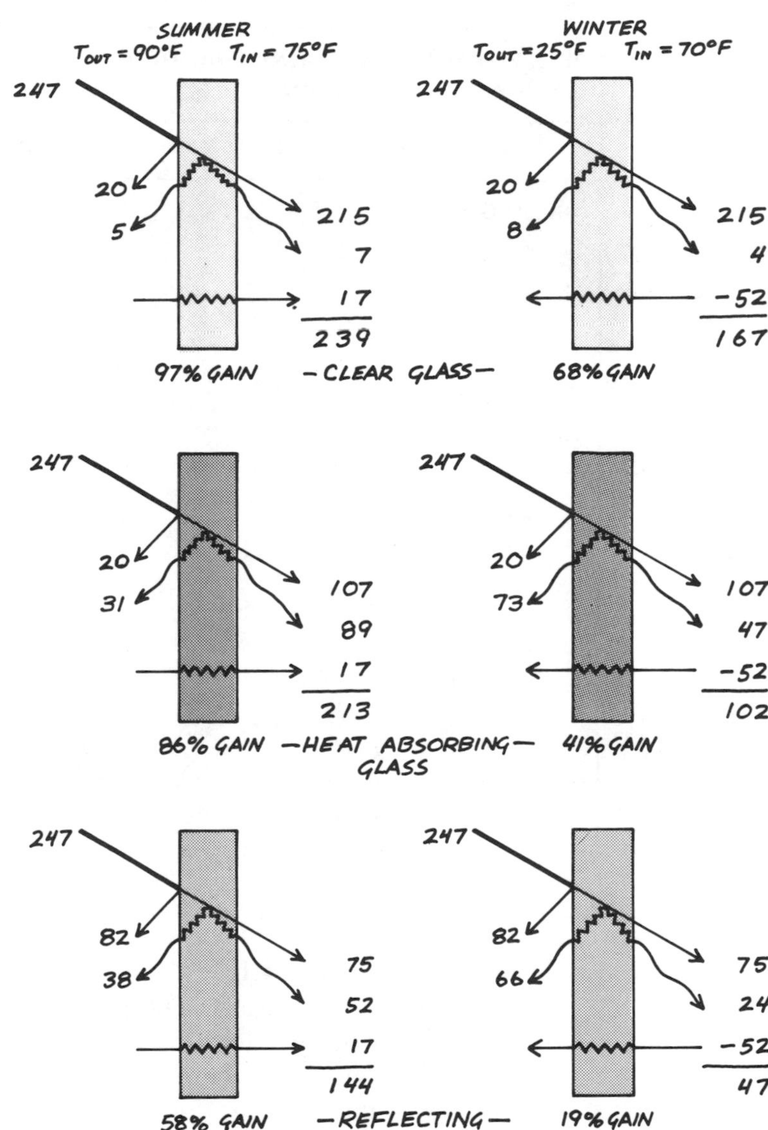

Solar heat gains through different types of single glass. Listed values are in Btu per hour.

PERCENTAGE HEAT GAINS THROUGH VARIOUS TYPES AND COMBINATIONS OF GLASS		
Glass Type	**Summer**	**Winter**
Single Glazing		
Clear	97	68
Heat-absorbing[1]	86	41
Reflective[2]	58	19
Double Glazing		
Clear outside Clear inside	83	68
Clear outside Heat-absorbing inside	74	52
Clear outside Reflective inside	50	42
Heat-absorbing outside Clear inside	42	28
Reflective outside Heat absorbing inside	31	17

[1] Shading coefficient = 0.50
[2] Shading coefficient = 0.35

heat gains for various combinations of single and double glass. The percentage of solar heat gain includes a contribution from heat conduction through the glass. The heat gains are approximate for the sunny day conditions shown, and no attempt has been made to account for the differing solar angles in summer and winter.

The percentage summer and winter heat gains of these combinations of glass are summarized in the table. To reduce summer heat gain, you might use reflecting glass on the outside and clear glass on the inside of two-pane windows facing into the sun. Unfortunately, this combination drastically reduces the winter heat gain, and is not recommended for south-facing glass. Two clear panes of glass is generally the best for these windows. You can replace the outer pane in summer with a single pane of reflecting glass. Or, use shading—both natural and artificial—to keep out the hot summer sun.

In many climates, keeping the sunshine out during warm weather is very important to human comfort. In such areas, the use of heat-absorbing or reflecting glass is one alternative, especially for the east and west sides. The important factors to consider in the use of specialized glass bear repeating:

- Such glass *does* reduce solar heat gain, which can be more of a disadvantage in the winter than an advantage in the summer
- Except for glare control, reflecting and heat-absorbing glass is almost always unnecessary on north, north-northeast, and north-northwest orientations
- In latitudes south of 40°N, heat absorbing and reflecting glass should not be considered for south-facing windows
- The use of vegetation or movable shading devices is a more sensible solu-

tion than the use of heat-absorbing or reflecting glass for south, southeast, and southwest orientations

The four (or more) sides of a building need not, and in fact should not, be identical in appearance. Substantial savings in heating and cooling costs will result from the use of well-insulated walls on the north, east and west. The few windows needed on these sides of the house for lighting and outdoor views should use the glazing methods advocated here. In most areas of the United States, double clear glass windows on the south sides provide the optimum winter heat gain.

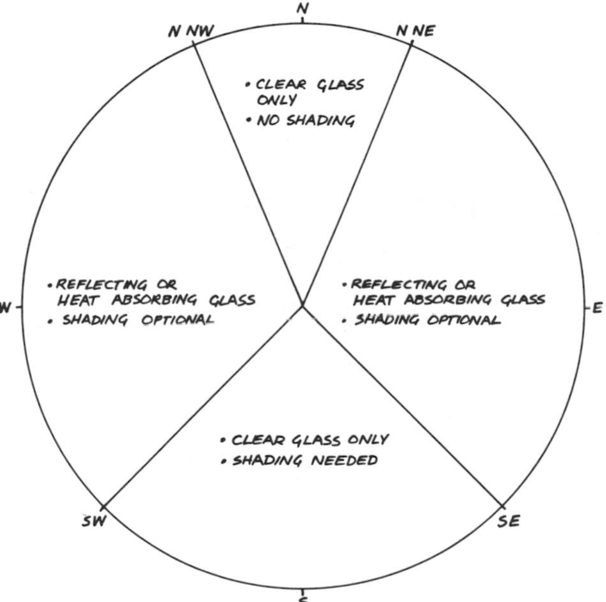

Different glass types are recommended for various window orientations. Use of shading can be helpful where indicated.

Shading

Through the intelligent use of shading, you can minimize the summer heat gain through your windows. Perhaps the simplest and most effective methods of shading use devices that are exterior to the house, such as overhangs or awnings. One difficulty with fixed overhangs is that the amount of shading follows the seasons of the sun rather than the climatic seasons. The middle of the summer for the sun is June 21, but the hottest times occur from the end of July to the middle of August. A fixed overhang designed for optimal shading on August 10 causes the same shadow on May 1. The overhang designed for optimal shading on September 21, when the weather is still somewhat warm and solar heat gain is unwelcome, causes the same shading situation on March 21—when the weather is cooler and solar heat gain is most welcome.

Vegetation, which follows the climatic seasons quite closely, can provide better shading the year round. On March 21, for example, there are no leaves on most plants, and sunlight will pass readily. On September 21, however, the leaves are still full, providing the necessary shading. Placement of deciduous trees directly in front of south-facing windows can provide shade from the intense midday summer sun. Even better is an over-

Shading a south window with a fixed overhang (at solar noon).

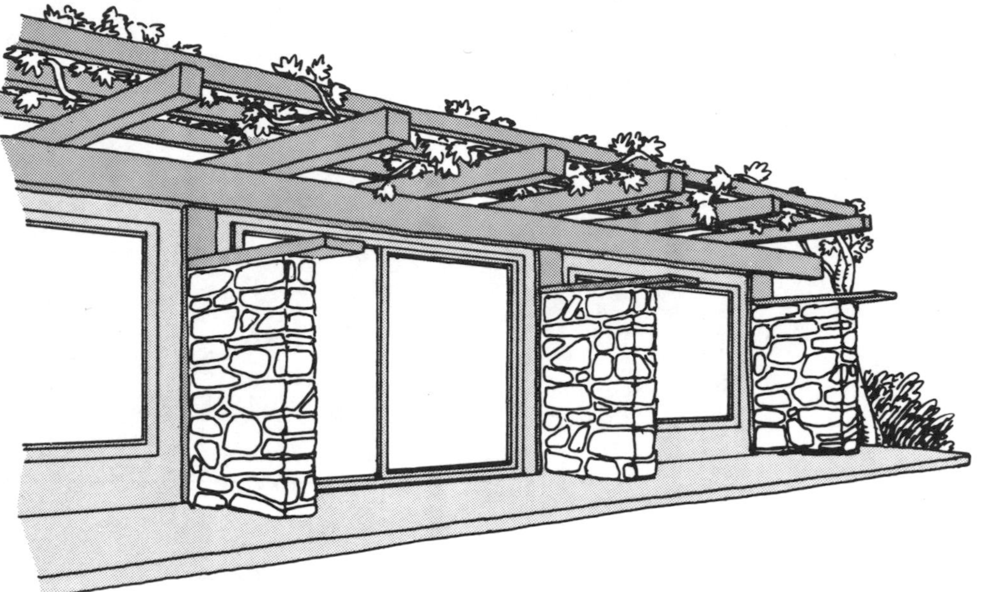

Shading a south window with an overhanging trellis.

Sun angles in the evening and morning.

hanging trellis with a climbing vine that sheds its leaves in winter.

Movable shading devices are even more amenable to human comfort needs than fixed overhangs or vegetation, but they have their own problems. Movable shading devices on the outsides of buildings are difficult to maintain and deteriorate rapidly. Efforts have been made to design them with long lifetimes, but success in these endeavors is unusual. Awnings are perhaps the simplest and most reliable movable shading devices, but their aesthetic appeal is limited. The requirement for frequent human intervention is often seen as a drawback, but perhaps designers should encourage people to participate more in providing a comfortable home for themselves.

Operable shading placed between two layers of glass is not as effective as an exterior device, but it is still more effective than an interior shading device. Venetian blinds between glass panes are often expensive, and difficult to clean or repair. Interior shading devices, such as roller shades and draperies, give the least effective shading but offer versatile operation by the people inside. And they do keep direct sunlight from bleaching the colors of walls, furniture, and floors.

East- and west-facing glass is extremely difficult to shade because the sun is low in the sky both early morning and late afternoon. Overhangs do not prevent the penetration of the sun during the summer much more than they do during the winter. Vertical louvers or extensions are probably the best means of shading such glass, but you might consider reflecting and heat-absorbing glass. For this purpose, you should be familiar with the values of the *shading coefficient* of the various glasses. A single layer of clear, double-strength glass has a shading coefficient of 1. The shading coefficient for any other glazing system, in combination with shading devices, is the ratio of the solar heat gain through that system to the solar heat gain through the double-strength glass. Solar heat gain through a glazing system is therefore the product of its shading coefficient times the Solar Heat Gain Factors listed on page 82 (and in the ASHRAE *Handbook of Fundamentals)* for clear, double-strength glass.

Sun Path Diagrams

It is usually necessary to describe the position of the sun in order to determine the size of a window shading device. Earlier, in Chapter 3, we described the sun's path in terms of the solar altitude angle θ and the azimuth angle ϕ. These can be determined for the 21st day of any month by using tables, or can be calculated directly from formulas. Another method for determining solar altitude and azimuth for the 21st day of each month is the use of *sun path diagrams*. A different diagram is required for each latitude, although interpolation between graphs is reasonably accurate. Diagrams for 24°N, 28°N, 32°N, 36°N, 40°N, 44°N, 48°N, and 52°N are provided in Appendix 1.

SHADING COEFFICIENTS FOR VARIOUS SHADING CONDITIONS

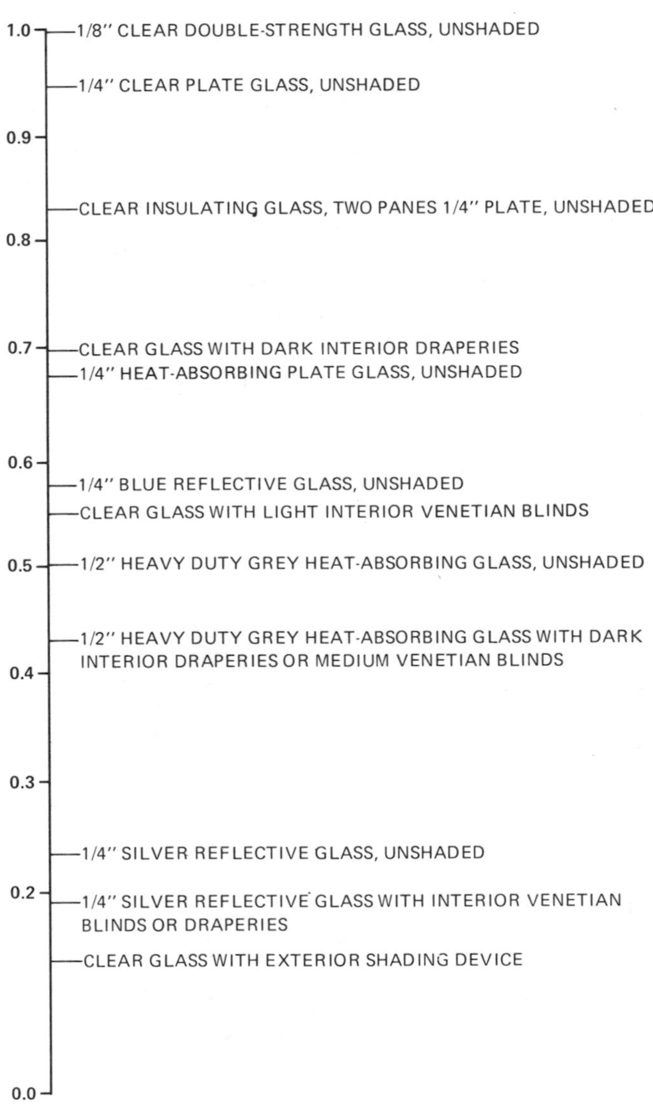

Use of Sun Path Diagrams

A sun path diagram is a projection of the sky vault, just as a world map is a projection of the globe. The paths of the sun across the sky are recorded as lines superimposed on a grid that represents the solar angles. Sun path diagrams can be used to determine these angles for any date and time. Different sun path diagrams are needed for different latitudes.

EXAMPLE: Find the solar altitude and azimuth angles at 4:00 p.m. on April 21 in Philadelphia (40°N). First locate the April line–the dark line running left to right numbered "IV" for the fourth month; and the 4:00 p.m. line–the dark line running vertically numbered "4". The intersection of these lines indicates the solar position at that time and day. Solar altitude is read from the concentric circles–in this case it's 30°. The solar azimuth is read from the radial lines–in this case it's 80° west of true south. If you trust your judgement, you can also use these diagrams to give you the solar positions on days other than the 21st of each month.

The shading mask protractor provided here will help you to construct masks for any shading situation. First determine the shading angle of the horizontal overhang or vertical fins, as shown on the opposite page. For a horizontal overhang, find the arc corresponding to the angle a in the lower half of the shading mask protractor. All the area above that arc is the segmental shading mask for that overhang. For vertical fins, find the radial lines corresponding to the shading angle b in the upper half of the shading mask protractor. All the area below these lines is the radial shading mask for those fins.

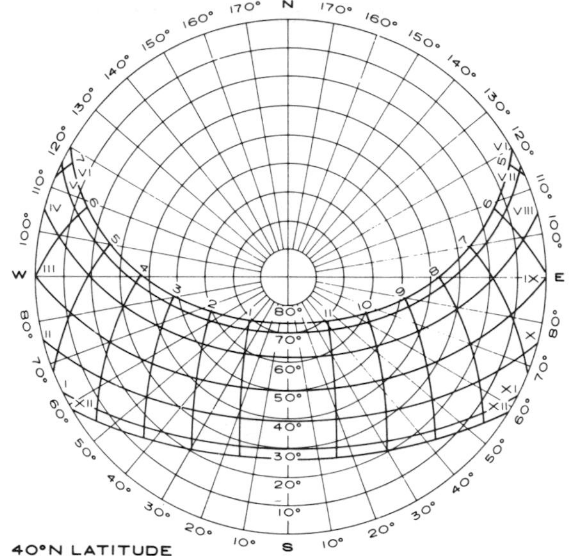

40°N LATITUDE

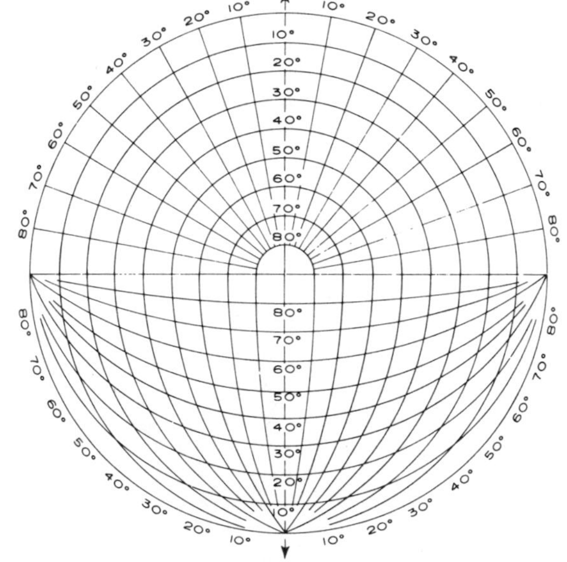

SHADING MASK PROTRACTOR

SOURCE: Ramsey and Sleeper, *Architectural Graphic Standards*, 1972.

The 40°N diagram is reprinted here for your convenience, along with an example of its use.

You can also use sun path diagrams to determine the effects of shading devices. There are two basic categories of shading–a horizontal overhang above the window or vertical fins to the sides. As shown in the diagram opposite, the shading angles *a* and *b* of these two basic obstructions are the two important variables available to the designer. The broader the overhang or fin, the larger the corresponding angle. Each basic shading device determines a specific *shading mask.* A horizontal overhang determines a "segmental" shading mask while vertical fins determine a "radial" one. These shading masks are constructed with the help of the shading mask protractor provided here.

These masks can then be superimposed upon the appropriate sun path diagram for your latitude to determine the amount of shading on a window. Those parts of the diagram that are covered by the shading mask indicate the months of the year (and the times of day) when the window will be in shade.

Sun path diagrams and the shading mask protractor can also be used to *design* shading devices. If you specify the times of year that shading is needed and plot these on the appropriate sun path diagram, you have determined the shading mask for your desired conditions. The shading angles *a* and *b* can be read from this mask using the shading mask protractor. From these angles you can then figure the dimensions of the appropriate shading devices.

Sun path diagrams, shading masks and the shading mask protractor are quick, convenient devices for organizing the rather complicated geometries of solar angles. For more detailed advice on their use, consult *Solar Control and Shading Devices,* by Aladar and Victor Olgyay, or *Architectural Graphic Standards,* by C. G. Ramsey and H. R. Sleeper.

THE HOUSE AS A HEAT STOREHOUSE

A vital component in a solar heated house is a "container" for storing heat. When the house is used as the solar collector, it too needs a method of "soaking up" or storing heat so that it doesn't become too hot when the sun shines upon it and retains some of this heat for use when the sun is not shining. One of the major problems Hutchinson encountered in his "solar house" was the wide fluctuation in temperature. From the discussion of the Purdue House in Chapter 2, you may recall that the inside temperature reached 80°F on January 15, while it plummeted to 25°F that very night. This drop was partly traceable to the lack of insulation on the windows at night, but more massive walls and interiors would have solved much of his problem.

Heat Storage in a House

Probably the most efficient heat storage container is the material of which the house

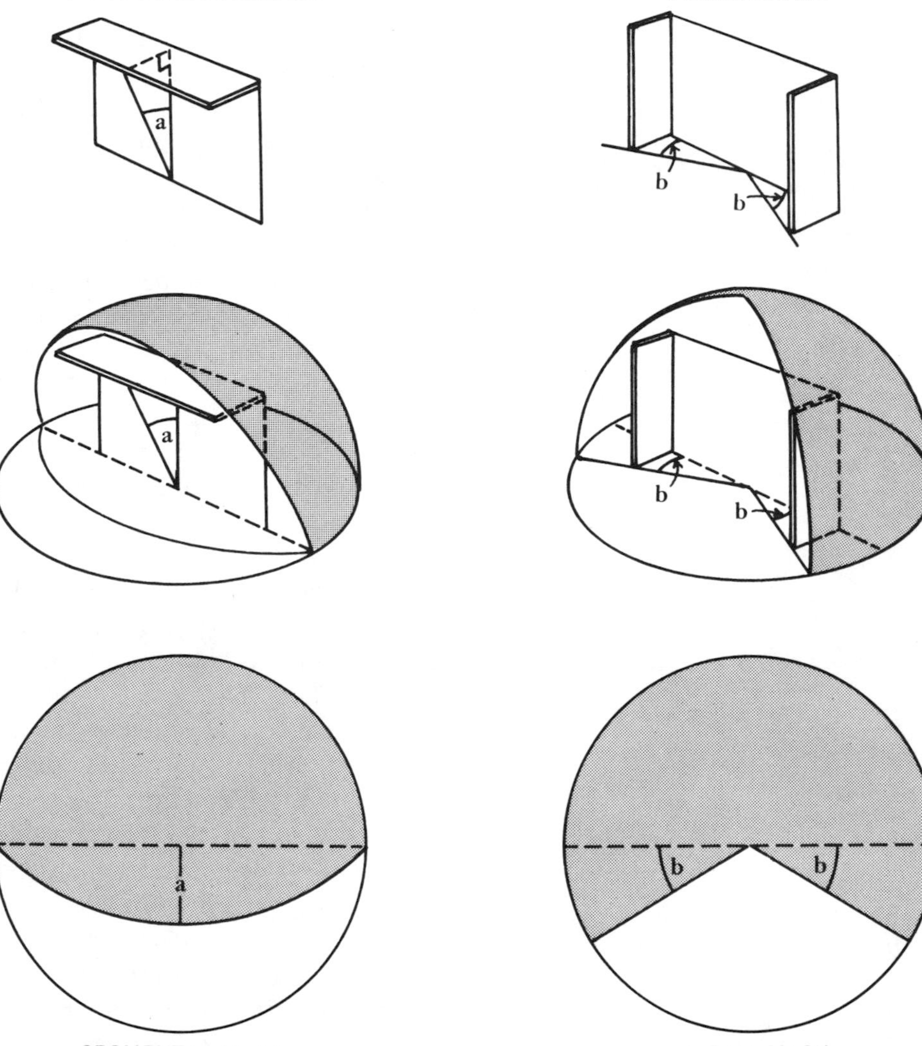

Determining masks for horizontal and vertical shading obstructions. Use the shading mask protractor to convert a particular shading angle into the corresponding mask.

itself is built—the walls, floors, and roofs. All materials absorb and store heat as they are warmed. For example, water or stone will absorb more heat for a fixed temperature rise than straw or wood. Heavy materials can store large quantities of heat without becoming too hot. When conditions around them become cooler, the stored heat is released and the materials themselves cool.

This *heat storage capacity* of various materials can be used to store the sun's heat for later use. As discussed in the previous section, solar energy penetrates through walls, roofs, and windows to the interior of a house. This solar heat is absorbed in the air and surrounding materials. The air in the house is likely to heat up first. It then distributes this heat to the rest of the materials. If they have already reached the temperature of the air or cannot absorb the heat quickly, the air continues to warm and overheats. The greater the heat storage capacity of the materials in the house, the longer it will take for the air to reach uncomfortable temperatures and the more heat that can be stored inside the house.

If it is cold outside when the sun sets, the house begins losing heat through its exterior skin, even if it is well insulated. To maintain comfortable temperatures, this heat must be replaced. In houses which have not stored much solar heat during the day, auxiliary heating devices must provide this heat.

If the interiors are massive enough, however, and the solar energy has been allowed to penetrate and warm them during the day, the house can be heated by the sun, even at night. As the inside air cools, the warmed materials replace this lost heat, keeping the rooms warm and cozy. Depending upon the heat storage capacities of the inside materials, the amount of solar energy penetrating into the house, and the heat loss of the house, temperatures can remain comfortable for many hours. Really massive houses will stay warm for a few days without needing extra heat from fires or furnaces.

During the summer, a massive house can also store coolness during the night for use during the hot day. At night, when outside air is cooler than it is during the day, ventilation of that air into the house will cool the air and all of the materials inside. Since they will be cool at the beginning of the next day, they will be able to absorb and store more heat before they become warm—thereby cooling the indoor air as they absorb heat from it. Thus, if the materials are cool in the morning, it will be a long time before they have warmed to the point that additional cooling is needed to remove the excess heat.

Temperature Fluctuations of Houses

The effects of varying outdoor temperatures upon the indoor temperatures can be very different for different types of houses. The first graph shows the effects of a sharp drop in outdoor temperature on the indoor temperatures of three types of houses. In a lightweight house, such as one with wood-frame construction, the temperature drops relatively quickly, *even if the house is well-insulated.* A massive structure built of concrete, brick, or stone maintains its temperature over a longer period of time if it is insulated on the outside of the walls. The heavy materials which store most of the heat are poor insulators, and they must be located within the confines of the insulation. A massive house set into the side of a hill or covered with earth has an even slower response to a drop in the outdoor air temperature. Ideally, the interior concrete or stone walls in this house are insulated from the earth by rigid board insulation, such as polystyrene or polyurethane. One or two walls can be exposed to the outside air and still the temperature will drop very slowly to a temperature close to that of the earth.

The second graph shows the effects of a sharp rise in outdoor temperature on the same three houses. Again, the lightweight house responds quickly to the change in outdoor temperature; in spite of being well-insulated, its temperature rises quickly. The heavy houses, however, absorb the heat and delay the temperature rise. The house set into a hill or covered with earth has the longest time delay in its response to the outdoor air change; if properly designed, it will never become too warm.

The effects of alternately rising and falling outdoor air temperatures on indoor air temperatures are illustrated in the third graph. Without any sources of internal heat, the inside air temperature of the lightweight house fluctuates widely, while that of the earth-embedded house remains almost constant near the temperature of the earth. We say that massive houses, whose indoor temperatures do not respond quickly to fluctuations of outdoor temperature, have a large *thermal mass,* or *thermal inertia.*

If a house responds slowly to outdoor temperature fluctuations, you don't need heavy duty auxiliary equipment to keep the place comfy. Although the furnace in a lightweight, uninsulated wood-frame house might not be used at all on a cold, sunny day, it might have to labor at full throttle to keep the house warm at night. The massive earth-embedded house, on the other hand, averages the outdoor temperature fluctuations over a span of several days—or even weeks. A bantamweight heating system (such as a wood stove, for example) could operate constantly to assure an even comfort level throughout the house.

Heat Storage Capacities of Materials

All materials vary in their heat storage capacity, or their ability to store heat. One measure of this ability is the *specific heat* of a material, which is the number of Btu required to raise one pound of the material 1°F. For example, water has a specific heat of 1.0, which means that one Btu is required to raise the temperature of one pound of water 1°F. As one gallon of water weighs

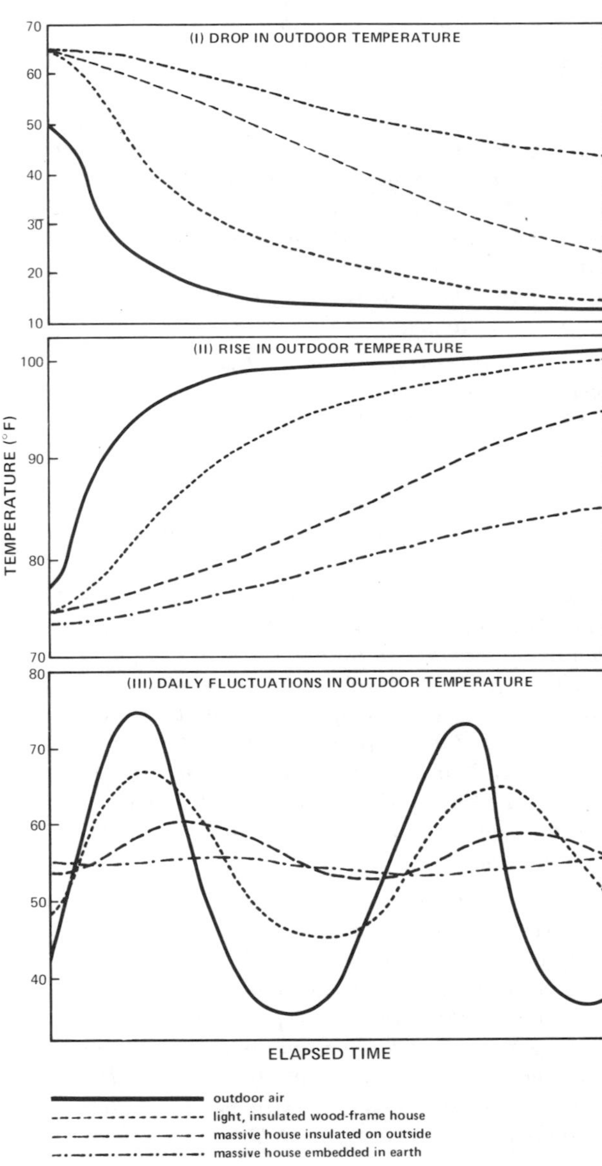

Effects of changes in outdoor air temperature on the indoor air temperatures of various houses.

Storing Heat in a Concrete Slab

Consider a 20 by 40 foot house with a well-insulated concrete slab 9 inches thick. By 5 p.m. on January 21, the slab has warmed up to 75°F from sunlight flooding in the south windows. From that time until early the next morning, the outdoor temperature averages 25°F, while the indoor air averages 65°F. If the house is well-insulated and loses heat at a rate of 300 Btu/hr/°F, and there is no source of auxiliary heat, what is the temperature of the slab at 9 o'clock the next morning?

Well, the total heat lost from the house during that period is the product of the (rate of heat loss) times the (number of hours) times the (average temperature difference between the indoor and outdoor air), or:

$$\Delta H = 300 \times 16 \times (65 - 25) = 192{,}000 \text{ Btu.}$$

With a total volume of 600 cubic feet (20 × 40 × ¾) and a heat capacity of 32 Btu/ft³/°F, the concrete slab stores 19,200 Btu for a 1°F rise in its temperature (600 ft³ × 32 Btu/ft³/°F = 19,200 Btu/°F). For a 1°F drop in its temperature, the slab releases the same 19,200 Btu. If the slab drops 10°F, from 75°F to 65°F, it will release just enough heat to replace that lost by the house during the night. So the slab drops to a temperature of 65°F by 9:00 the next morning.

In reality, things are a bit more complicated. But this exercise helps to give a rough idea of how much heat you can store in a concrete slab. If the house has 200 square feet of south windows, and a solar heat gain of 1000 Btu/ft² is typical for a sunny January day, the slab can store the 200,000 Btu of solar energy with a temperature rise of about 10°F. The stored solar heat is then released at night to keep the house warm as the inhabitants slumber.

SPECIFIC HEATS AND HEAT CAPACITIES OF COMMON MATERIALS

Material	Specific Heat (Btu/lb/°F)	Density (lb/ft³)	Heat Capacity (Btu/ft³/°F)
Water (40°F)	1.00	62.5	62.5
Steel	0.12	489	58.7
Cast Iron	0.12	450	54.0
Copper	0.092	556	51.2
Aluminum	0.214	171	36.6
Basalt	0.20	180	36.0
Marble	0.21	162	34.0
Concrete	0.22	144	31.7
Asphalt	0.22	132	29.0
Ice (32°F)	0.487	57.5	28.0
Glass	0.18	154	27.7
White Oak	0.57	47	26.8
Brick	0.20	123	24.6
Limestone	0.217	103	22.4
Gypsum	0.26	78	20.3
Sand	0.191	94.6	18.1
White Pine	0.67	27	18.1
White Fir	0.65	27	17.6
Clay	0.22	63	13.9
Asbestos Wool	0.20	36	7.2
Glass Wool	0.157	3.25	0.51
Air (75°F)	0.24	0.075	0.018

8.4 pounds, it requires 8.4 Btu to raise a gallon of water 1°F.

Different materials absorb different amounts of heat while undergoing the same temperature rise. While it takes 100 Btu to heat 100 pounds of water 1°F, it takes only 22.5 Btu to heat 100 pounds of aluminum 1°F. The specific heat of aluminum is therefore 0.225.

The specific heats of various building materials and other common materials found inside buildings are listed in the accompanying table. The *heat capacity,* or the amount of heat needed to raise one cubic foot of the material 1°F, is also listed along with the density of each material. Although the specific heat of concrete, for example, is only ¼ that of water, its heat capacity is more than half that of water. The density of concrete compensates somewhat for its low specific heat, and concrete stores relatively large amounts of heat per unit volume.

To store large amounts of heat in a given volume of floor or wall, use materials which have high heat capacities—those near the top of the table. As heat storage devices, concrete or stone walls insulated on the *outside* are superior to wood-framed walls having a plywood exterior and a gypsum wallboard interior with fiberglass insulation stuffed between them. You can use the heat capacities listed in this table and in Appendix 2 to calculate the total heat storage capacity, or thermal mass, of your home. The larger its thermal mass, the slower its response to outdoor temperature fluctuations.

Building With Thermal Mass

Thermal mass is one of the most underrated aspects of current building practice. Unfortunately, heavy buildings are hardly the favorite children of architects and building contractors. The so-called creative designers are trying to "do more with less," and the architectural genius is the person who uses the least material to enclose a space. Such thinking usually treats notions of durability and energy conservation with a shabbiness that would make Machiavelli blush. The visual weight of buildings is the important aesthetic consideration and it's trendy to design and build a structure that appears to be light in weight. Perhaps todays designers would be more prone to design massive buildings if the popularity of concrete didn't soar and plummet with a predictability akin to the New York Stock Exchange. Cost, availability, ease of handling, and *weight* are among the factors which influence its popularity.

However, the task of adding thermal mass need not be very difficult or costly, particularly at the scale of a single family home. Placing containers of water within the building confines, especially in front of a window, is a simple but excellent solution—a fortunate thing indeed for tropical fish enthusiasts! Massive fireplaces, interior partitions of brick or adobe, and even several inches of concrete or brick on the floor can greatly increase the thermal mass of a house.

Putting insulation on the outside of a house is not standard construction practice and involves some new problems. Insulation has customarily been placed between the inner and outer surfaces of a wall, or inside the house itself. Insulation on the outside of a concrete or masonry wall requires protection from the weather and contact with people or animals.

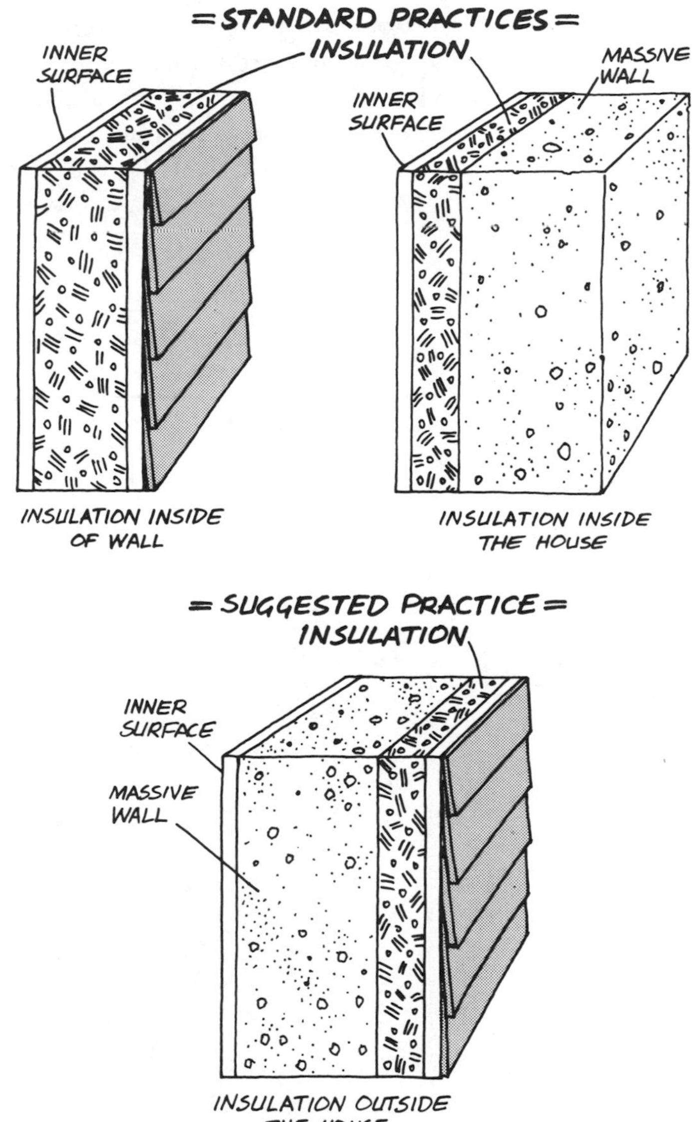

Insulation placed on the outside of a masonry wall increases thermal mass. Conventional insulation methods do not.

In the example below, 3 inches of rigid polystyrene board covers the outside surface of a poured concrete foundation. Above the surface of the ground, this insulation must be protected from rain, physical abuse, and solar radiation—particularly the ultra-violet rays. Below ground level, it must be protected from the unmerciful attacks of moisture and vermin. The polystyrene can be placed inside the formwork before the concrete is poured, and the bond between the two materials is extremely strong. But it is still difficult to protect the insulation above ground level. One alternative is to plaster the insulation with a "cementitious" material such as fiberglass-reinforced mortar. Another alternative is to fasten a rigid sheet of moisture-treated plywood or cement-asbestos board onto the insulation.

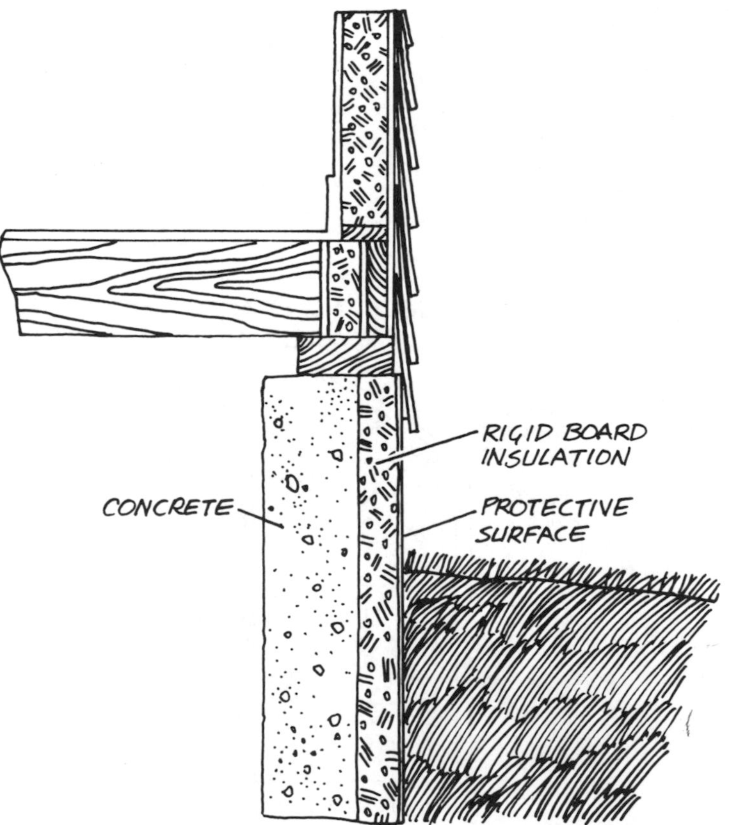

Insulation on the exterior of a house must be protected from weather and vermin.

Lamentably, the effect of thermal mass on the overall energy consumption of houses is rarely understood or used as a design tool in the same way as other options such as insulation and storm windows. Victor Olgyay, in *Design With Climate,* and Baruch Givoni in *Man, Climate and Architecture* give more insight into this topic. These other options for reducing energy consumption are discussed in the following section.

THE HOUSE AS A HEAT TRAP

If you design a house to collect and store solar heat, you should also take steps to retard the escape of that heat. The escape of heat from a house during winter is usually called its "heat loss." In addition, houses also absorb heat through their walls and windows during summer – their "solar heat gain." Retarding this movement of heat both into and out of a house is the essence of energy conservation in housing design.

Fortunately, most efforts to reduce winter heat loss will also reduce summer heat gain.

Energy conservation goes hand in hand with the use of solar energy for heating and cooling houses. By decreasing the demand for heat, you reduce the required size and cost of a solar heating installation and its backup system. This is true whether you use direct solar heating methods, integrated systems, or indirect systems.

Lowering the thermostat is the easiest way of reducing your winter heating costs (but perhaps the most difficult for many of us to accept). As explained in the last chapter, the heat loss through walls and windows is proportional to the difference between indoor and outdoor temperatures. Reducing this difference can definitely reduce your heat loss. You can do this without undue discomfort by wearing more and heavier clothing while awake, or by using more blankets while sleeping. The accompanying table shows that lowering the thermostat at night *does* save energy. A nightly 10°F setback reduces energy consumption by at least 10 percent in every city listed.

There are so many ways to save energy in a house that discussing them all would require an entire volume. What I will discuss here are the fundamental methods of keeping heat inside a house. These methods include insulation, choices of windows, window shutters, entrances, weatherstripping, and wind protection.

There are three primary ways that heat escapes from a house:

- by conduction through walls, roofs, and floors
- by conduction through windows and doors
- by convection of air through openings in the exterior surface

As discussed earlier, conduction works together with radiation and convection—both within the walls and floors and at the inner and outer wall surfaces—to produce the overall heat flow. The third mode of heat loss includes air infiltration through open windows, doors, or vents, and through cracks in the skin of the house or around windows and doors.

Depending upon the insulation of the house, the number and placement of windows, and the movement of air within, each of these three modes can contribute 20 to 50 percent of the total heat loss. If the total heat loss is divided evenly among these modes, and any one mode is reduced by half, the total heat loss is reduced by only one sixth. Clearly, you should attack *all three modes* of heat loss with the same vigor if you want to achieve the best results. Focusing your efforts on only one aspect of heat loss can only result in incomplete solutions.

Insulation

The only way to reduce the heat loss through air-tight walls, floors, and roofs is to add more resistance to this heat flow.

PERCENT FUEL SAVINGS WITH NIGHT THERMOSTAT SETBACK FROM 75°F
8 Hour Setback—10 p.m. to 6 a.m.

City	Setback		
	5°F	7½°F	10°F
Atlanta	11	13	15
Boston	7	9	11
Buffalo	6	8	10
Chicago	7	9	11
Cincinnati	8	10	12
Cleveland	8	10	12
Dallas	11	13	15
Denver	7	9	11
Des Moines	7	9	11
Detroit	7	9	11
Kansas City	8	10	12
Los Angeles	12	14	16
Louisville	9	11	13
Milwaukee	6	8	10
Minneapolis	8	10	12
New York City	8	10	12
Omaha	7	9	11
Philadelphia	8	10	12
Pittsburgh	7	9	11
Portland	9	11	13
Salt Lake City	7	9	11
San Francisco	10	12	14
St. Louis	8	10	12
Seattle	8	10	12
Washington, D.C.	9	11	13

SOURCE: Minneapolis-Honeywell Data, 1973.

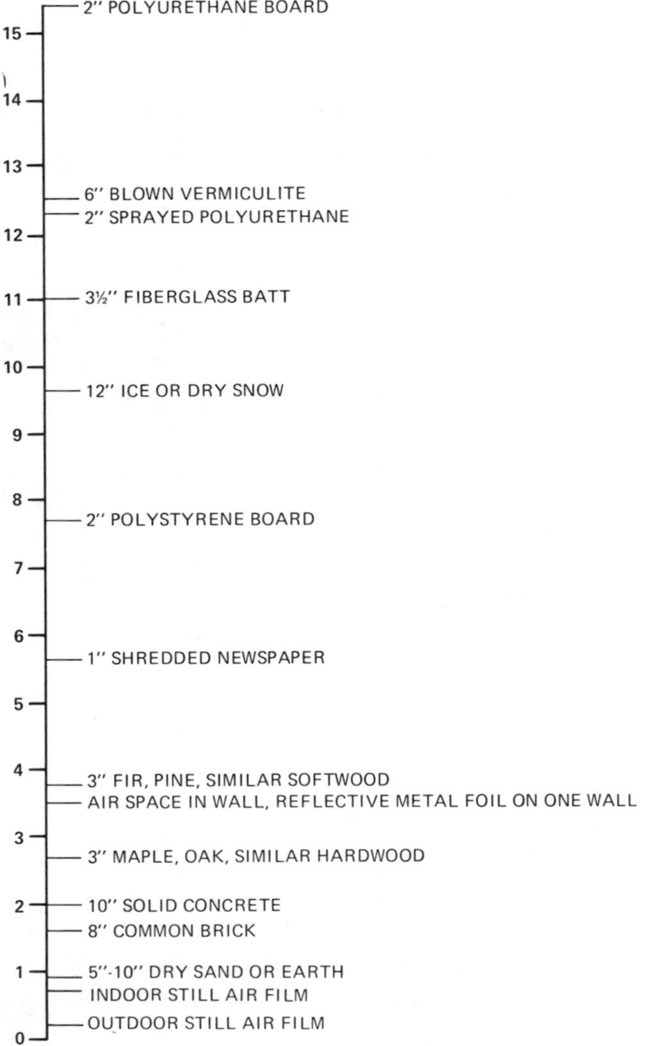

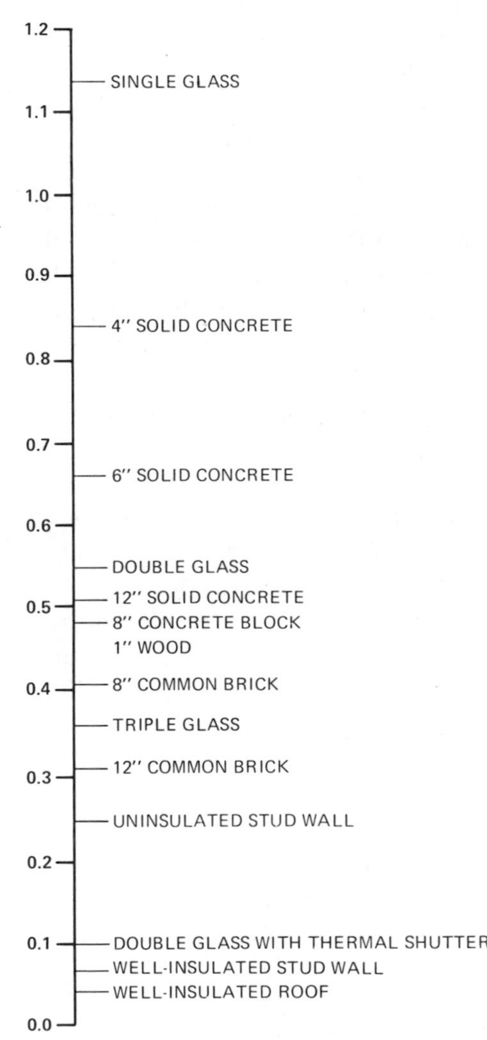

Insulation retards the flow of heat, keeping the interior surfaces warmer in winter and cooler in summer. Because of radiation heat transfer from your body to the walls (which can be 8°F to 14°F colder than the room air during winter), you can feel cold and uncomfortable even when the room air is 70°F. Eliminate this "cold-wall effect" by adding insulation and you will feel comfortable at lower thermostat settings.

Well-insulated buildings also foster more uniform distribution of air temperatures. The air adjacent to cold, uninsulated walls cools, becomes more dense, and falls to the floor, displacing the warm air. These "ghost" drafts are considerably reduced in well-insulated houses.

You probably gained some familiarity with the language of insulation in the preceding chapter. A knowledge of the R-values of materials and the U-values of exterior surfaces is crucial for an understanding of the heat flow through walls, floors and roofs. The resistance R of a building material indicates the difficulty with which heat flows through that material. Typical R-values of common building materials and insulation are compared in the first line chart. The *total* resistance R_t, of an exterior surface is the sum of the R-values of all materials in it. The greater the value of R_t, the smaller the overall heat flow through that surface.

The U-value is a measure of how readily heat flows through an exterior surface. Mathematically, the U-value is the reciprocal of the total resistance, $U = 1/R_t$. U-values for typical building surfaces are provided in the second line chart; they range from 0.05 for a well-insulated roof to 1.15 for a single pane of glass.

You can reduce the U-value of an exterior surface, and consequently its heat loss by adding more insulation. However, your investment for insulation will quickly reach the point of diminishing returns. For example, by adding two inches of polystyrene board insulation to the exterior of a concrete wall, you can reduce its U-value from 0.66 to 0.11—an 83 percent decrease in heat loss. Adding another two inches of polystyrene lowers the U-value to 0.06—an additional savings of only 7½ percent. Your money is better spent on window shutters and weatherstripping.

The *placement* of insulation is also important. First, you should insulate roofs and the upper portions of walls, primarily because heat is conducted vertically much more easily than horizontally. Also, warm air collects at the ceilings of rooms, producing a greater temperature difference there between indoor and outdoor air.

Six inches of fiberglass batt insulation (R-19) for roofs and 3½ inches (R-11) for walls have been the standards for cold climates. These standards are being quickly accepted in mild climates and greatly upgraded in cold. Polystyrene or polyurethane board insulation built into the exterior of conventional wood-framed walls gives even better R-values. Construction details for adding such insulation are shown in the diagrams. If you have a ventilated crawl space under your uninsulated floor, you should tack on 3½-inch batts of fiberglass insulation to the underside. The increased comfort and energy savings will be well worth the added cost.

There are many more insulation techniques than we could possibly cover here. One of the best references for insulating dwellings is the *Insulation Manual: Homes, Apartments* by the National Association of Home Builders (NAHB) Research Foundation, Inc., of Rockville, Maryland. It may be available at your local library.

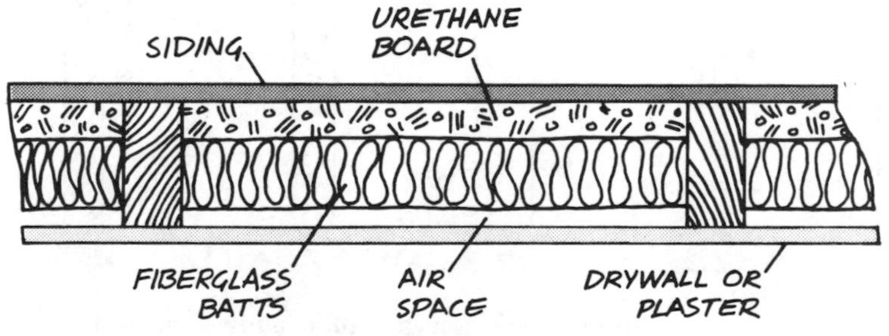

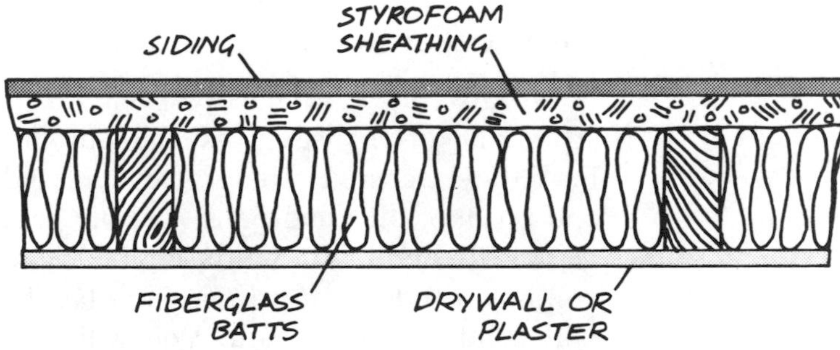

Adding rigid board insulation to exterior stud walls—plan view.

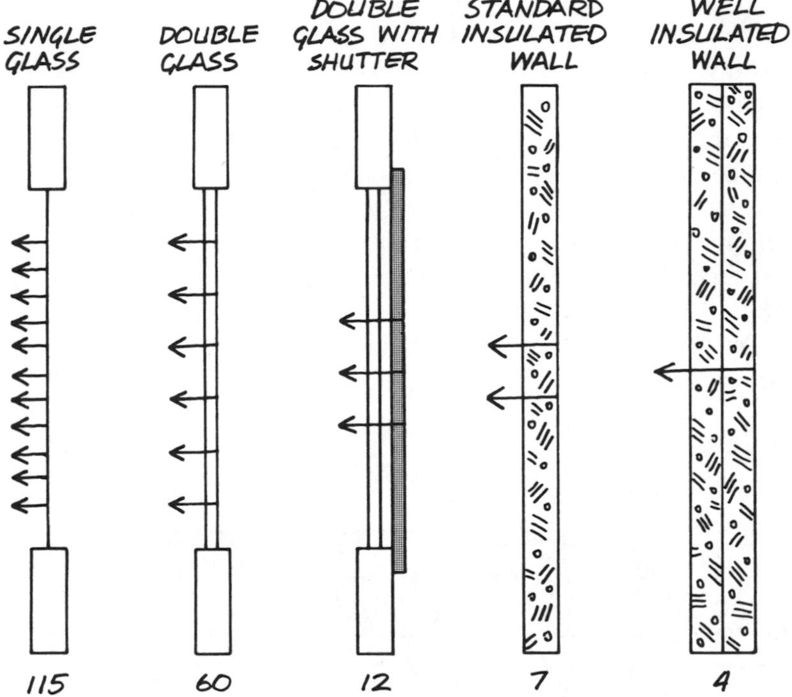

Relative heat losses through various types of windows and walls. These values represent only conduction heat loss.

Window Shutters

Various types of window and wall construction differ widely in the amount of heat they transmit. Under the same indoor and outdoor air conditions, a single pane of glass will conduct 115 Btu, double glass will conduct 60 Btu, and a well insulated wall will conduct only 4 Btu. You will lose the same quantity of heat through a well-insulated wall 30 feet long and 8 feet high as through a single glass window 2 feet wide and 4 feet high!

As indicated in the diagram, however, the use of insulating shutters for closing off windows greatly reduces this heat loss. The rate of heat loss through a shuttered, double-glazed window is comparable to that through a standard insulated wall. Depending upon its thermal resistance, an insulating shutter can reduce conduction heat loss through a window by a factor ranging from two to ten. Shutters also reduce radiation heat transfer from warm bodies to the cold window glass and, depending upon construction, can practically eliminate air leakage.

To make the best use of these shutters during the winter, you should close them only at night. They should be opened during the day, particularly on south-facing windows, to admit as much solar radiation as possible. If shutters are opened only during daylight hours (that is, between 40 percent and 55 percent of the time during the heating season), heat loss during the remaining 45-60 percent will be greatly reduced. In a 5000 degree-day climate, a single glass pane with a U-value of 1.15 has a heat loss of 138,000 Btu per square foot of glass per season. If you apply a shutter that gives an overall U-value of 0.12 for the combination of window and shutter, and if the shutter always covers the window at night, you will save about 60 percent of this energy—80,000 Btu (the equivalent of about one gallon of heating oil burned at 60 percent efficiency). A single 10 square foot window shutter can save you 10 gallons of heating oil, or four to five dollars at 1976 prices, for each heating season. The value of shutters is unquestionable, and their use is strongly advised.

The necessity of opening and closing shutters can be an encumbrance to people with busy schedules, but you may feel occasional Franklinesque glee once this thrifty habit becomes a part of your daily routine. Closing your shutters when the sun goes behind clouds at midday may prove more difficult. The shutters also have to be stored somewhere when open, and hinged interior shutters generally require large amounts of open space to swing through. One possible solution, developed by Zomeworks Corporation, is the shutter arrangement shown in the photograph. Or you can use thick insulating cloth held against the window frame by Velcro liners (a design developed by Sunshine Designs, Inc.). The cloth is rolled up above the window in the morning and rolled down again at night.

You can fashion a cheap, effective window shutter from one- or two-inch thick polystyrene board insulation. That gallon of oil from which the polystyrene is made will go much further on your window than in your furnace. But first, do a bit of measuring. It is essential that shutters close tightly to block the flow of cold air from the window surface to the room. And you can save yourself a lot of money, fuss and bother by dismissing that old wives' tale about the insulating value of heavy drapes. Because they don't block cold air movement, they make poor window

Manually operated shutter developed by Zomeworks Corporation.

insulators. After you've made your measurements, simply cut the polystyrene boards so they fit snugly into the window frame. Then push them into place, and—Voila!—you've begun to acquire a modest fortune.

Skylid and Beadwall

A shutter device that doesn't require frequent owner intervention is the Skylid, developed and marketed by the Zomeworks Corporation of Albuquerque, New Mexico Skylids are insulated louvers that can be placed inside the house behind skylights or vertical windows. The louvers pivot open during sunny weather but close at night or during cloudy periods. When closed, the Skylid is an effective thermal barrier, with an R-value of about 3 or 4. It would be an even *better* insulator, were it not for small leakage of air around the seals.

Skylids used as shutters in an Albuquerque home.

The louvers are opened and closed by the movement of freon (a liquid used in refrigerators) between two cannisters mounted on the inside and outside surface of one louver. When the sun heats the outer cannister, the freon inside evaporates and flows to the inner one, where it condenses to counterbalance the louvers and causes them to fall open. A reverse flow at night closes the louvers, and a manual override permits the owner to close them and keep out solar heat during summer days.

Another window shutter design from Zomeworks is a revolutionary device called Beadwall. Polystyrene beads are blown between two panes of glass or plastic to prevent undesirable heat loss (or gain). Three inches of beads provide an R-value of about 10. The beads are sucked out during sunny winter weather by small vacuum cleaner motors, which operate only a few times a day to fill and empty the Beadwall.

Both Skylid and Beadwall have striking visual impacts upon the adjoining rooms. Soft, diffuse lighting is possible when Skylids are used with skylights or clerestories. The same room can be radiantly warm during winter and refreshingly cool in summer. When full, Beadwall provides a soft glow to the interior space. Unlike Skylid, however, it requires owner intervention or expensive automatic controls to operate it at the appropriate times. Several applications of

Skylid and Beadwall are discussed in the next chapter.

Air Infiltration

Heat loss by infiltration, or the movement of air through the exterior skin of a house, can be a large part of the overall loss. All the cold air that enters the house through window cracks, door jambs, and other openings must be warmed to room temperature if a semblance of comfort is to be maintained. We confess it's difficult to separate infiltration losses from conduction losses through walls and windows. The addition of storm windows or shutters to existing windows cuts the conduction loss drastically and restricts the air leakage—thereby licking part of the infiltration problem. Caulking and weatherstripping are straightforward, traditional solutions to air infiltration. You can also cut down on infiltration by installing tighter-fitting windows or planting trees for wind protection, but these measures involve more planning and expense.

People require *some* outdoor air for ventilation and a feeling of freshness, and the penetration of air through the cracks in the surface of a house usually satisfies this need—particularly if cigarette smoking is avoided or done outside. You should make every effort, however, to reduce such uncontrolled air infiltration. As you reduce other heat loss factors, the penetration of outdoor air becomes a larger part of the remaining heat loss. If you reduce air infiltration to a minimum and open a window to bring in occasional fresh air, you will save large amounts of energy.

One of the main reasons for spreading building paper between the plywood sheathing and the exterior siding of houses is to reduce air infiltration through cracks in the walls. Good trim details on a house exterior also reduce air penetration. Mortar joints in brick and concrete block facades should be tight and complete.

More important than cracks in the wall surfaces, however, are those around the windows and doors. Different types of windows vary greatly in their relative air infiltration heat losses, as the line chart indicates. Weatherstripping improves the performance of any window, particularly in high winds. Fixed windows save the most energy. You need operable windows for ventilation, but how many do you really need? In an existing house some windows can usually be caulked shut for the winter. All others should be weatherstripped, preferably with durable metal weatherstripping. Forget about that cheap, spongy stuff. Its effectiveness deteriorates rapidly and so does its appearance.

If you plan to install new windows, you should be careful in selecting them. Operable windows should be chosen for their tight fit when closed—not only when first installed, but also after being used hundreds of times over a period of decades. Pivoted and awning windows are the loosest, and casement

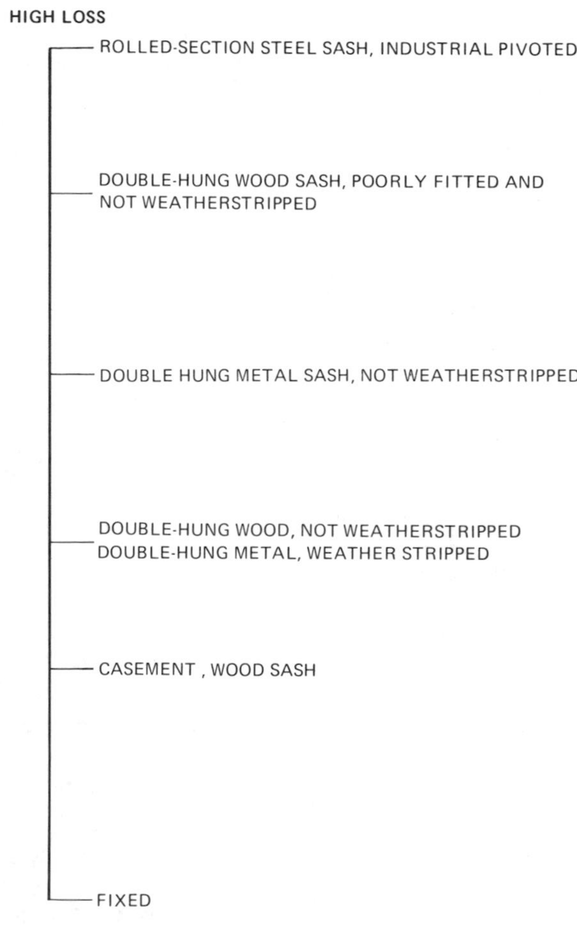

windows are among the closest fitting.

Wind Control

Wind is the arch-culprit in the moment-to-moment variation in the amount of air which penetrates a house. Olgyay reports in *Design With Climate* that a 20 mph wind doubles the heat loss of a house normally exposed to 5 mph winds. He also notes that the effectiveness of a belt of sheltering trees increases at higher wind velocities. With good wind protection on three sides, fuel savings can be as great as 30 percent.

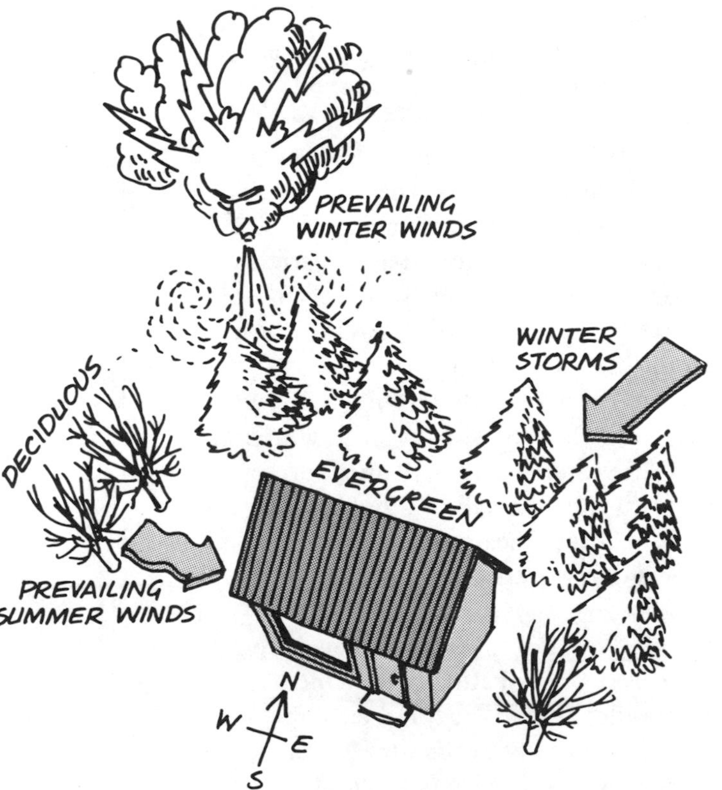

Proper orientation and vegetation shields protect a house from the wind.

Houses should be oriented away from prevailing winter winds or screened by natural vegetation in order to block heat-pilfering air flows around windows and doors. Man-made windscreens, such as the baffles on the Saunders House discussed in Chapter 2, can also be very effective (while generously providing mural artists with a chance to display their talents!)

Winter winds blow, whistle, and wail from the north and west in most locales. Entrances should not be located on these sides, and the number of windows (the smaller the better!) should be kept to a minimum. Double doors and storm windows are especially desirable there. Wind directions do vary, however, with locality and season. Monthly maps of the "Surface Wind Roses," available in the *Climatic Atlas of the United States*, can be most helpful in the layout of windows and doors. These maps give the average wind velocity at many weather stations and show the percentage of each month that the wind blows in various directions.

The conduction heat loss through the surfaces of a house also increases with wind velocity. And the higher the U-value of a surface, the more you need to protect it from wind. A single-pane window needs much more wind protection than a well-insulated wall, because the air film clinging to its exterior surface contributes a much larger part of its overall thermal resistance. As the thickness of this air film decreases with an increase in the velocity of the air striking a surface, the effective insulating value of the film also decreases. This decrease is large for windows but almost negligible for well-insulated walls.

A storm window added to existing windows almost halves conduction heat loss for single-pane windows and reduces air infiltration similarly. A two-window sash—the standard single-pane window in combination with a storm window—is superior to a single sash containing insulating glass. A standard window of insulating glass in conjunction with a storm window is an even better solution.

You will find many more excellent suggestions for limiting air infiltration and conduction heat losses in the recent book, *Low-Cost Energy-Efficient Shelter,* edited by Eugene Eccli. This book also has many helpful comparisons of products and building materials.

PUTTING IT ALL TOGETHER

The foregoing sections of this chapter introduced you to the basic principles of building houses that utilize the energy flows in nature. You were shown how to allow sunlight to penetrate at the proper time, how to store solar heat inside the house, and how to retard the escape of that heat. Including all three criteria in a coherent

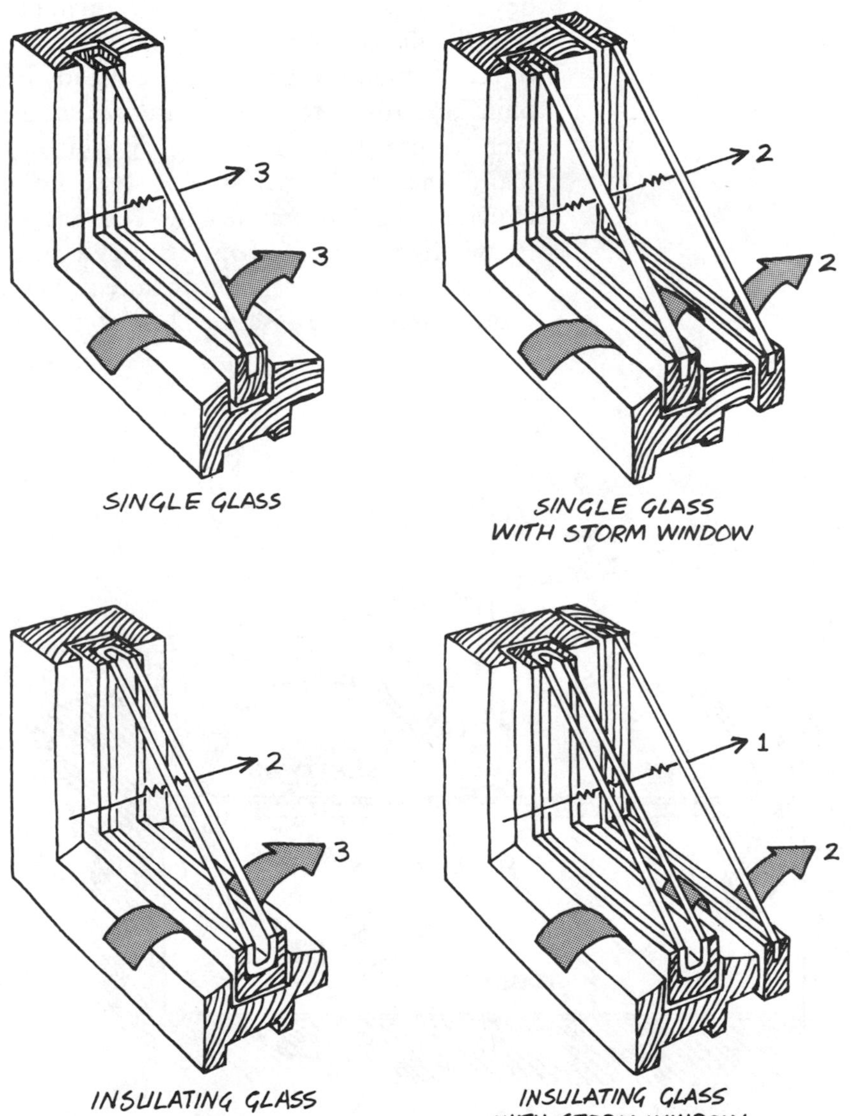

Relative conduction and air infiltration heat losses from various windows. An added storm window cuts both kinds of heat loss.

design that satisfies many other human needs is difficult—but not impossible. Thousands of very habitable dwellings have already been built according to these criteria. Primitive architecture throughout the world offers many shelter designs suited to both local climates and cultural needs. The Saunders house discussed in Chapter 2 achieves this synthesis in a more modern context. Several more examples are described here to illustrate further the possibilities of building with the sun.

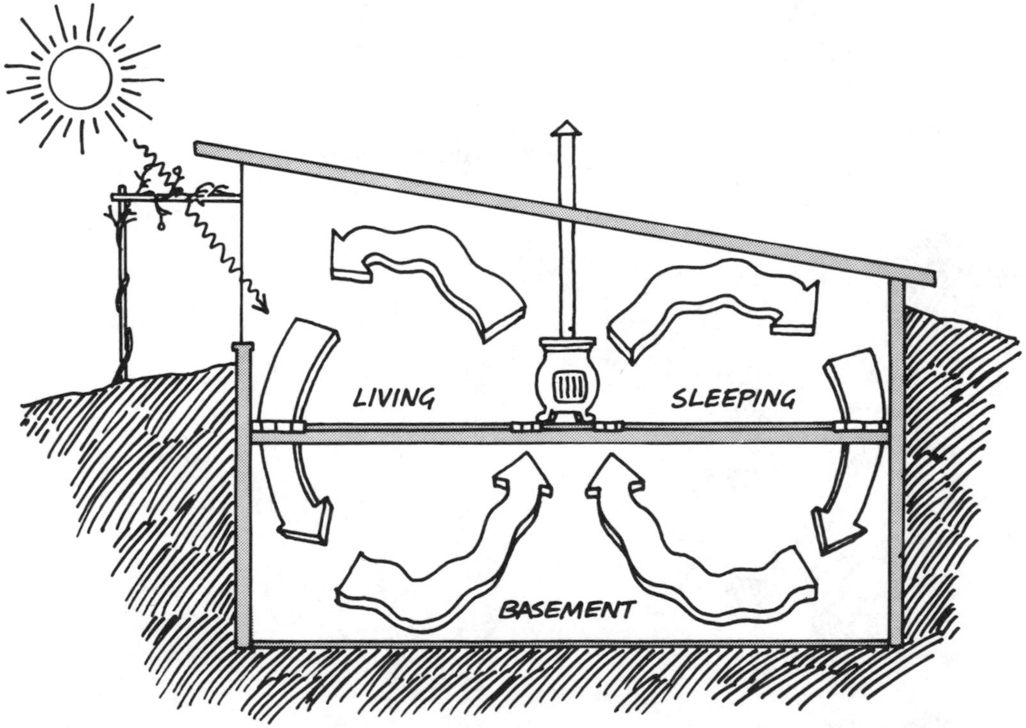

LIVING
SLEEPING
BASEMENT

Air circulation in Sunnycave. Slots in the floor drain cool air from the walls into the cellar below. Sunlight and a small stove keep this house warm throughout the winter.

Sunnycrest and Sunnycave

In the 1940's and 50's, Mr. and Mrs. Wendell Thomas built two small houses in the Celo Community located in the mountains of western North Carolina. Finished in 1948, the first house was a 32 by 24 foot shed-roof dwelling called "Sunnycrest." This house had a cellar set deeply into the ground, and earth was partially banked around the north, east and west walls. A 2-inch slot in the floor at the base of the walls drained cool air from the walls and windows down into the completely dry, sealed cellar. There the cool air was warmed slightly by the earth. A central heater on the main floor induced this air to rise through vents around it—warming the air further and distributing it evenly throughout the house. There were few windows on the east, west, and north walls, but there was plentiful glass on the south to allow for extra solar heat gains in winter.

In 1957, Wendell Thomas built a slightly smaller house called "Sunnycave." Earth was banked almost to the roof on the north and west and to the window sills on the south and east. The west wall of Sunnycave protrudes 4 feet beyond the south wall to form a baffle that protects the south windows from the cold west wind in winter. About half the south wall is glass, but there are few windows on the east, west and north. All windows are double or triple glazed and only a few can be opened for ventilation. Thomas sealed the inside panes but left the outer ones unsealed to prevent fogging. At night and on cold dark days, he covers the inside surface of the windows with aluminum-painted insulation boards and pulls heavy drapes behind them. Two large triple-glazed windows are left uncovered for lighting.

To shade the south windows in summer, a thick growth of grapevines and woodbine covers a trellis stretching parallel to the south wall. Their leaves block the sunlight in summer but drop off to allow its penetration in late fall and winter, when solar heat is much more welcome. As in Sunnycrest, a central wood-burning heater warms air from the cellar and drives the natural circulation loop that distributes this warm air to the rooms.

In Sunnycrest, the temperature fell to 50°F on the coldest winter morning, and on the hottest summer afternoon it rose to 85°F. In Sunnycave, however, the temperature never falls below 60°F and never rises above 75°F.

Raven Rocks

A house designed by Malcolm Wells in Barnesville, Ohio, carries the ideas of Wendell Thomas even further. Called "Raven Rocks," the house is oriented from east to west on a south-facing slope. Extensive south windows and a sweeping array of tilted glass along the full length of the roof provide most of the heating requirements. The several hundred yards of concrete in the walls, floors and ceilings store the solar heat that accumulates inside. Excess hot air from the topmost heights of the house is ducted to a gravel bed located under the concrete floor slab.

Except for the south-facing glass, the building is completely buried. Rigid polyurethane board insulation isolates the concrete walls and gravel bed from the surrounding earth.

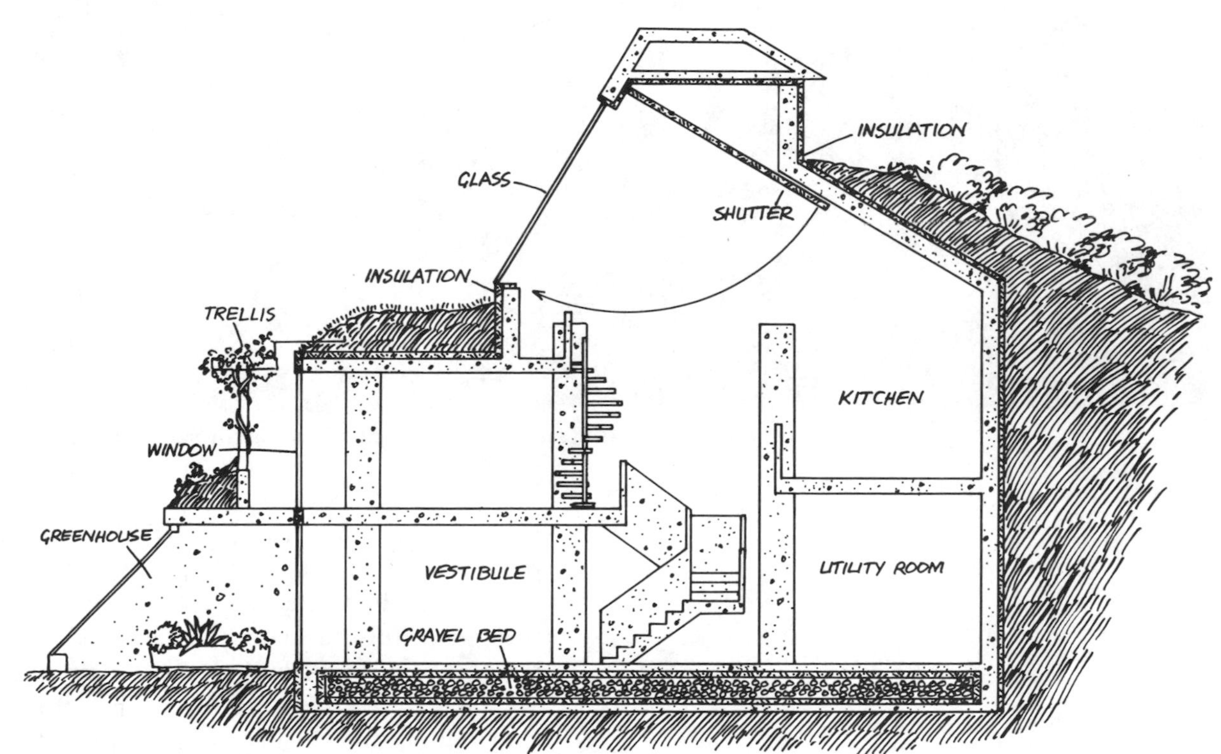

Raven Rocks in Barnesville, Ohio, is almost completely buried in a south facing hill.

Almost the entire south face of David Wright's house is covered with glass.

Heat loss through the glazing is drastically reduced by shutters that are closed at night. Because of the tremendous thermal mass and the plentiful insulation, the indoor air temperature drops only 3°F per day during average January weather—even when the sun isn't shining and the backup heater is off.

Wright House

David Wright, an environmentalist and architect, designed his own solar home in Santa Fe, New Mexico. Using only the direct methods we have been studying, he achieves more than 90 percent solar heating in a 6200 degree day climate. Because temperatures are allowed to fluctuate between 60°F and 80°F in this massive house, it can absorb very large amounts of solar heat—and very simple designs are possible.

Completed in 1974, this semi-cylindrical house faces the southern horizon with 384 square feet of insulating glass. Only a few windows can be found on the other facades. The 13-17 inch thick adobe walls and the 2 foot thick adobe floor are insulated on the outside by 2 inches of polyurethane foam. The house loses about 13,000 Btu per degree day or 32,000 Btu/hr at the design temperature of 7°F.

On a clear January day, as much as 483,000 Btu of solar energy enters the house through the south window and is stored in the massive floor and walls. Several adobe-covered, water-filled drums sitting directly behind the glass provide additional heat storage. The drums, walls, and floors can store enough heat to keep the house comfortable for 3 or 4 sunless winter days. Because of the spacious, open interiors, this stored heat can easily flow to the rest of the house by radiation and convection. David Wright claims the house is very comfortable—even at 60°F indoor temperatures. The insulated walls remain at room temperature—eliminating the cold-wall effect discussed earlier.

At night, the large expanses of south glass are insulated by folding shutters made from canvas and 2-inch polyurethane. By day, these clever shutters are cranked up and stored like an accordion near the ceiling. The only drawback is their loose fit. They do not fit snugly enough against the glass to prevent small cold air drafts from the inside surfaces. Eaves projecting 4 feet from the front wall exclude most of the summer sun from the south glass. Vents allow any accumulated hot air to escape through the roof.

The performance of this house was studied during a 2 week period in January of 1976. Without any backup heating from the fireplace, the house remained comfortable for that entire period. Indoor temperatures swung a maximum of 20°F in any one day, reaching a high of 80°F and a low of 58°F. Solar heat provided about 90 percent of the house's annual heating needs.

Jackson House

Many of the energy conserving measures discussed in this chapter are embodied in a recently built house near the town of Jackson in Western Tennessee. Designed by architect Lee Porter Butler, this house is extremely well adapted to the Tennessee climate with its mild (3200 degree day) winters and hot, humid summers. But according to Butler, indoor temperatures would remain above 68°F during the severest winter weather on record.

Window shutters in the Wright House. These canvas-and-polyurethane shutters are cranked up and stored near the ceiling during the daytime.

Built onto the south side of this 1440 square foot house is a large greenhouse or "sun room" that doubles as a solar collector. With 360 square feet of single glass in the roof and south wall, this greenhouse can collect upwards of 500,000 Btu of solar heat on a sunny winter day. Some of the solar radiation penetrates all the way into the house through a glass dividing wall, while the rest is absorbed in the greenhouse. Glass doors can be opened to admit solar heated air into the main part of the house.

Jackson House collects the sun's heat in a south-facing greenhouse.

Once inside the house, this solar heat is trapped by an airtight insulating envelope. Walls and roofs have R-11 and R-19 fiberglass insulation. And there are no windows on the north, east or west facades. Entry to the house is gained via two long vestibules whose doors are weatherstripped to insure a tight "air-lock." Because of these measures, air infiltration into this house is extremely small.

The solar heat is stored in the fabric of the home—mostly in the 4 inch concrete slab and in the 18 inches of dirt sitting beneath it. Insulated by a 2 inch layer of urethane foam on the underside, the concrete and dirt alone can store about 500,000 Btu for a 10°F rise in temperature. Natural circulation of the heat through the house is aided by vents in the floors and sheet metal ductwork in the ceiling, floor, and north wall.

The house is equal to the task of summer cooling, which in this locale is more important than winter heating. Deciduous trees shade the greenhouse in summer, and roof overhangs prevent the penetration of sunlight directly into the house. Excess hot air from the house and greenhouse escapes through vents at the roof peak. To replace this hot air, earth-cooled outdoor air flows into the house through an underground pipe that feeds into the subfloor ductwork. The warm, moisture-laden air drops some of its water as it cools, thereby lowering the humidity indoors.

According to Butler, the performance of the house exceeded expectations during the first year of operation. Indoor temperatures never rose above 75°F during the summer of 1975. In winter, the sun provided almost all the heating needs of the house. The total cost of the house and lot was $38,000—well within reach of the middle-income homeowner.

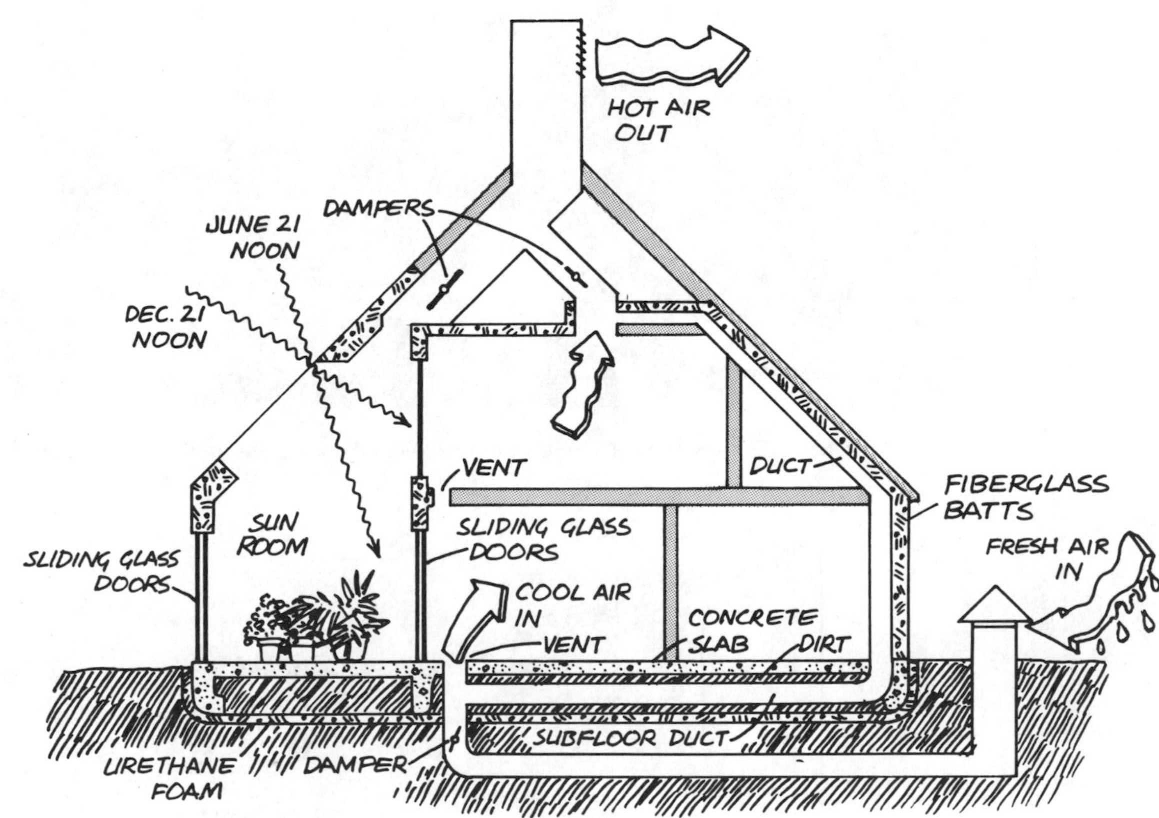

Cross section of the Jackson House—summer operation.

St. George's County Secondary School

A final example of direct solar heating is the annex to St. George's School in Wallasey, England, near Liverpool. It was built in 1961 for the use of 300 students. Designed by the late Emslie A. Morgan to gain as much heat as possible from direct solar radiation, this building can store this heat for days at a time. Its massive concrete floors and brick walls prevent rapid fluctuations of indoor temperature, and the exterior building insulation traps the solar heat so well that little other heat is needed. In fact, the only other sources of heat are the electric lights and the students themselves – even though winters in Wallasey feature fierce winds and frequent cloudiness.

The annex is a long two-story structure facing 9° west of south. The ground floor is a 6-inch concrete slab sitting over 4 inches of crushed rock, and the intermediate floor is made of 9-inch concrete. The roof has 7 inches of concrete insulated on the top by 5 inches of expanded polystyrene. Partitions between the classrooms are made of 9-inch brick, as are the exterior walls. The thermal

The solar-heated annex to St. George's County Secondary School in Wallasey, England.

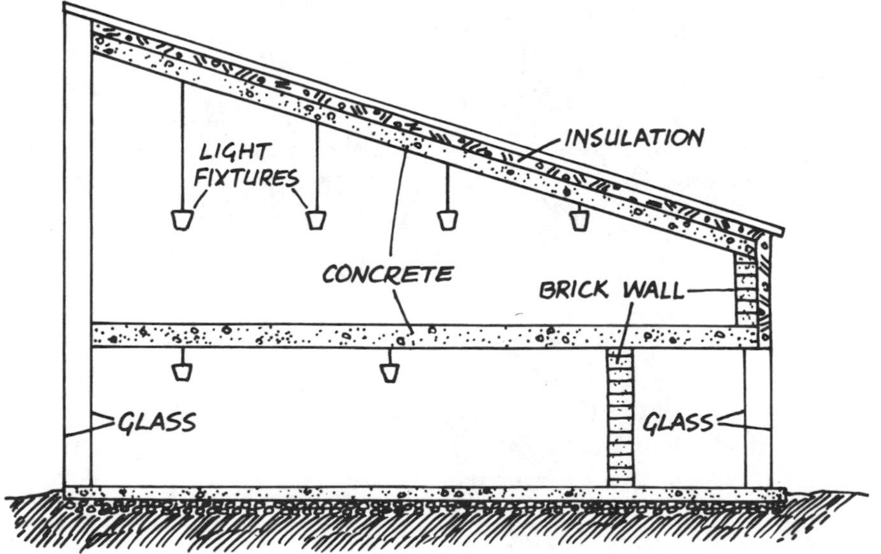

The massive interiors of St. George's School prevent rapid fluctuations in temperature. Solar energy provides about half the heat needed inside, while the students and electric lights provide the rest.

mass of all the interior brick and concrete is great enough to prevent the daily temperature swing from exceeding 6° F on a normal winter day. The north, east, and west walls have few windows and are also insulated with 5 inches of expanded polystyrene.

By contrast, the 230 foot long by 27 foot high south wall is mostly glass—two fixed panes separated by a 2 foot air gap. The inner pane has a ripply, textured surface to scatter the incoming sunlight in many directions—this cuts glare and insures that solar radiation is striking almost all the brick and concrete directly.

A research team from the Department of Building Science at Liverpool University has made extensive studies of this remarkable building. For most weather conditions, solar energy is the principal source of heat. On a sunny day, for example, solar energy accounted for six times the heat gain from the students and three times the heat gain from the electric lights. Over an entire year, solar energy supplied 50 percent of all the heating needs, while the lights and students provided the rest. But some stiff upper lip was needed to accomplish this feat. During December and January, indoor temperatures averaged between 59 and 61° F, occasionally falling as low as 52° F. The classrooms were ventilated only sparingly during these months and became quite stuffy as a result.

These examples of environment-sensitive buildings help to clarify the major points of this chapter. Properly designed homes can take the best advantage of solar energy when they act as solar collectors, heat storehouses, and heat traps. The design principles discussed here are universal – they remain valid for the solar homes discussed in the rest of this book.

FURTHER READING

ASHRAE. *Guide and Data Book.* New York: American Society of Heating, Refrigerating and Air Conditioning Engineers, 1970.

ASHRAE. *Handbook of Fundamentals.* New York: ASHRAE, 1972 (also 1967).

Aronin, J. E. *Climate and Architecture.* New York: Reinhold Publishing, 1953.

Conklin, Groff. *The Weather Conditioned House.* New York: Reinhold Publishing, 1958.

Dubin, Fred S. "Energy for Architects." Reported by Margot Villecco in *Architecture Plus,* July 1973.

Eccli, Eugene, ed. *Low-Cost Energy-Efficient Shelter for the Owner and Builder.* Emmaus, Pennsylvania: Rodale Press, 1976.

Givoni, Baruch. *Man, Climate, and Architecture.* Barking, Essex, U.K.: Applied Science Publishers, 1969.

Hutchinson, F.W. "The Solar House: Analysis and Research." *Progressive Architecture,* Vol. 28, May 1947.

Kern, Ken. *The Owner-Built Home.* Oakhurst, California, 1961 (available for $5.00 from Ken Kern Drafting, Sierra Route, Oakhurst, CA 93644).

NAHB Research Foundation, Inc. *Insulation Manual.* Rockville, Maryland, 1971 (available for $4.00 from NAHB Research Foundation, Inc., P.O. Box 1727, Rockville, Maryland 20850).

National Bureau of Standards. "Technical Options for Energy Conservation in Buildings." NBS Technical Note 789. Institute for Applied Technology, Washington, D.C. July 1973.

Olgyay, Aladar and Victor. *Solar Control and Shading Devices.* Princeton: Princeton University Press, 1957.

Olgyay, Victor. *Design with Climate: A Bioclimatic Approach to Architectural Regionalism.* Princeton: Princeton University Press, 1963.

Ramsey, C. G. and H. R. Sleeper. *Architectural Graphic Standards.* New York: John Wiley & Sons, 1972.

Robert, Rex. *Your Engineered House.* New York: M. Evans & Co., 1964 (available for $8.95 from J. P. Lippincott, E. Washington St., Philadelphia, PA 19105).

Rudolfsky, Bernard. *Architecture Without Architects.* New York: Museum of Modern Art, 1964 (available for $4.50 from Doubleday & Co., Garden City, NY 11530).

Severns, William H. and J. R. Fellows. *Air Conditioning and Refrigeration.* New York: John Wiley & Sons, 1966.

Thomas, Wendell. "The Self-Heating, Self-Cooling House." *Mother Earth News,* Issue No. 10, 1970.

5
Soft Technology Approaches

Solar Heating and Cooling Systems Built into the Fabric of a Dwelling

After a long journey, the Aztec god Xochipilli warms himself against a stone wall at the Inca citadel of Machu Picchu.

I believe the ground rules can be transformed so that technology simplifies life instead of continually complicating it.

Steve Baer

The most sensible ways of using solar energy are the *simple* ways. They are more efficient, more reliable, and less disruptive to the environment. You have already encountered the simplest of such *low impact* or *soft technology* approaches—south facing windows, insulating shutters, and thermal mass. Solar heat is trapped right inside the house, and no pumps or fans are needed to distribute it elsewhere.

In the methods to be described here, the fabric of the house is still used to collect and store solar heat, but the sun's rays do not penetrate to the interior of the house. They are intercepted and absorbed by the walls or roof, which usually double as the heat storage container. The water wall of the second MIT solar house, the concrete south walls of the Odeillo houses, and the roof ponds used in the Winters House typify these methods. The solar "apparatus" is an intrinsic part of the house—not a separate appliance. Such approaches extend our concepts of a house and generally require departures from standard building practice. But materials are kept simple and little or no electrical power is used to distribute the heat—and the resulting systems are cheap and reliable.

Such integrated systems are commonly called *passive* systems by workers in the trade. Natural convection or thermal radiation, rather than pumps or fans, are used to move the solar heat from storage volumes to the living areas. Passive systems have few complex moving parts or controls and therefore little potential for breakdown. Designers of such systems recognize that human control is often more reliable than mechanical control in providing human comfort.

The personal effort and involvement required of the inhabitants of a passively-heated house are often critical issues. For people weaned on the thermostat, even the simple activity of opening and closing shutters may be too much of an imposition on their lives. Others welcome this opportunity to participate in tempering their indoor climates. Hopefully you are warming up to the attitude changes dictated by changing environmental imperatives. Passive systems for solar heating and cooling make the most sense within the context of such changes.

THERMOSIPHONING AIR COLLECTORS

A grasp of the principles of thermosiphoning—where the natural bouyancy of heated air or water is used to circulate heat—is crucial to an understanding of the passive uses of solar energy. When heated, air expands and becomes lighter than the surrounding air. The heated air drifts upward and cooler air moves in to replace it.

You have all observed the process of

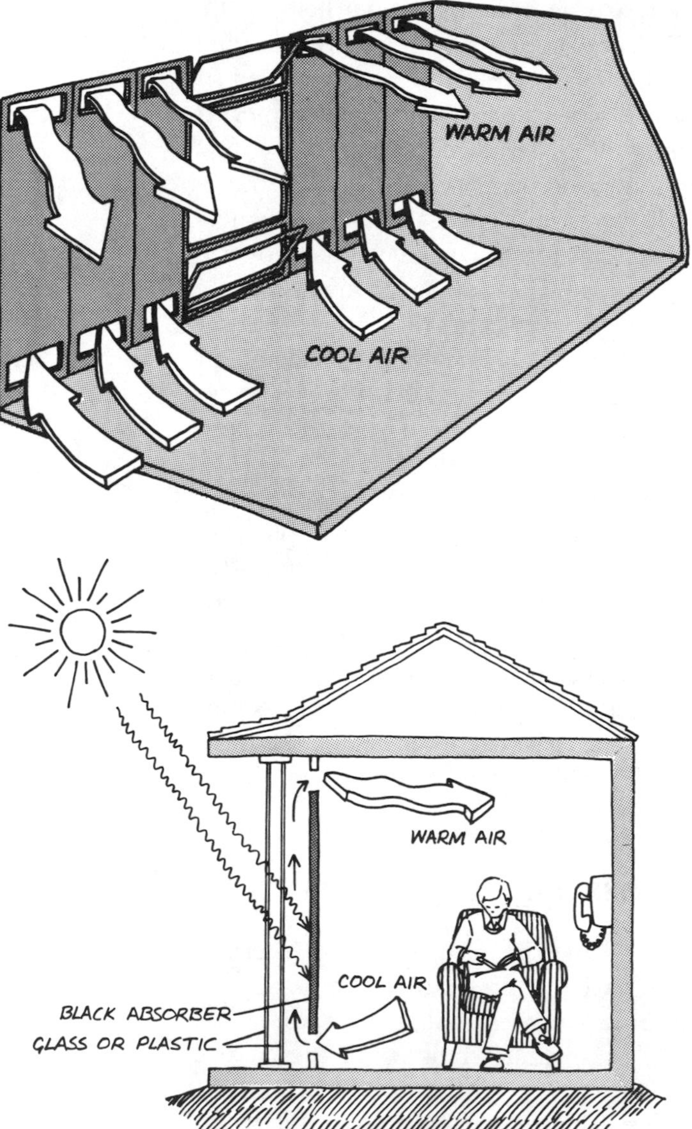

A thermosiphoning air collector—two views.

thermosiphoning, also called natural or gravity convection, at work in a fireplace. Because the hot air just above the fire is much lighter than the surrounding air, it rises rapidly up the chimney. Cooler, heavier room air replaces it—bringing more oxygen to maintain the flames. Most of the fire's heat is delivered to the outdoors by this "chimney effect."

The simplest form of a thermosiphoning air collector is illustrated in the diagrams. As the air in the space between the glass and the blackened absorber wall is heated, it expands and becomes lighter, rises through the collector, and flows into the room from a vent at the top. Cool room air is drawn through another vent at the base of the wall, heated in turn, and subsequently expelled at the top. This process continues as long as there is enough sunlight to push the temperature of the absorber wall above the room air temperature.

CNRS Wall Collectors

Some of the most significant work in thermosiphoning air collectors is being done at the *Centre National de la Recherche Scientifique* (CNRS) in Odeillo, France. Under the direction of Professor Felix Trombe, this laboratory has developed several low-technology approaches to solar heating. The main building, which houses the world-famous solar furnace, is an excellent example of the passive use of solar energy. The south, east, and west walls of this building are a composite of windows and thermosiphoning solar collectors, which supply about half of the building's winter heat.

A cross-sectional view of these collectors is shown in the accompanying diagram. Blackened corrugated metal panels are located behind a single pane of glass. Solar radiation passes through the glass and is absorbed by the panels, which are contained entirely within the volume defined by the glass and duct. As the metal heats, so does the air between the absorber panel and glass. The heated air flows upward through vents into the rooms. Simultaneously, cooler room air falls through the same vent and sinks down between the back of the absorber and the duct wall. This air returns to the face of the absorber—where it, too, is heated and expelled into the rooms.

Exterior and interior views of the windows and collectors are shown in the photographs. The heated air rises into the room through the perforated stainless steel window sill. Such collectors are mounted between floors on three facades of the building.

No provision has been made to store the solar heat, other than the thermal mass of the building – particularly the reinforced concrete slab floors. Consequently, the system is most effective when the sun is shining—almost 90 percent of the daytime hours in Odeillo. The air temperature in the offices and laboratories remains relatively constant during the day. Even during February, auxiliary heat is required only at night and on overcast days. Outdoor temperatures are

Wall collectors in CNRS Laboratory Building.

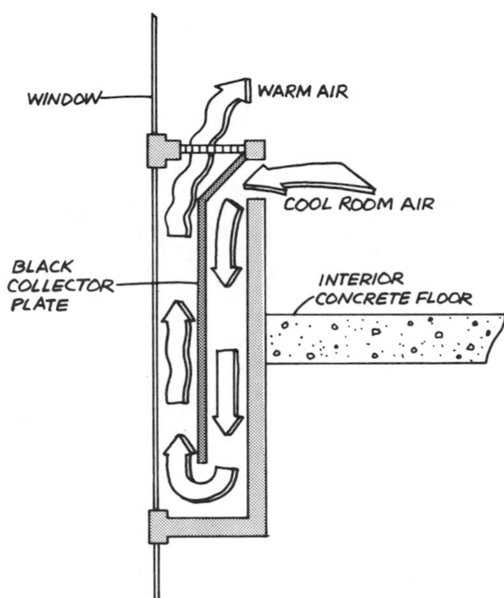

Cross-section of CNRS wall collector.

relatively cool in summer, allowing the use of east and west facing collectors, which would overheat most buildings in hot climates.

Thermosiphoning Collector Variations

A number of variations on the basic design of thermosiphoning air collectors can improve their performance. These variations include insulation, improved absorber surfaces, and dampers and fans to regulate the flow of air.

During a sunny winter day, no insulation is needed on the wall between the absorber surface and room. To reduce room heat losses on cloudy days or at night, however, the wall should be adequately insulated behind the absorber. Or you can apply insulating shutters on the outside of the glass.

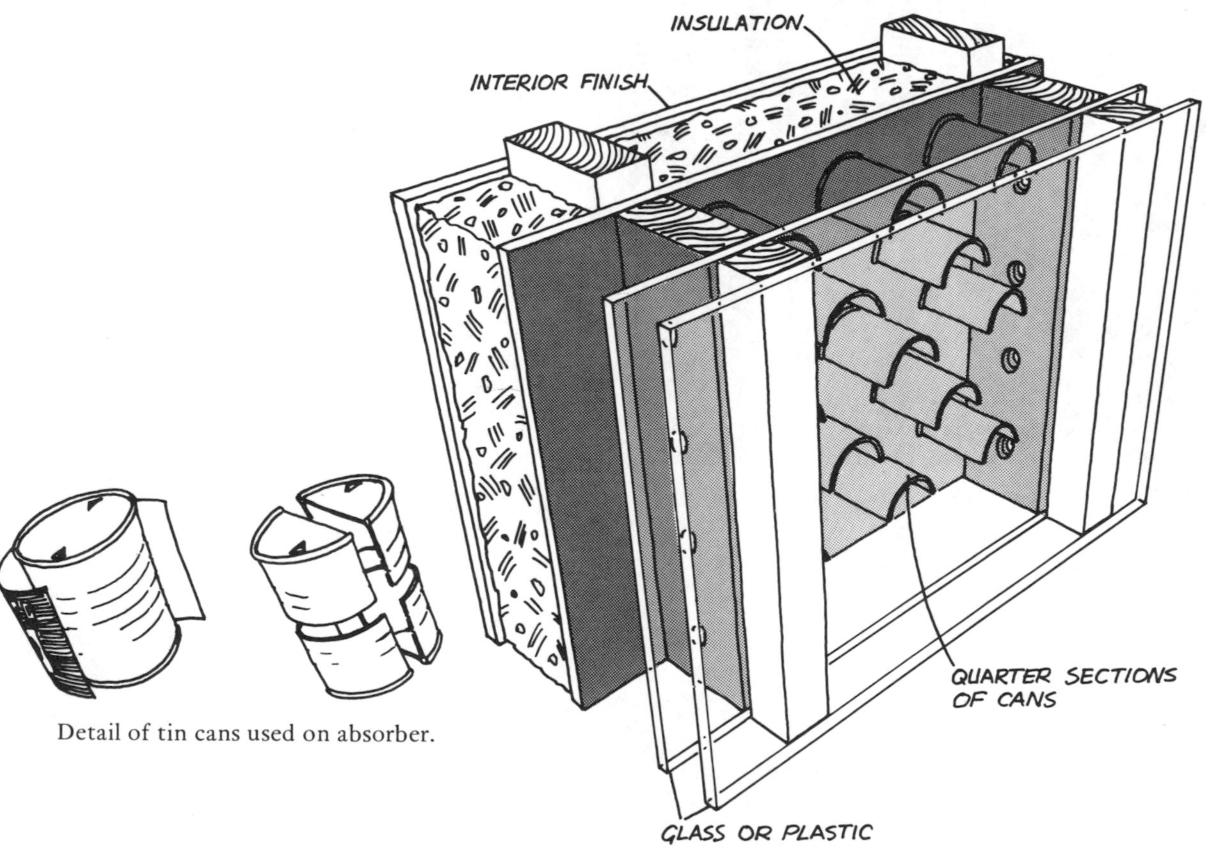

Detail of tin cans used on absorber.

Low-cost thermosiphoning air collector built onto an exterior wall.

The easiest way of including exterior insulation is to use sliding shutters. But be careful to insure a tight seal between adjacent shutters. If the collector is double-glazed, you can fill the space between the two panes with styrofoam beads. In this adaptation of Beadwall, the beads are withdrawn in the morning and replaced at night.

A metal absorber plate isn't an absolute necessity for a thermosiphoning air collector. Since the temperature of the collector wall does not get extremely high, blackened masonry or wood surfaces are also possible, and costs need not be excessive. Alternatives which increase the total absorber surface can be particularly effective, if they do not hinder the natural convection air flow. Rough surfaces make better absorbers than smooth ones. Pebbles cast in a blackened concrete wall are a good example of such an absorber surface.

Another option was developed by Jim Peterson and Marc Thomsen of Boulder and Jerry Plunkett of Denver. They used aluminum beer cans cut to two inches in height and attached in a hexagonal array to a sheet of plywood. With about ten cans per square foot of collector, the whole assembly is painted black and covered with transparent glazing to form one wall of a house. Another variation on this theme (shown in diagram) has tin cans cut into quarters and mounted on the standard plywood sheathing of conventionally framed houses. Many other alternatives to flat metal sheets are discussed in the next chapter—in the section on air-type collectors. Be sure to check the design criteria given there.

Venetian-blind type louvers can be positioned between two layers of glass to provide summer shading, nighttime insulation, and thermosiphoning solar-heated air. One side of the louver is black (as shown), the other reflective silver, and the inside core rigid board insulation. Here are some of the many possible positions and functions of the louvers.

1. To obtain direct solar heat gain, they can be either pulled up to the top for maximum glass exposure, or left down but turned parallel to the sun's rays.
2. For controlled solar heat gain, the louvers are down and slightly tilted, with the black side toward the sun to absorb some of the sunlight.
3. While in position 2, air circulates between the panes of glass and absorbs heat from the black surfaces. This hot air can be stored or fed directly to the rooms on the north side of the house.
4. To prevent all sunlight from entering the space, the louvers can be turned vertically to form a continuous surface. With the black sides facing out during the winter, solar heat can be collected as in position 3, but the louvers form a more effective heat-retaining barrier, especially if they are thick. The silvered surface facing inwards reflects thermal radiation back to the room.

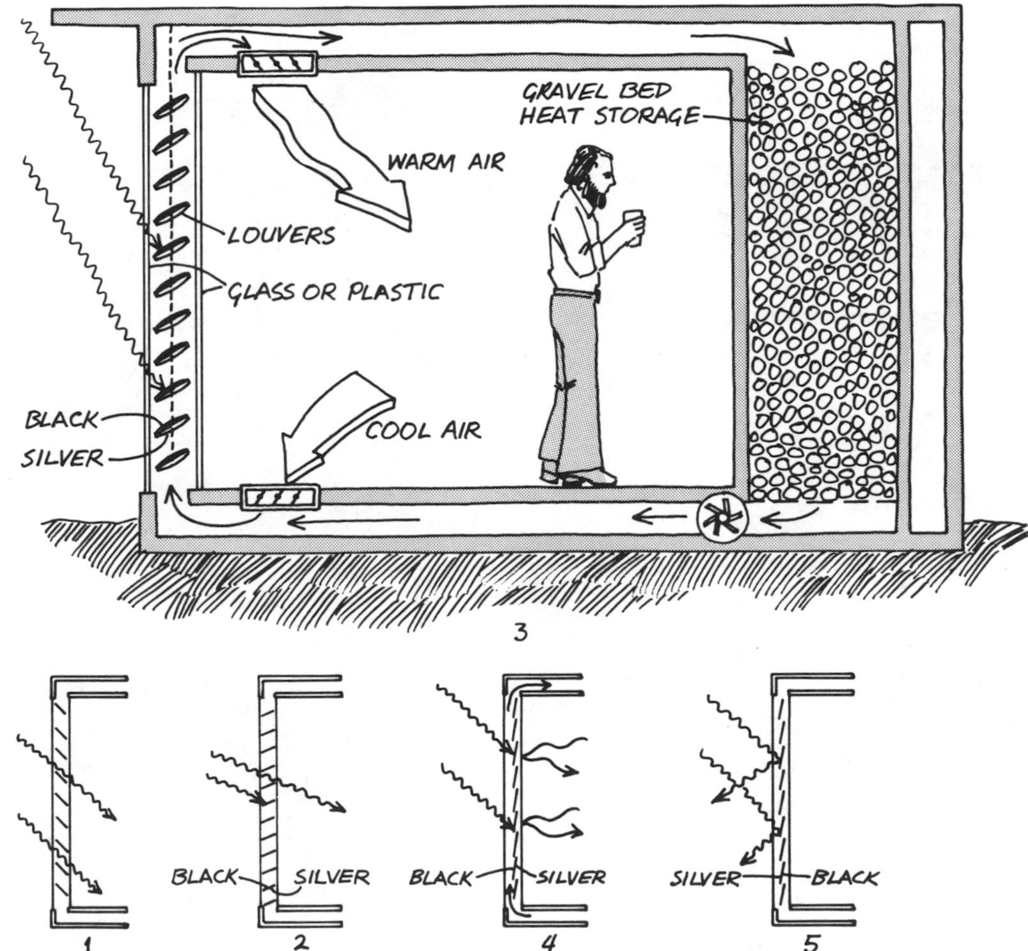

A "Venetian blind" solar collector and heat control device—various operating modes.

5. During hot sunny weather, the silvered side can be turned facing outward, reflecting the sun's rays and keeping the interior cool.

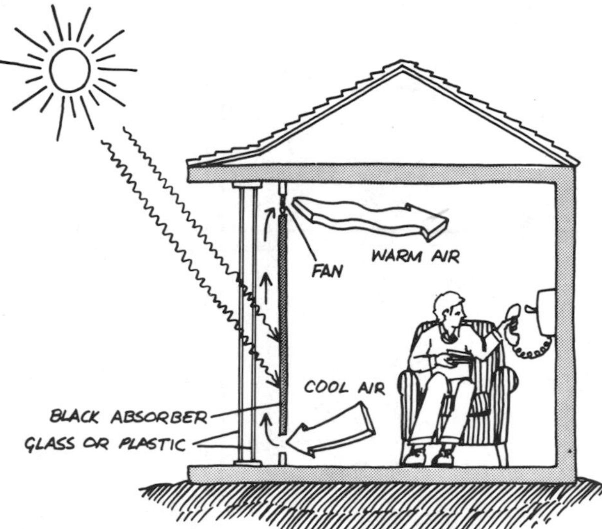

Added fan provides heat control.

Reverse thermosiphoning cools the room.

Damper prevents reverse thermosiphoning.

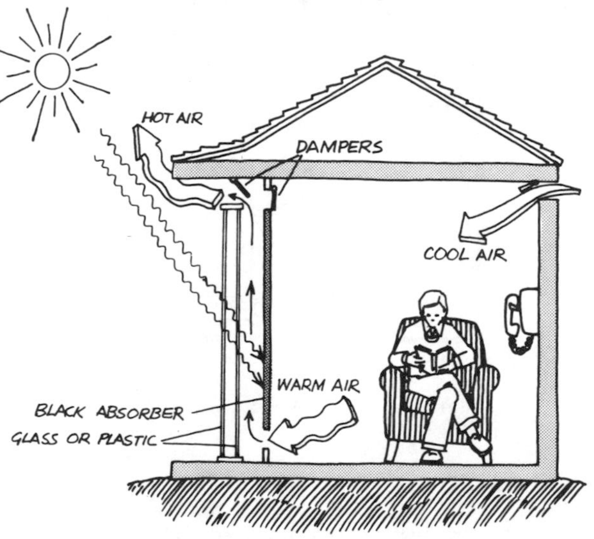

Chimney effect induces natural ventilation.

To provide greater control of air flows, you can add a fan to the exhaust or supply duct of the solar collector. Faster movement of air across the absorber surface boosts the collector efficiency and allows the use of a smaller air gap between absorber and glass. A fan can also deliver warm air to other parts of the house, such as north rooms, or heat storage bins. Using fans with a proper combination of windows and wall collectors, you can simultaneously heat the rooms exposed to the sun and those in the shade.

Dampers help to control the air flow and prevent the cooling effect of *reverse* thermosiphoning. When the sun isn't shining, the air in the collector loses heat by conduction through the glass and radiation to the outside. As this air cools, it travels down the absorber face and flows out into the room. Warm room air is drawn in at the top to the collector and cooled in turn. Although this reverse thermosiphoning can be a benefit in summer, it is most undesirable in winter. It can be prevented by shutting dampers at night.

You can design the dampers to operate manually or automatically. Natural air currents or fan pressure can open or close them. You can also use the dampers to prevent overheating by inducing natural ventilation through houses. Cool air can be drawn into the house from the north side and warm air expelled by the "chimney" exhaust system shown here. As with all air-type solar heating devices, the dampers should be simple in design and operation.

They should close tightly and there should be as few of them as possible.

CONCRETE WALL COLLECTORS

Heat storage capacity can be added directly to these vertical wall collectors. The overall simplicity of this synthesis of collector and heat storage is compelling. Large cost reductions are possible by avoiding heat transport systems of ducts, pipes, fans, and pumps. Operation and maintenance are far simpler, and comfort and efficiency generally greater, than with systems having separate collectors and storage.

The schematic diagram shows a concrete wall used as a solar collector *and* heat storage device. When sunlight strikes the rough blackened surface, the concrete becomes warm and heats the air in the space between wall and glass. Some of the solar heat is carried off by the air, which rises and enters the room, but a large portion of this heat migrates slowly through the concrete. The wall continues to radiate heat into the house well into the night—after the thermosiphoning action has ceased.

With good exterior insulation, such concrete wall collectors can be sized to maintain comfort for 2 or 3 days of sunless weather. Rigid insulation on the interior wall will reduce the amount of direct radiant heat from the wall and keep the rooms from overheating. The solar heat stored in the wall is then retrieved primarily through the thermosiphoning air circulation. In cold climates (5000+ degree days), no insulation is needed on thick walls (10+ inches) because the interior surface rarely becomes too warm.

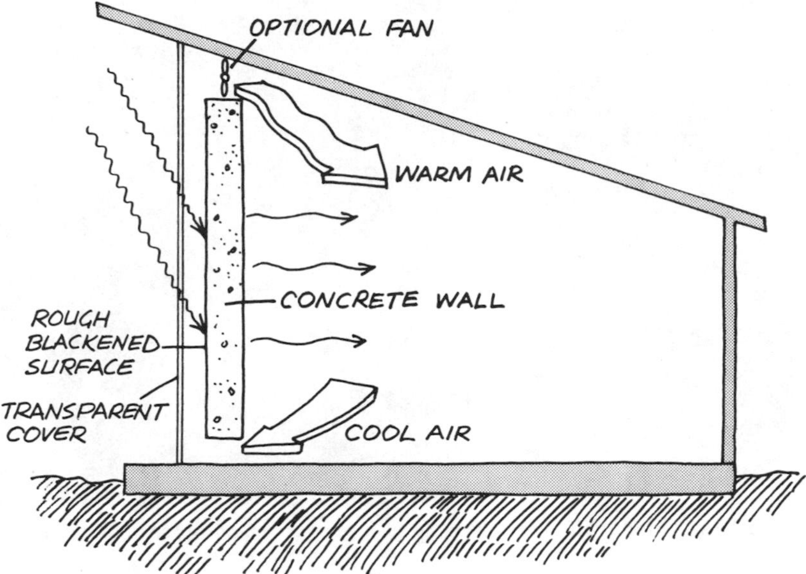

A concrete south wall collector. Solar collection, heat storage, and heat distribution are combined in one unit.

The 1967 Odeillo Houses

The pioneering work in concrete wall collectors was done at Odeillo under the direction of Professor Trombe and architect Jacques Michel. The first actual dwelling to make significant use of concrete wall collectors was built there in 1967. The south

One of the 1967 Odeillo Houses. Close-up of concrete wall collector shown below.

face of the house is a 2-foot thick concrete wall covered by double panes of glass 4 inches away. The outside surface of the concrete is rough in texture and painted black to absorb the solar radiation that penetrates the glass. The 4-room house has a floor area of 818 square feet and a collector area of 516 square feet.

This collector operates in a fashion similar to the one in the previous diagram. After solar radiation penetrates the glass, it is absorbed in the black coating—heating the concrete to temperatures as high as 150°F. As the longwave thermal radiation cannot penetrate back through the glass, the air in the gap between the wall and glass becomes hot and rises. This warm air passes through slots in the top of the wall and into the room, simultaneously drawing cool room air in through ducts at the bottom. This thermosiphoning action continues until the outside wall surface temperature returns to 70°F—about 2 to 3 hours after sunset. To prevent reverse thermosiphoning, the lower ducts are located above the bottom of the collector. Cool air settling to the bottom at night is trapped there.

A large portion of the solar heat is stored directly in the concrete and migrates through the wall with a time delay of 10 to 15 hours. At night, the rooms are warmed by thermal radiation from the inside surface. On a clear day, as much as 70 percent of the solar heat eventually reaching the rooms enters by this process. The rest of the heat is brought in by the thermosiphoning air circulation.

Two of these houses have been occupied since 1967 by engineers of the *Laboratoire de l'Energie Solaire.* Because they are not very well insulated, the houses lose about 22,000 Btu per degree day. Nevertheless, the concrete wall collectors supply 60 to 70 percent of the heat needed during an average Odeillo winter—where temperatures frequently plummet to 0°F. From November to February, the collectors harvest more than 30 percent of the sunlight falling upon them. Over a typical heating season, this passive system supplies about 200,000 Btu (or the usable heat equivalent of 2 gallons of oil) per square foot of collector. The collectors, which replace the entire south wall of these houses, cost between 11 and 12 dollars per square foot.

In 1971, another house was built in the Meuse—near the Belgian border. With 1080 square feet of floor area, this house is larger and better insulated than the Odeillo prototypes. Its concrete wall collector is 16 inches thick and triple glazed. While the Odeillo houses had about 1 square foot of collection surface for every 2 cubic feet of house volume, this dwelling has only 400 square feet of collector—or 1 square foot for every 2.5 cubic feet. The winters are warmer but cloudier here, and the concrete wall supplies 40 to 45 percent of the heat needed.

Scale model of the 1974 Odeillo Residences.

The 1974 Odeillo Residences

A three-unit dwelling in Odeillo, also designed by Jacques Michel, uses a Trombe wall solar heating and cooling system. Windows are incorporated into the collector walls of this building, and the concrete is painted dark blue, red or green instead of black. Other major differences include additional air ducts at the top of the collector and dampers at these ducts to direct the flow of heated air either into or out of the dwelling.

In summer, the dampers are positioned (as shown) so that solar heated air is exhausted at the top of the collector. This "chimney effect" draws warm room air to the base of the collector and cool outdoor air into the house through a vent in the north wall. In this way, the sun is used to drive a natural

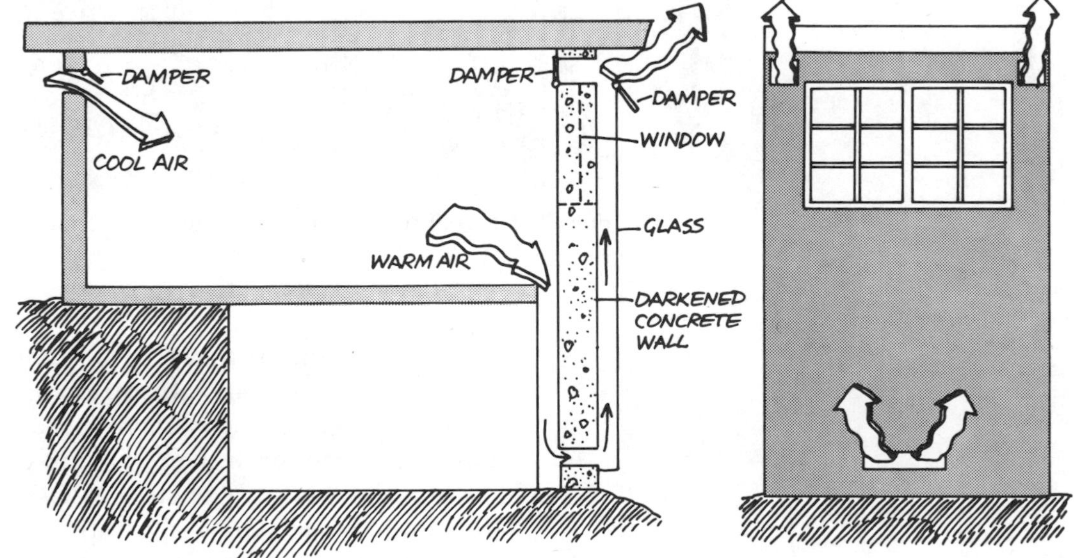

Cross-section and front view of solar heating and cooling system in the 1974 Odeillo Residences—summer operation.

South wall of Tyrrell House near Manchester, New Hampshire.

ventilation system that provides summer daytime cooling.

In winter, the dampers are positioned so as to direct the solar heated air *into* the house—just as in the earlier houses. This gravity convection continues to supply warm air at night, but at a reduced rate. Reverse thermosiphoning during the early morning hours or on cloudy days is prevented because cool air from the absorber face settles into the "kitchen-sink trap" at the collector bottom. Migration of solar heat through the 16 inches of concrete takes 8 to 12 hours. For a well-insulated dwelling such as this, Michel recommends about 1 square foot of collection surface for every 3.5 cubic feet of house volume.

Tyrrell House

Nightly heat losses through the glazing can degrade the performance of a concrete wall collector. A house near Manchester, New Hampshire, licks this problem by incorporating Beadwall insulation into the collector design. The 1500 square foot house was designed and built for Ralph Tyrrell and Holly Anderson by Total Environmental Action, Inc., a research, education, and design firm located in Harrisville, New Hampshire. The exterior walls are concrete, and those on the north, east, and west are banked with earth, sometimes up to the eaves. The exterior surfaces of the concrete walls are insulated with two inches of polyurethane; and the ceiling with 8 inches of fiberglass batts. The only auxiliary heat is provided by two wood stoves; thus, the concrete slab floor was purposely left uninsulated to prevent winter freeze-up when the Tyrrells were away for several days.

As shown in the drawing, the south wall of the house is a composite of windows and vertical wall collectors. The foot thick concrete walls are covered by two sheets of Sun-Lite (a fiberglass-reinforced polyester manufactured by the Kalwall Corporation of

Manchester, New Hampshire) spaced 2¾ inches apart. On cloudy days and winter evenings, styrofoam beads are blown from storage drums to fill this gap. On sunny winter mornings the beads are sucked out to expose the blackened concrete surface to the sun. Some difficulty was originally encountered in getting the Beadwall to fill completely and evenly, and the collectors worked throughout the first winter without any insulation. The source of this difficulty is not yet understood, but it is not believed to be an inherent fault of Beadwall itself.

Summer and winter operation of these collectors is quite similar to that of the Trombe-Michel solar walls used in the 1974 Odeillo residence. As shown in the photograph (taken before the Kalwall was installed), a series of air ducts run along the top and bottom of the collectors. Warm air in the 3½ inch space between concrete and the inner Kalwall pane rises and flows into the room via the top ducts. Additional heat migrates through the concrete with an 8 to 10 hour time delay. With the Beadwall insulation in place, natural convection continues unabated well into the night. A set of dampers allows the Tyrrells to control the airflow manually in order to provide either winter heating or summer cooling. In summer, the collectors are exhausted to the outdoors—drawing cool air into the house from vents in the north wall.

In addition to the 300 square feet of concrete wall collectors, another 150 square feet of south windows admit solar heat directly into the house. At night, these double-glazed windows are covered inside with 2-inch styrofoam shutters that prevent excess heat loss. In the first winter of operation, which was abnormally cold, the solar heating system performed well—without any help from the Beadwall. The Tyrrells claim they needed only one cord of wood beyond what they normally used in the wood cooking stove.

Tyrrell House collectors under construction.

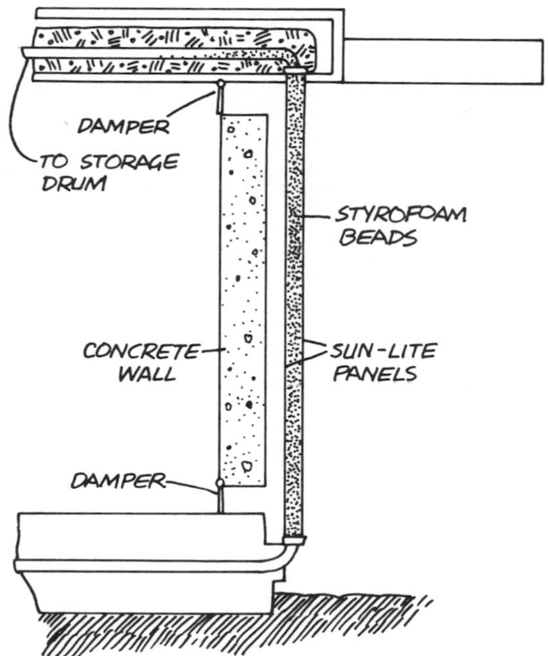

Beadwall insulation used with a concrete wall collector—nighttime operation.

Concrete Wall Variations

A number of other variations in the design

of concrete walls are possible. You can construct the wall from hollow masonry blocks and fill the voids with sand, earth or concrete. Empty voids can be used as air ducts for thermosiphoning. You can also fill the voids with vinyl plastic bags of water, as discussed in the following section on water wall collectors. The water stores twice as much heat as the same volume of concrete and delivers it to the rooms far more rapidly. Brick or adobe can also be used instead of concrete and need not be painted black—if they are dark enough.

There are advantages in making the space between the concrete wall and the glass covers wide enough for human use. The space can be used as a porch or vestibule—or even as a greenhouse. But the thermosiphoning heat flow to the interior does not work very well for such large spaces because the air does not get quite as hot. Fortunately, there will still be large heat flows by conduction through the wall and radiation to the rooms.

The Hofman house has a concrete south wall combined with a greenhouse.

The Hofman house in Canterbury, New Hampshire, was designed by Dan Scully of Total Environmental Action to use a greenhouse as a solar collector. The interiors of this 2-story house are designed to encourage natural circulation of heated air. The concrete south wall has two layers of Kalwall Sun-Lite covering it, with a curved third layer forming a greenhouse. Heated air from the concrete wall can be directed into either the ground floor or the floor above by dampers and registers. Heat lost from the wall warms the greenhouse before passing to the outdoors, thereby extending the short New Hampshire growing season. At the same time, the concrete wall keeps the greenhouse from becoming overheated on clear sunny days. Excess heat is absorbed in the concrete and held there until needed by the house or greenhouse.

Biosphere

A further exploration of this theme is the Biosphere—a conceptual integration of house, greenhouse, solar heater, and solar still. As conceived by physicist Day Chahroudi, the space between the transparent surface and the concrete wall has become large enough for full scale gardening. Solar energy is absorbed by the concrete wall, the ground, and the plants themselves. Plastic water-filled tubes lying in troughs between the rows of the garden provide even more heat storage. A rubber liner several feet below the surface of the ground keeps water from draining away and provides some insulation. The transparent cover of the greenhouse consists of several layers of plastic, including a protective layer and a layer of clear acrylic bubble-foam.

Excess heat that builds up in the enclosure is vented into the house as needed. The 400 square foot garden sits in a tropical environment and provides food for an average family the year round. Very little water is lost from the enclosure, and soil nutrients are replaced by composting the kitchen wastes with plant remains.

Together with Sean Wellesley-Miller of MIT, Chahroudi is now developing a thin plastic membrane whose solar transmittance varies with temperature. When cool, it transmits up to 95 percent of the solar radiation striking it, but it is almost totally opaque when warm. This membrane would provide high solar heat gain into the greenhouse during sunny winter weather and almost none during excessive summer heat. Except in the outer layers of the human skin, this combination of opacity and transparency has never before been achieved without the use of mechanized controls and moving parts. Although the Biosphere itself remains only a concept, it clearly demonstrates the efficiencies that come with an integration of life functions in solar home design.

Wall, Window, and Roof Collectors

Ease of construction is perhaps the most important reason for the emphasis upon vertical wall collectors rather than sloping roof collectors. Glazing is much easier to install, weatherproof, and maintain in a vertical orientation. The cost difference between windows and skylights is testimony to this fact. It is so much easier to keep weather out of vertical surfaces than tilted or horizontal ones. There are fewer structural

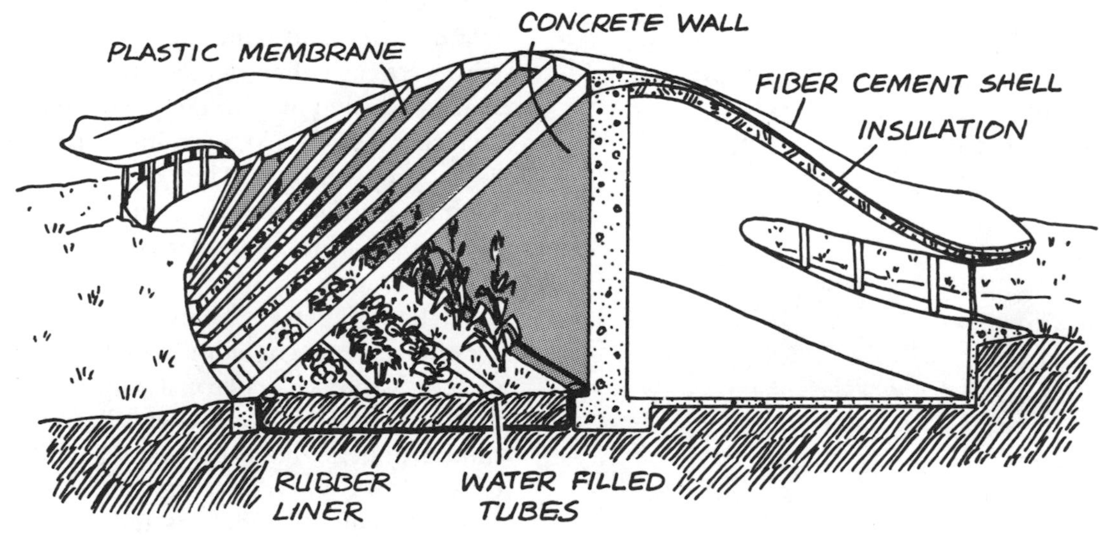

Day Chahroudi's Biosphere produces its own food, heat, and fresh water.

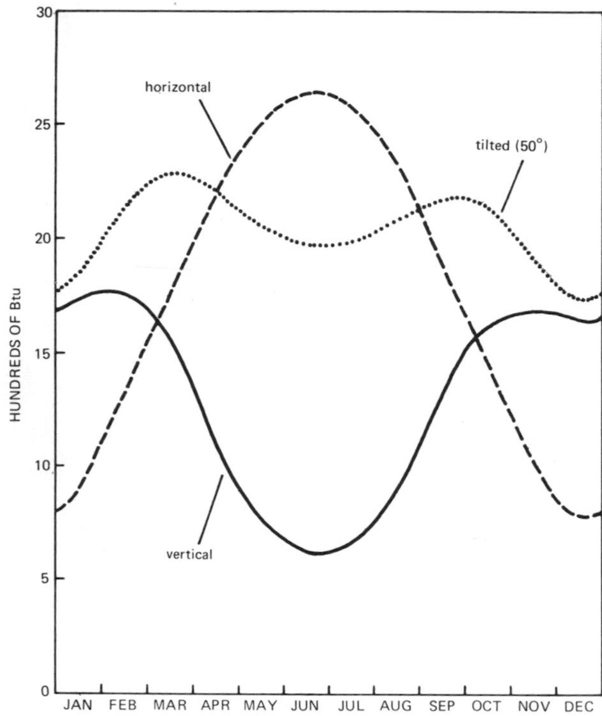

Clear day insolation on horizontal surfaces, and on south-facing vertical and tilted surfaces. Reflected radiation not included.

complications with walls than with roofs, and you needn't worry much about hail or snow build-up. Another important architectural constraint of large, steeply pitched roofs is that interior space under such roofs is difficult to use.

The total amount of clear day solar heat gain on south walls follows the seasonal need very closely. In most of the United States, the greatest heat gain on vertical south walls occurs in December and January, the coldest months, and the least occurs in June and July. The mid-winter clear day insolation on vertical south walls is only about 10 percent less than that on tilted roofs facing south. With an additional 10-30 percent more sunlight reflected onto vertical surfaces from fallen snow, they can actually receive *more* solar heat gain than tilted surfaces. Other types of reflectors, such as swimming pools, white gravel, and concrete walks, work best with vertical collectors. South walls can be shaded easily in summer, preventing the collector surface from attaining dangerously high temperatures. The small amount of summer sun striking a south wall can be used to aid natural ventilation as in the Odeillo houses.

At first glance, it seems foolish to remove a window which admits light and heat directly, only to replace it with an opaque wall solar collector. By now, however, you should understand the advantages of a mix of windows and collectors, as used in the 1974 Odeillo Residence and the Tyrrell house. Interior wall surface is lost if the entire south facade is glass, and excessive sunlight can damage furniture, floors, and fabrics. A section of wall provides an interior space in which you may place delicate objects that could not take direct sunlight.

People can be very uncomfortable when the sun shines directly upon them. Overheating is often a problem with an all-glass wall, even with massive floors and partitions. But with solar collectors and heat storage in the south walls, the excess heat can be transported to cool parts of the house or trapped and stored for later use.

WATER WALLS

Water is an even better heat storage medium than concrete. A cubic foot of water can store 62 Btu for a 1°F temperature rise, while the same volume of concrete can store only 32 Btu. The convection currents in water rapidly transfer the solar heat from the collection surface to the rest of the volume of water and thence to the rooms. Under similar operating conditions, the collection surface of a water storage wall will remain much cooler than that of a concrete wall. Consequently, water walls lose less heat to the outdoors than concrete walls and are more efficient collectors of solar energy. And water is cheap and plentiful virtually everywhere in the United States.

Being a liquid, however, water would rather lie flat than stand erect. This propensity is not a problem during arctic winters, and the igloo shows a remarkable use of a plentiful local material for shelter construction. For use as a solar heat storage medium, however, water must have temperatures in excess of 70°F—where it is not quite as rigid! The structures or containers necessary to support and confine water are often the major expense involved in the construction of a water wall collector. They must be strong and corrosion-resistant.

Early Water Walls

The use of water wall collectors began in 1946 with the second MIT Solar House, discussed briefly in Chapter 2. Six different wall configurations were tested simultaneously under the same operating conditions. High nocturnal heat losses and the resulting low collection efficiencies achieved in these tests led the MIT team to abandon this avenue of research. Insulation materials and techniques have been vastly improved since 1946, however, and water walls are being tried once again.

A solar house built at Odeillo in 1962 used a combination of windows and water wall collectors on its south facade. The primary difference between this house and the others at Odeillo is that water instead of concrete was used as the thermal storage medium. The solar collectors for this house (see diagram) were water-filled radiators painted black and located between the south-facing glass and the interior rooms. The inside surfaces of the radiators were insulated to control heat flows to and from the rooms. Solar heated water rose behind the absorber surface into the tank above, while cooler water sank down to replace it. Heat flowed to the rooms by thermal radiation from the bottom surface of the tank.

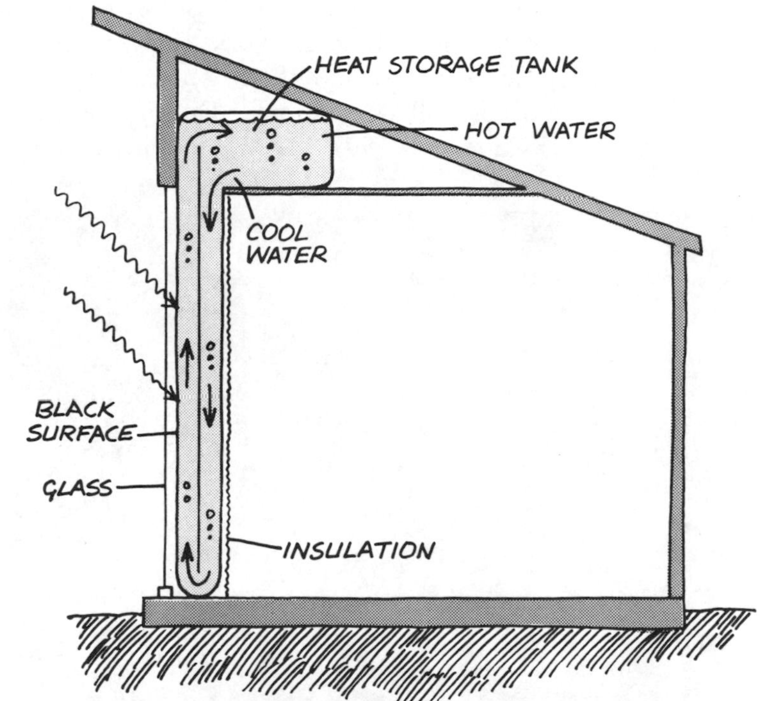

Water wall collector in 1962 Odeillo house.

Baer House

Steve Baer, one of the most outspoken advocates of passive systems, built a unique structure that uses south-facing walls of water to collect and store solar heat. Together with Zomeworks Corporation, Baer developed a modified dome with vertical walls called a "zome." During the fall and winter of 1971-72, he built his own zome dwelling in Corrales, New Mexico. This structure is a cluster of "10 exploded rhomic dodecahedra stretched and fused to form the different sized rooms." The walls and roofs are made of a cardboard and paper honeycomb filled with urethane foam, and covered by an aluminized surface.

South face of Steve Baer's zome. Insulating shutters are lowered on winter days to expose the drumwalls.

The four south-facing walls of this house are arrays of 55-gallon oil drums filled with water and stacked horizontally. The drums are oriented north-south with the outward facing ends painted black. The 20 to 25 drums in each wall sit behind single panes of ordinary glass. Each wall has an area of about 100 square feet, for a total collector area of 400 square feet. The drums themselves have a total blackened area of 260 square feet facing the sun.

The collector surfaces are covered at night or on cloudy days by insulating shutters which are hinged at the base of each wall. Each shutter lies flat on the ground during sunny days with its aluminum inner surface reflecting additional sunlight onto the drums. At night the shutter is hoisted back into the vertical position by a hand crank and nylon rope. The shutters are made from the same aluminized, urethane-filled honeycomb that is used for the walls. Each shutter weighs about 150 pounds and costs about 2 dollars per square foot. Steve admits that the shutters do not attain the predicted R-value of 10 because of air leaks around their edges. But reflected sunlight from the aluminum inner surface compensates for these heat losses when the sun shines.

On sunny mid-winter days, these Drumwalls absorb up to 1400 Btu of solar energy per square foot of glass. Convection currents within the drums rapidly distribute this solar heat to the remaining volume of water—keeping the solar collection surfaces relatively cool. Steve claims that water temperatures remain below 100°F and that 60 to 70 percent collection efficiencies are common. The drums distribute the heat to the rooms by radiation and natural convection, controlled somewhat by a sliding curtain located behind the Drumwall.

Although the house has only 400 square feet of collection surface for 2000 square feet of floor area, the temperature inside usually remains between 63°F and 70°F during the winter. The 5000 gallons of water in the drums, together with the concrete slab floor and adobe partitions, moderate the extremes of temperature. During cloudy winter weather, the indoor temperature drops 2 to 3°F per day. Daily temperature swings of more than 5°F are rare. If swings of 10°F are allowed, the house can be 75 percent solar heated by the Drumwalls alone. Auxiliary heat is supplied by two wood stoves, but the Baers use less than a cord of wood each winter.

That the Drumwall can be an aesthetically pleasing contribution to interior living spaces is evident from the photograph. The drum surfaces facing inward are painted white, and several plants are suspended from the metal support structure. One of the most striking features of this design is the soft and seemingly mystical lighting that emanates from the spaces between the drums.

During the summer, the shutters are lowered at night and the Drumwalls release

Interior view of Drumwall in the Baer House.

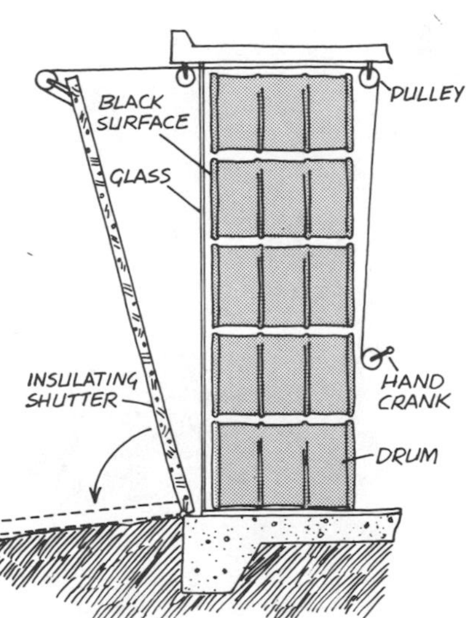

Hand-operated insulating shutter and drumwall.

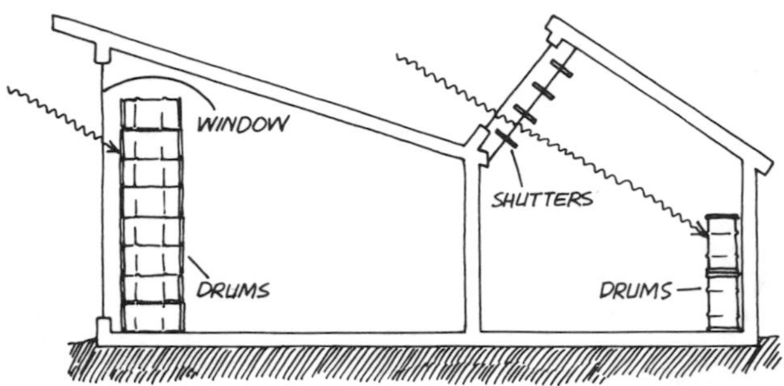

Two locations for Drumwalls.

stored heat to the cool desert air. With the shutters closed during the day, the cool water in the drums acts as a sink for excess interior heat. The reflective aluminum skin of the entire zome cluster also helps to shed the hot desert sun.

Other features of the house include two thermosiphoning solar water heaters and a windmill that pumps well water into a 5000 gallon tank above the home. Additional solar heat enters the building through Skylids and other apertures in the roof. Altogether, the house has 588 square feet of glass on its south, southeast, or southwest facades.

The Drumwalls have been in routine use since 1972 with no major failures. Two zinc-bunged drums have sprung leaks, but there have been no problems with the steel-bunged drums. Baer acknowledges that all the drums will rust out eventually, but he expects lifetimes of 20 to 30 years. With his innovative yet simple design, Steve seems to have solved the problem of high cost of a water support-and-containment structure. At a cost of $2 to $4 for a used oil drum ($9 to $10 for a new one), an owner-builder can keep the total costs of a Drumwall below $6 a square foot. Drumwall plans are available for $5 from Zomeworks Corporation.

Southwest view of the Baer House in Corrales, New Mexico.

Water Wall Variations

There are many other ways of getting large masses of water inside of your house. Drums or other containers can be stacked against interior walls under skylights or Skylids. Or for those who have trouble embracing the aesthetics of Drumwalls, the drums can be hidden under window seats. As discussed in the last chapter, the Wright house has water-filled drums embedded in adobe sitting at the base of its enormous south window. In the Winters house (discussed in Chapter 2), a single row of drums is located on the floor, just behind the south windows. Other drums further inside the rooms provide extra thermal mass not directly exposed to the sun.

Steve Baer suggests stacking and bracing the drums vertically—in order to have fewer pounds of water per square foot of collector surface. In fact, recent studies made at Los Alamos Scientific Laboratories indicate that a 6 inch thickness may be the optimum for water walls. The storage walls of the second MIT house were close to this in thickness.

The Kalwall Corporation is now marketing moulded cylinders of Sun-Lite, their fiberglass-reinforced polyester, that are ideal for storage wall collectors. The tubes can sit directly in the rooms behind a south window. But some designers worry that the tubes might leak or even topple (during an earthquake!), with disastrous results for floors and furniture. In one solution to this problem, a thin wall is placed between the tubes and the room. Heat is delivered to the room by forced or natural convection via vents in the wall. At a cost of $25 for an 8 foot cylinder 1 foot in diameter, the total costs of such a system could be kept below $8 per square foot of collection surface.

A final variation, suggested by Harold Hay of Sky Therm Processes and Engineering in Los Angeles, combines these water walls with concrete wall collectors. His ingenious idea is to build hollow masonry-block walls as shown and to fill the cavities with vinyl plastic bags containing water. The entire wall is then faced with glass as in a normal concrete wall collector. In addition to increasing the heat storage capacity of this wall over that of a normal concrete wall, the water bags aid conduction heat flow to the interior, and the cooler outer surface makes a more efficient collector of solar energy. This idea will be tried in Hay's Solarchitecture House, discussed in the following section on roof pond collectors. You might want to consult Hay before building such a collector wall; he owns a patent on the idea and can

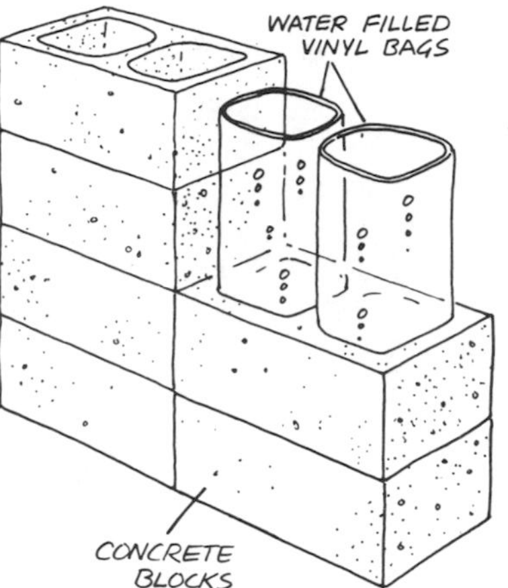

Concrete-and-water wall.

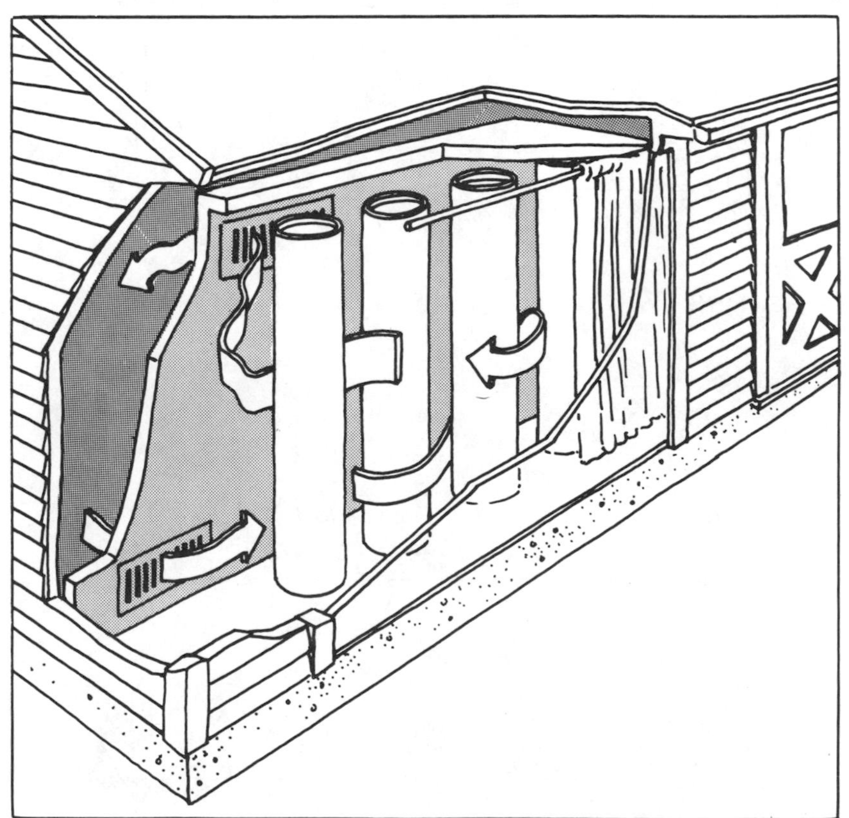
Water-filled Sun-Lite cylinders used for integrated collection and storage.

SUMMER COOLING

WINTER HEATING

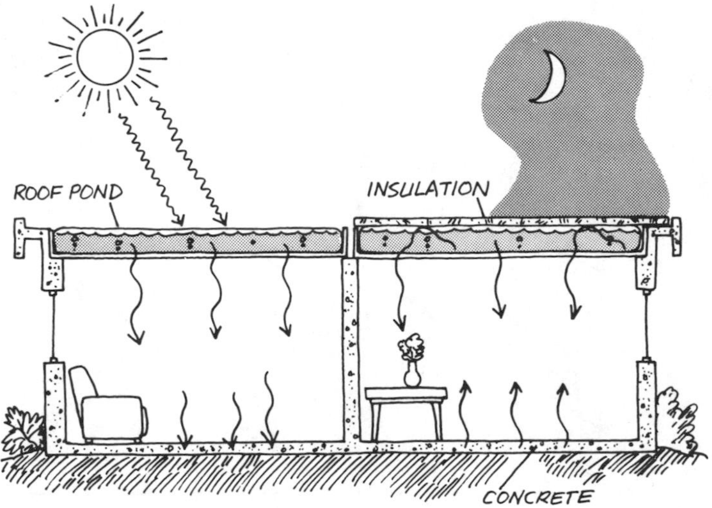

Roof pond collectors—summer and winter operation.

furnish you with helpful suggestions.

ROOF PONDS

Containers of water on the roof of a house can also be used to collect and store the sun's energy. A container tilted at 45°-60° would have the best exposure to the winter sun, but supporting and confining a large mass of water at such an angle is a formidable task. More feasible are horizontal rooftop collectors, which confine the water by exploiting its propensity to lie flat. The water can be kept in shallow pans or in plastic bags upon sheet metal decking and supported by concrete walls or thick wooden beams. The solar heat collected in such roof ponds is radiated directly to the rooms below. Very even heating occurs, and water temperatures as low as 70°F can still be used to heat the house.

Unfortunately, roof pond collectors also have a number of drawbacks. As the sun is low on the horizon during winter, the daily insolation on a horizontal surface is the least when it's most needed. On a clear December day in Philadelphia, for example, the sunlight striking a horizontal roof is less than half that hitting an equal area of a concrete south wall. To compensate for this deficiency, the ponds must cover most of a roof. Stratification of the heat within the roof pond itself is another problem—though a minor one. The warmer, lighter water remains near the top of the pond at night—

losing heat to the outside air. The cooler, denser water falls to the bottom, just above the rooms, where heat is needed the most. And potential snow accumulation limits the use of roof ponds in colder climates.

Roof pond collectors are much better suited to lower latitudes—those between 35°S and 35°N. At these latitudes, the sun climbs higher in the sky on a winter day, and snow build-up is not a threat. Also, roof ponds are extremely well suited to the purpose of summer cooling, which is more important than winter heating in these climates. Insulated panels are used to cover a roof pond on summer days—as in the Winters House. The water absorbs excess heat from the rooms below and radiates this heat to the sky at night, when the panels have been removed. Heat stratification within the roof pond aids this cooling because the coldest water lies just above the rooms. During the winter, the movable insulation is operated in reverse. It is removed on sunny days to permit solar collection and replaced at night or on cloudy days to trap the collected heat.

The Phoenix Test Building

For the most part, the development of roof pond solar collection and nocturnal cooling systems is the work of Harold Hay. A chemist and building materials specialist, Hay has toured much of the world on behalf of the United Nations and the U.S. Department of Housing and Urban Development. He conceived the idea of movable insulation in India, when he witnessed the human discomfort that occurred when native thatched roofs were replaced by sheet metal. Stripped of their natural insulating layer, these houses subjected the poor inhabitants to intense summer heat and chilling winter cold. Without any recourse to Western methods of heating and air conditioning, they shivered through winter nights and sweated through summer days.

Returning to Phoenix, Hay performed an experiment to test his ideas about movable insulation. He filled a plastic-lined styrofoam box with water and set it out in the open. During the summer months, he put a styrofoam lid on by day and removed it at night. During the winter, he reversed this cycle, opening the lid by day and closing it at night. The water remained cool in summer and warm in winter. A little puzzled about the significance of his results, Hay showed them to John Yellott, an engineer and perennial advocate of solar energy who told him, "What you have there, Harold, is an effective natural air-conditioning system!"

Together, the two men set out to test this principle in an actual structure. In 1967, they built a one-room, single-story building with an adjoining carport. On the roof of this tiny building was a series of ponds with a total area of 170 square feet and two movable insulating covers above them. Open troughs of water were used at first, but these had to be covered with a transparent plastic sheet because of evaporation. In the final

The Phoenix Test Building, built by Harold Hay and John Yellott.

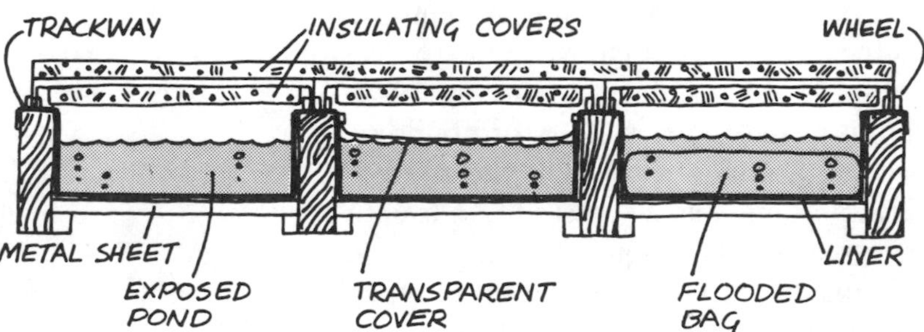

Cross section of roof ponds in the Phoenix Test Building. Several water containment methods were studied.

version, the water was contained in plastic bags supported on corrugated steel decking lined with black polyethylene sheets. Hay and Yellott positioned the rigid polyurethane panels over the ponds or stacked them above the carport by pulling them along aluminum trackways with a rope. During the summer, they removed the covers at night and replaced them by day, just as with the insulated box. They reversed this cycle during the winter, with the roof ponds exposed on sunny days and covered at night or on cloudy days.

Conversion of solar energy into heat occurred in the black plastic liner. The heat flowed readily through the metal decking and radiated into the room below. Excess solar heat was stored in the roof ponds where it was trapped by the insulation until needed in the room.

With *no* auxiliary sources of heating or cooling, the building remained comfortable throughout that year. The extra cooling needed when the temperature rose above 100°F was obtained by flooding the area above the bags with a thin layer of water and letting it evaporate at night. While outside air temperatures ranged from 30°F to 115°F in a normal year of Phoenix weather, the roof ponds kept room temperatures between 66°F and 82°F. During the winter, the pond water often reached a temperature of 85°F by the end of a sunny day.

Later, while traveling and working in Iran, Hay observed that nocturnal cooling is an ancient art. For centuries, the desert people there have used a device known as a *yakh-*

chal to produce ice by thermal radiation to the night sky. A trough is dug in the earth and a mud-brick wall built just to the south of it to provide shade. Water in the trough would freeze hard when the temperature of the night air was as high as 48° F!

The Solarchitecture House in Atascadero, California, looks like any other suburban house when viewed from the street.

Solarchitecture House

In 1973, a full scale house based upon the principles borne out by the Phoenix tests was built in Atascadero, California. The single story house was designed by Hay and architect Ken Haggard. It has an adjoining carport and patio and includes about 1100 square feet of heated floor area. An approximately equal area of roof pond collectors covers the entire roof of the main part of the house. Nine movable insulated panels slide along aluminum tracks. They can be positioned over the roof ponds or stacked three-deep above the patio and carport.

Water is contained in 8 foot by 38 foot transparent plastic bags that sit in four troughs running north to south on the roof. With a total volume of 7000 gallons, the 8 inch depth of water has the heat capacity of about 16 inches of concrete but the weight of only 4 inches. The waterbed-like bags sit on an earthquake-proof steel deck supported by interior and exterior concrete block walls. A black plastic liner between the water bags and the roof deck protects against leaks and rainwater seepage. Just above the water bags, an inflatable transparent plastic cover protects them from the sun's ultraviolet radiation and helps trap the solar heat.

The nine insulated panels slide on wheels along 5 sets of aluminum guiderails parallel to the 4 troughs. These panels are made from 2-inch rigid polyurethane foam (R=15) insulation. They can be moved manually or automatically in response to a signal from thermostats. A ¼-horsepower electric motor requires only 3 minutes to move the panels from the carport to the roof ponds or vice-versa. During the first winter of operation, there were large heat losses around the panel edges, but these were stopped by installing better seals.

Operation of these roof pond collectors is similar to that of the Phoenix Test Building.

Insulating roof panels on the Solarchitecture House are positioned by a ¼ horsepower electric motor.

Massive interiors of the Solarchitecture House provide additional heat storage capacity.

During the winter, solar heat is absorbed in the black plastic liner, travels through the metal ceiling of the house, and radiates uniformly into the rooms. Excess solar heat frequently warms the ponds to 85°F during the winter. At this temperature, enough heat can be stored to keep the house warm over a period of 4 cold cloudy days in January. During the summer, excess heat from the rooms travels upward through the ceiling and is absorbed in the roof ponds. At night the bags radiate their heat to the night sky—providing cool water for the next day. The plastic is protected from the most intense summer sun by the insulating panels, thereby prolonging its lifetime.

The concrete block walls and interior partitions provide additional thermal mass for the house. The cavities of the interior partitions are filled with sand, while those of the north wall are filled with vermiculite. Hay plans to fill the cavities of the south wall with vinyl plastic bags of water, both to give added thermal mass and for use in a south-wall solar collector.

A research group from the School of Architecture at the California Polytechnic Institute in nearby San Luis Obispo has evaluated the architecture, construction, maintenance, thermal performance, and economics of the house and the tenant reaction to it. During 1973 and 1974, the solar heating system worked well, providing *all* of the heating needs of the house. Only 20 percent of the sunlight hitting the roof was actually used to heat the rooms, but it was enough to supply all the needs. On a typical February day when the outdoor air temperature fluctuated from 32°F to 60°F, the room temperature stayed between 68°F and 72°F. During the entire test period, the room temperature remained between 66°F and 74°F. The tenants were extremely pleased with the cooling system, which they rated as "far superior" to conventional air conditioning. Lack of blowing air and noise, and the uniformity of temperatures throughout the house made it "one of the most comfortable houses" they'd ever experienced.

The cost of the house and land was about $40,000, but Hay attributes $7000 of this to the fact that the house is a prototype. With some hindsight, the research group estimates that the cost of the house alone could have been kept as low as $25,500 (1973 prices)—only a little more expensive than a custom-built house of the same floor area with conventional heating and cooling. The extra costs of the concrete block walls and partitions, the roof ponds, and the insulation panels is offset by the usual cost of roofs, ceilings, and normal heating and cooling equipment. Harold Hay calculates the cost of his system at $5.20 per square foot, which with mass production might be as low as $2.70. By comparison, a typical tile or shake roof in the area would cost $3.50 per square foot—including the cost of a heating and air-conditioning system.

Hay's Sky Therm nocturnal cooling and solar heating system has many obvious advantages over other solar energy systems. It can be installed, operated, and maintained by relatively unskilled people with a minimum of technical supervision. It does not need pumps, compressors, piping, or ducts. Nor does it require any materials other than those common to modern construction practice. The Sky Therm technique provides gentle and uniform comfort without noise or drafts. "It is ironic that a highly developed country is developing a system having greatest merit for developing countries," Hay remarks. "Where a narrow comfort zone is not required, movable insulation can be used without roof ponds in simpler and cheaper ways. Such means could improve the health and productivity of people with low incomes."

Roof Pond Variations

Under a contract from the Department of Housing and Urban Development, a team of Mexican-American farm workers are building five Sky Therm houses in a single city block in Selma, California. The contractor, Self-Help Enterprises of nearby Visalia, hopes to demonstrate that Sky Therm can be built *on site* and installed by properly supervised laborers who are otherwise unskilled in the building trades. The five houses differ in orientation, floor plan, and appearance. Three of the houses have standard frame construction and two have concrete block walls. Built in an earthquake-prone region, each house supports a 12-inch depth of water on its roof. Because they operate at low temperature and face a large portion of the sky, the roof ponds can collect diffuse solar radiation during the foggy winter weather common to California's San Joaquin Valley.

Movable insulation can also reflect additional sunlight onto the roof ponds, as in the Winters House. Similar to the shutters used in Steve Baer's house, the inside surface of these insulating lids is a reflecting aluminum surface. The panels are hinged on

Harold Hay examining the attic ponds in the Freese house.

one end and pivot up to the optimum angle for solar collection at each time of year. The additional sunlight reflected onto the roof ponds compensates for the low angle of the winter sun—allowing the use of smaller ponds. Harold Hay remarks, however, that non-uniform room temperatures can be expected as a result of the decreased area of these ponds. And he worries that the wind will eventually snap the lids off during an unguarded moment.

Several other modifications of roof pond collectors can make them more suitable for colder climates—where freezing is a problem and winter heating is more important than summer cooling. One example is a house built for Dr. Jackson Freese near Concord, New Hampshire. The "attic ponds" in this house were married with Beadwall to provide solar collection in a winter climate where the temperature frequently falls to −10°F. Originally a rambling farmhouse, this building was extensively renovated under the supervision of a local architect. Total Environmental Action provided advice and consultation on the solar heating system. Sitting in the attic of the two story addition to this farmhouse are 14 black plastic waterbed bags holding a total of 3500 gallons of water. With plastic liners underneath them to protect against leaks, the bags rest on the plywood floor of the attic, which is supported by thick wooden beams and insulated from the rooms directly below.

The attic roof is insulated with 9½ inches of foil-faced fiberglass batt insulation, except for a 280 square foot portion of the south-facing roof covered by a double layer of Kalwall Sun-Lite. These panes slope at an angle of 60° to the horizontal. The 2¾ inch space between them is filled with styrofoam beads (R=10) at night and on cloudy days. When the sun is shining, the beads are sucked out by vacuum cleaner fans and stored in a series of recycled oil drums.

Sunlight passes through the Kalwall panes and heats the water in the attic ponds to temperatures between 80°F and 90°F. Because the house is two stories high, thermal radiation cannot be used to heat the bottom floor. And the heavy insulation just below the attic prevents any significant thermal radiation heat flow to the second story. Instead, a large blower feeds the warm air from the attic space above the roof ponds to the rest of the house. Dr. Freese estimates that this attic pond system will provide half his heating needs, or $600 of an annual $1200 bill for fuel oil. When he combines this savings with a $500 deduction on his real estate taxes, he should recoup the $8000 initial investment in less than 8 years.

Another approach to roof ponds in cold climates is called Sky Therm North by Harold Hay, who invented and patented the idea. Instead of Beadwall insulation, he suggests using an insulated shutter behind a glazed roof section that sits above a series of attic ponds. The shutter is hinged at the top and pivots backward to open when the sun is shining—just like the Zomeworks shutter shown in Chapter 4. Its outer surface is a layer of aluminum foil to reflect extra sunlight onto the ponds when the shutter is open. These attic ponds are *not* insulated underneath, and thermal radiation distributes solar heat to the rooms below.

Harry Thomason, whose innovative contributions to solar home heating were discussed earlier, proposes yet another variation of roof pond collectors. Instead of movable insulation to cover the roof ponds, he suggests that the water be drained to a basement reservoir filled with fist sized stones. During the winter, water is pumped to the roof only when the sun is shining. The solar heated water returns to warm the stones, and together they warm the house by conduction and radiation heat flow through the concrete floor. The warm stones continue to heat the house after the water has been pumped back to the roof. With this clever modification that allows heat stratification and natural convection to augment the heat flows into the house, the roof pond system is better suited to solar heating than to nocturnal cooling.

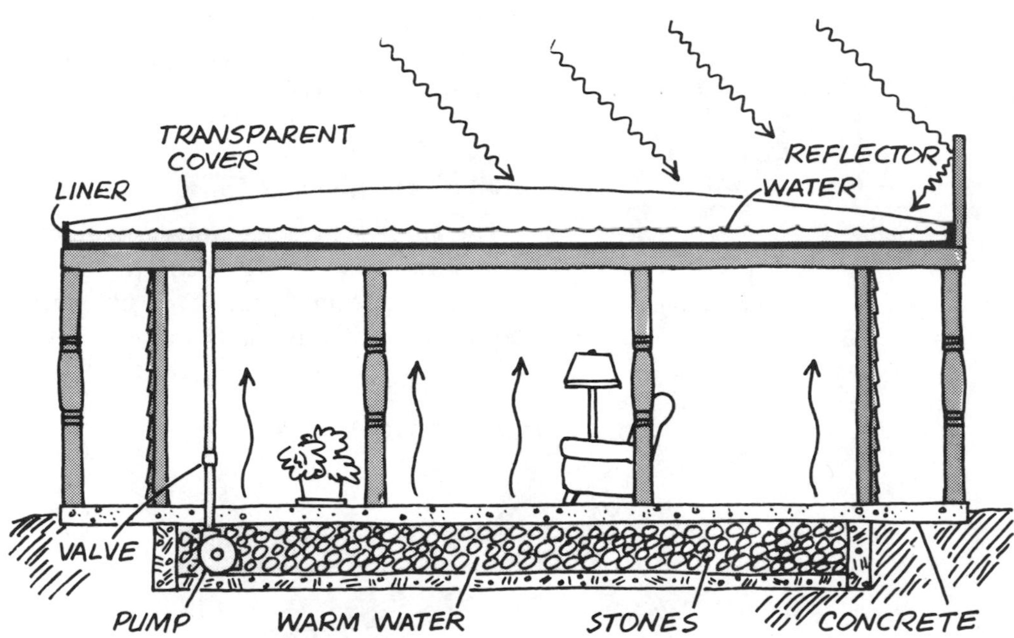

Harry Thomason's novel idea for a roof pond solar collector.

HEAT STORAGE METHODS

Thomason's idea illustrates that not all the heat storage need be built directly into the solar collector itself. Even if some heat storage *is* built in, it need not provide *all* the heat storage for a house. As emphasized in Chapter 4, the walls and floors can store the excess solar heat. If their storage capacity is too small and the collector does not have enough itself, more heat storage can be added elsewhere in the house.

But as you separate the heat storage from the collector, you begin to require mechanical power to move the solar heat around. For example, a fan may be needed to move excess warm air from a concrete wall collector to a subfloor heat storage bin. But most of the heat flow to the room still occurs by radiation and natural convection. Only when more solar energy is collected than needed does the fan spring

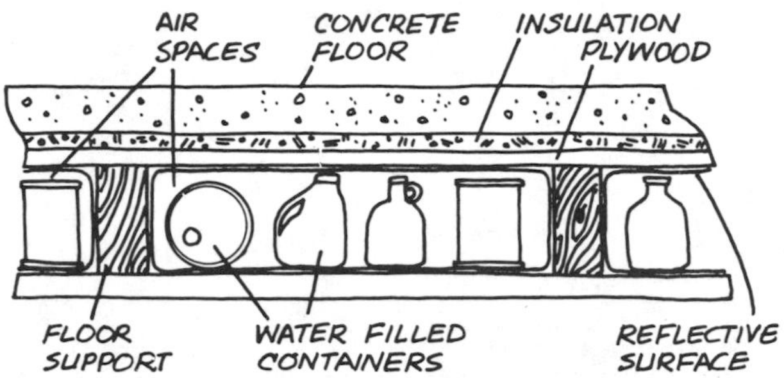

Old gallon jugs and containers used for subfloor heat storage.

into action. The rest of this chapter examines a few simple methods of heat storage that require little or no mechanical power to move the solar heat.

Subfloor Heat Storage

The most convenient place to put extra heat storage is directly beneath the floor. Heat stored there percolates up into the rooms without any mechanical assist. The rock beds beneath Peter van Dresser's house in Santa Fe are an excellent example of subfloor heat storage. Warm floors that result from putting the solar heat there are a special luxury for those of you who like to walk about barefoot.

Waterbed-like plastic bags can be used for subfloor heat storage.

Another place to put extra heat storage is in the spaces between floor joists. Solar heated air is blown through the spaces, which are lined with reflective foil to reduce heat loss and smooth the air flow. The heat is transferred to gallon containers of water made from plastic, glass, or metal and arrayed in an irregular pattern. This is an excellent way to recycle your many used containers but be sure to use some chlorine bleach to prevent algae. If desired, a concrete slab above the joists can be used to store direct solar heat gains from the windows. Rigid board insulation between the concrete and the subfloor storage bins reduces uncontrolled heat losses to the rooms above.

Day Chahroudi suggests another subfloor heat storage scheme that operates on similar principles. Warm air from the collector is circulated through the voids in masonry blocks lying on their sides under the floor. The warmed blocks transfer this heat to water in large plastic bags resting on top of them. Warm air from the top of the bags travels to the house by natural convection, while cool room air sifts down through vents in the floor and is heated again. Yet another alternative is to store water in large oil drums sitting in an insulated crawl space. Solar heated air circulates around and through them, while cool room air picks up this heat and carries it back into the house.

Interior Heat Storage

In similar fashion, you can also store heat in the ceiling, closets, and interior walls. Just as the water containers were arranged between the floor joists, so too can they be stacked on shelves inside the walls. But providing an unobstructed air flow past these containers is more difficult. You can also use tall containers of rocks for heat storage, as Dr. George Löf did in his Denver house. Rocks are better at extracting heat from air and easier to handle than hundreds of small bottles of water. Because of the large voids needed between these bottles, rocks can store almost as much heat in the same amount of space.

If possible, the supplementary heat storage should be located within the rooms themselves. For example, the heat storage tank shown in the diagram holds warm water from a solar collector. The surrounding rocks are heated by the warm tank. Cool room air is warmed by the rocks, rises, and enters the room through the louvers—drawing more room air in at the base. When heat isn't needed, the louvers on the top of the bin are closed. Movable insulating panels can also be used to regulate heat flow to the room. The heat is trapped by the insulation when not needed in the room. When heat is required, the panels are removed so that thermal radiation and natural convection from the surface of the storage bin warm the room.

Air Loop Rock Storage

Solar heat from a concrete south wall or other air collector can be stored in rock bins located inside the house. With clever designs, little or no mechanical power is needed to move the heated air. The accompanying design by Jonathan Hammond illustrates the basic principles of an air loop rock storage system. The heated air rises across the collector face and descends through a vertical bin of rocks in the center of the house. This air cools as it passes through the rocks—becoming heavier and falling faster. Air flow from the storage bin back to the collector passes directly through the living room. The fireplace and flue are embedded in the storage bin to provide supplementary heat during periods of cold cloudy weather.

The temperature difference between the solar collector and the rock storage bin leads to a pressure difference that drives such an air loop rock storage system. Hot air has a higher pressure than cold, and the air will move in a direction that eliminates this pressure difference. Circulation occurs whenever the pressure difference can overcome the resistance to this flow posed by the rocks and other obstructions. And unless the dampers are closed, a reverse air flow may cool the storage bin at night, when the temperature of the rocks exceeds that of the collector.

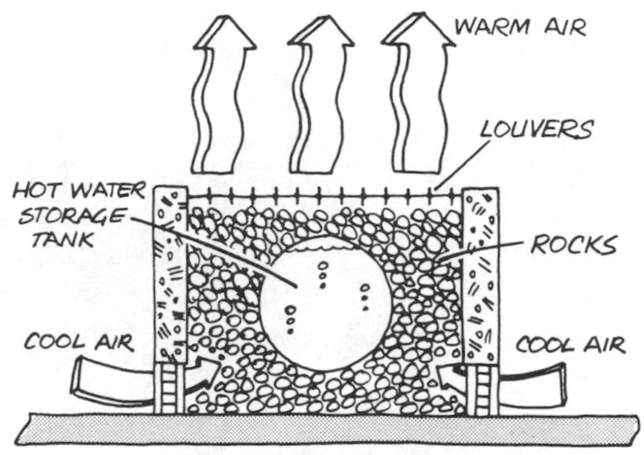

Solar heat storage directly inside the room.

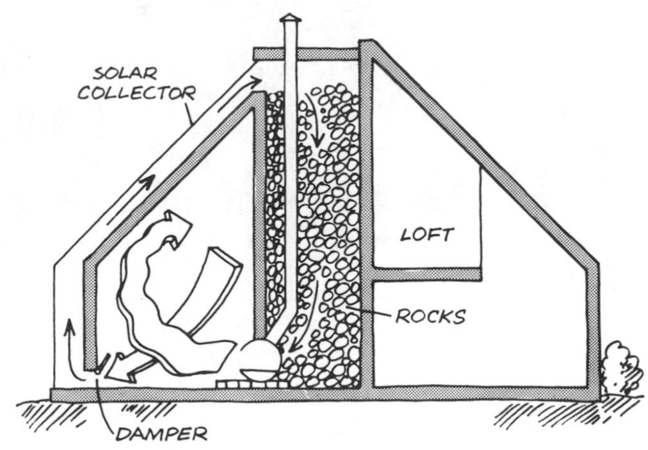

A basic air loop rock storage system for solar heating a house.

Solar collection and heat storage are integrated into the south porch of the Davis House.

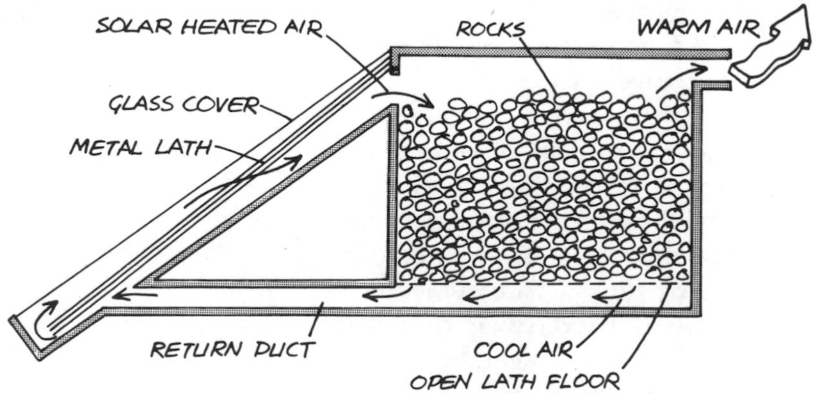

Davis House air loop rock storage system.

The trick in designing air loop systems lies in minimizing the resistance to the air flow. When this resistance is small, high collector temperatures (which would lead to large heat loss from the collector) are not needed to move the air. Steve Baer, who has done most of the seminal work in air loop rock storage systems, offers a number of helpful suggestions in his recent book *Sunspots.* He suggests the use of multiple layers of expanded metal lath as a solar collection surface. The lath readily transfers its heat to the passing air stream but offers little resistance to its flow. Pages 66-72 of this book are essential reading for anyone contemplating building such a solar heating system.

Davis House

Together with his associates at Zomeworks, Steve Baer has designed a number of houses that use air loop rock storage systems for their heating. One of these houses, built in 1972 by his neighbors Paul and Mary Davis, is shown in the photograph with Steve's zome in the background. The house is molded into a south facing slope, with the living quarters located above an air loop rock storage system built into the base of an expansive porch. This ingenious solar heater uses no mechanical power for its operation, yet it provides about 75 percent of the heat needed in the 1000 square foot house.

Solar heat gathered in the 320 square foot collector is stored in a rock bin located directly below the porch. Sunlight penetrates a single sheet of glass and is absorbed in 6 layers of blackened metal lath. Warm air rises through the lath and flows to the top of a 4 foot deep storage bin containing 43 cubic yards of cobblesized-rocks. As the air warms the rocks, it cools and falls through an open lath floor into a duct that returns the air to the base of the collector. On a sunny day, as much as 1000 Btu per square foot can be collected and stored in this manner. By mid-afternoon, the rocks have been warmed to 90° F—even though the outdoor air is a frigid 20° F. Reverse thermosiphoning during winter nights is prevented by dampers, but such a cooling process is encouraged during the summertime.

When warm air is needed in the house, other dampers are opened to permit cool room air to travel through a crawl space to the duct under the rock storage bin. This air rises as it is warmed by the rocks and re-enters the house through vents located in the floor just behind the south windows. The open interior of this two-story house permits this warm air to circulate up into sleeping lofts and down along the back wall before returning to the base of the rock storage bin.

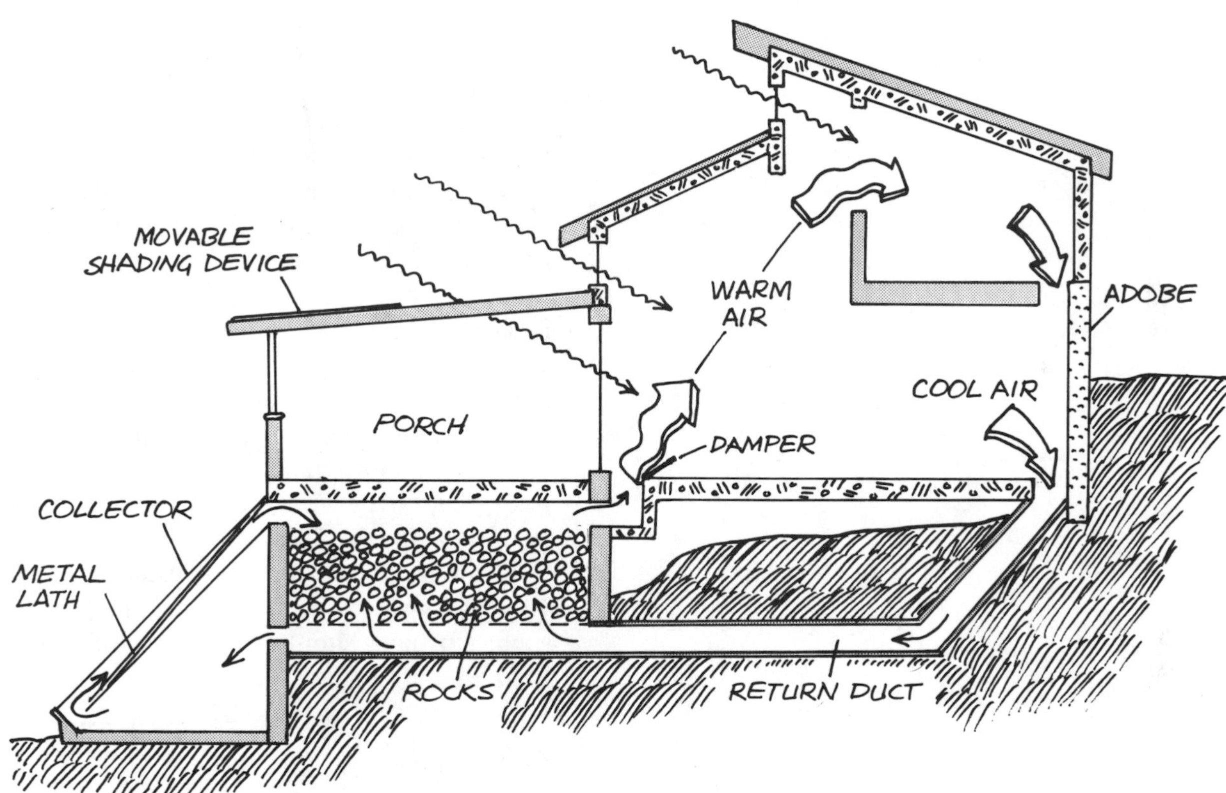

Solar heating system in the Davis House. There are two distinct flow loops—collector-to-storage and storage-to-house.

The house has a number of energy-conserving features that help it to retain the solar heat it gathers. The north and south walls of the bottom story are made from 10-inch adobe with 2 inches of styrofoam insulation on the outside. The east and west ends have double adobe walls—each 10 inches thick—with 4 inches of fiberglass in the cavity between them. In winter, sunlight floods in through the south windows and clerestories to provide direct solar heat. But in summer, fixed and movable overhangs above these windows block the summer sun. On summer nights, the clerestory windows can be opened to discharge warm air from the house while cooler desert air is drawn in at the north wall.

Paul Davis, a professor of English at the

University of New Mexico in nearby Albuquerque, is the most eloquent spokesman for the performance of this home. For him, "The natural aesthetic of the thermosiphon house produces a sense of being in a warm place rather than a warmed place. Its simplicity, depending on the most obvious of scientific principles and having no parts to break, no plumbing to leak, make it an ideal system for the non-mechanic." Given another chance, he would make two improvements in the design–he would put the rock storage under the house to keep heat losses "in the system," and he would add a greenhouse onto the collector to help raise the humidity inside the house. The design of such air loop rock storage systems is still a very tricky matter and anyone contemplating such a project should get some good advice before beginning.

Passive and Active Systems

In the Davis house, solar collection and heat storage occur *outside* the house–in distinctly separate units. In retrospect, this solar heater is a hybrid between a passive system and the indirect systems to be examined in the following chapter. Heat flows by natural convection, but large ducts and dampers are needed to direct and control this flow. You can identify *two* distinct heat flow loops–from collector to storage and from storage to house. The system has most of the earmarks of an active solar heating system except the most crucial–a source of mechanical power to move the heat.

As the three functions of solar heat collection, storage, and distribution become even more differentiated, external mechanical power must be summoned to move the heat around. Large ducts quickly become too expensive, and a pump or fan is a better alternative. With a few exceptions like the Davis House, indirect solar energy systems are also active systems. The real beauty of passive solar designs lies in their ability to function without external power sources and to liberate the inhabitants from such a dependency. People can return to a closer relationship with their environment, which they discover to be much more benign and much less fearsome than they had imagined. To quote Paul Davis once again, a passive solar heating system "is part of a living system which makes our world a microcosm again and reestablishes our bonds with the dynamics of our natural surroundings."

FURTHER READING

Baer, Steve. *Sunspots*. Albuquerque, New Mexico: Zomeworks Corp., 1975 (available for $3.00 postpaid from publisher, P.O. Box 712, Albuquerque, NM 87108).

Baer, Steve. "Solar House." *Alternative Sources of Energy,* Vol. 10, 1973, page 8.

Baer, Steve. "Zomes", in *Shelter.* Bolinas, California: Shelter Publications, 1973, pages 134-135.

Chahroudi, Day. "Biosphere." Albuquerque, New Mexico: Biotechnic Press, 1973.

Dietz, A. H. and E. L. Czapak. "Solar Heating of Houses by Vertical South Wall Storage Panels." *Heating, Piping and Air Conditioning,* March 1950, pages 118-125.

Haggard, Kenneth L. et al. "Research Evaluation of a System of Natural Air Conditioning." Report by California Polytechnic Institute, San Luis Obispo, 1975. (available for $10.00 from National Technical Information Service, Springfield, VA 22151; use order number PB-243498)

Hammond, Jonathan L., et al. "A Strategy for Energy Conservation." Report to the city of Davis, California, 1974. (available for $5.00 from Living Systems, Route 1, Box 170, Winters, CA 95694; see Chapter VI esp.)

Hay, H. R. and J. I. Yellott. "International Aspects of Air Conditioning with Movable Insulation." *Solar Energy,* Vol. 12, 1969, pages 427-38.

Hay, H. R. and J. I. Yellott. "A Naturally Air-Conditioned Building." *Mechanical Engineering,* Vol. 92, No. 1, 1970, pages 19-25.

Hogan, Ian. "Solar Building in the Pyrenees." *Architectural Design*, Vol. 44, January 1975, pages 13-17.

Moorcraft, Colin. "Solar Energy in Housing." *Architectural Design,* Vol. 42, October 1973, pages 634-658.

Walton, J. D., Jr. "Space Heating with Solar Energy at the CNRS Laboratory, Odeillo, France," in *Proceedings of the Solar Heating and Cooling for Buildings Workshop.* Washington, National Science Foundation, 1973. (a report by the Department of Mechanical Engineering, University of Maryland, under NSF-RANN Grant GI-32488)

Wellesley-Miller, Sean, and Day Chahroudi. "Bio Shelter." in *Architecture plus*, November-December 1974.

Yellott, J. I. "Utilization of Sun and Sky Radiation for Heating and Cooling of Buildings." *ASHRAE Journal,* December 1973.

6
Indirect Solar Energy Systems

Devices and Appliances that Trap and Store the Sun's Energy

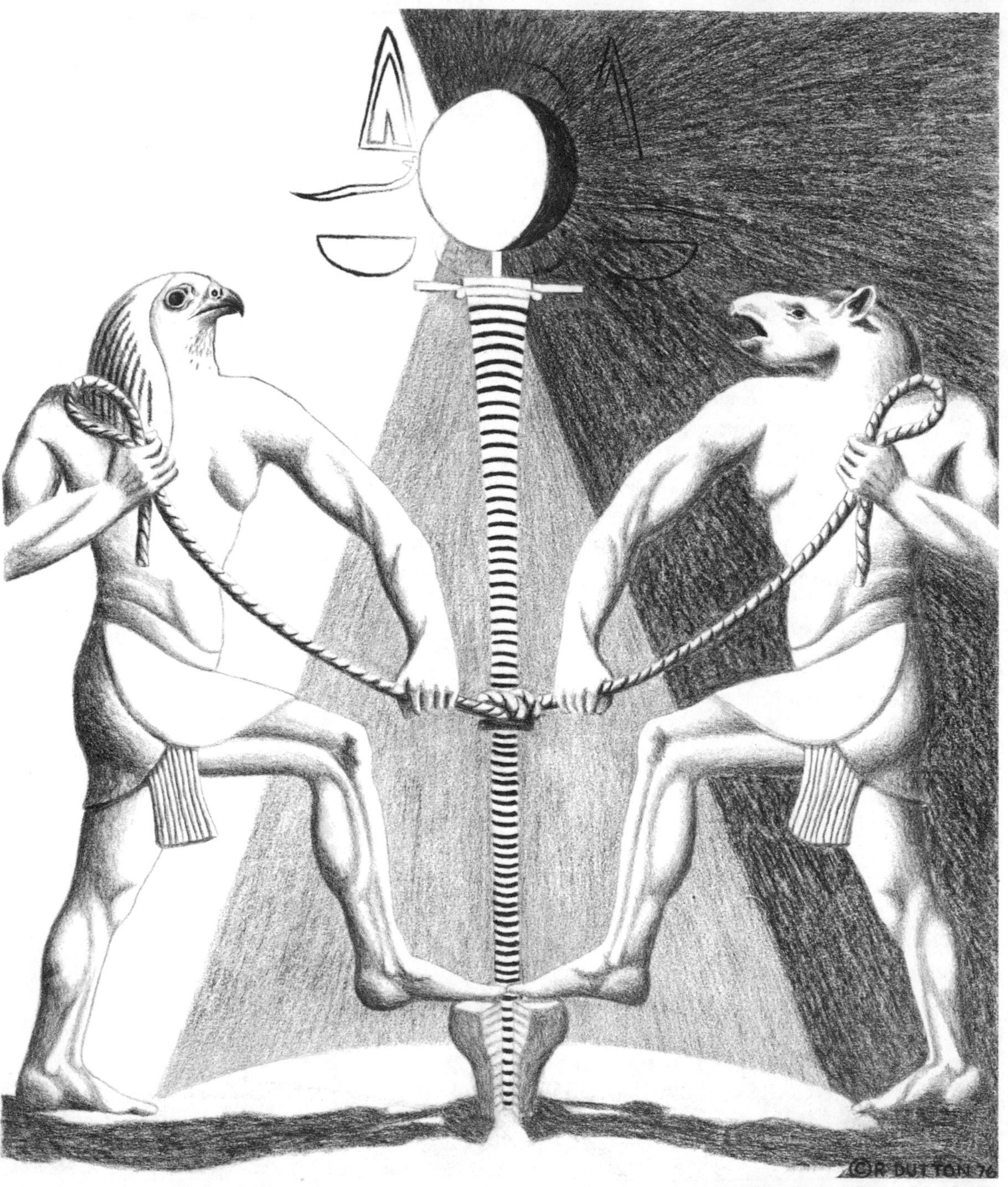

Ancient adversaries, the Egyptian gods Ra and Set turn the axis of the world, producing night and day.

Who does not remember the interest with which when young he looked at shelving rocks, or any approach to a cave? It was the natural yearning of that portion of our most primitive ancestor which still survived in us. From the cave we have advanced to roofs of palm leaves, of bark and boughs, of linen woven and stretched, of grass and straw, of boards and shingles, of stones and tiles. At last, we know not what it is to live in the open air, and our lives are domestic in more senses than we think.

Henry David Thoreau,
Walden

Indirect solar heating and cooling systems are very much in the public eye these days. They use large expanses of tilted, glass-covered surfaces to collect solar energy and convert it to heat. A fluid—either air or a liquid—carries this heat through pipes or ducts to the living areas or to storage units. As opposed to the direct methods and integrated systems discussed in the previous chapters, indirect or active systems involve many complex and interdependent components. Their elaborate collectors, fluid transport systems, and heat storage containers require a network of controls, valves, pumps, fans, and heat exchangers. They are generally more appropriate for apartment buildings, schools, and office buildings than for single family dwellings. Close coordination between the owner, the architect, the engineer, and the contractor is required to build an indirect solar energy system if undue complexity is to be avoided and cost effectiveness achieved.

HEAT TRANSPORT FLUIDS

When designing an indirect solar system, you must choose a fluid for transporting the heat energy. There are two primary heat transport loops: one links the solar collector to the heat storage container; the other delivers the heat from storage to the house. Liquids or gases may be used as the heat transport fluid in either loop. To date, liquids including water, oil, and solutions of ethylene glycol and propylene glycol have predominated. Air is the only gas that has been used. The following criteria influence the selection of a heat transport fluid:

- personal needs and comfort
- compatibility with the building design
- compatibility with the backup heating system
- compatibility with other mechanical devices
- climate (notably freezing)
- relative cost (initial, operating, maintainance)
- relative complexity
- long term reliability

When personal comfort requires only space heating, air-type transport systems are favored because of their relative simplicity and long lifetimes.

But when domestic hot water must also be provided, the choice between liquid and

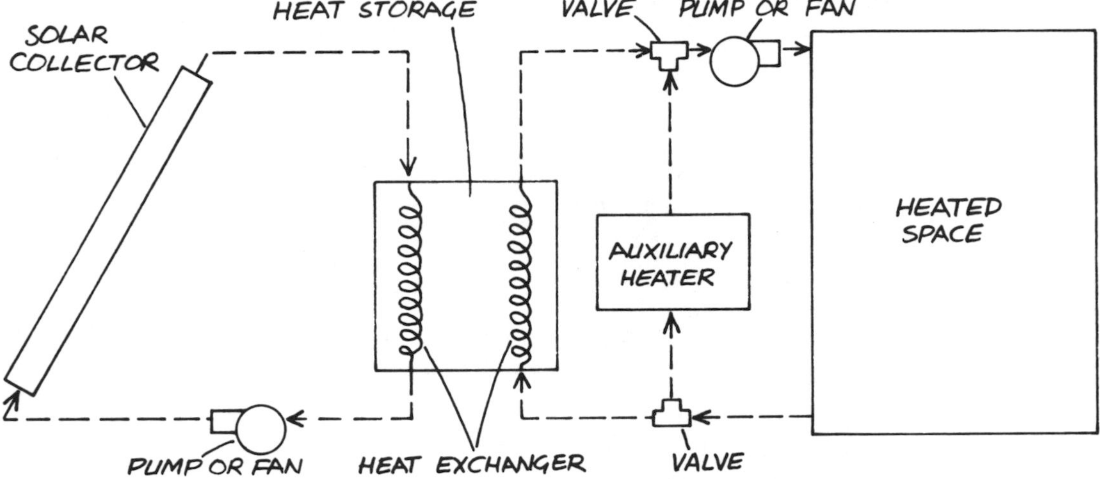

Basic components of an indirect solar heating system. There are two primary heat transport loops—from collector to storage and from storage to the rooms.

air systems becomes more difficult. Cold inlet water can be *preheated* before reaching the domestic hot water heater where it is then raised to its final temperature. This preheating is accomplished by passing the cold water supply through a heat exchanger in contact with the solar heated fluid in the storage bin or tank. Liquid systems are more compatible with such an arrangement than air systems. A possible alternative with air systems is to bury a small tank, say 20 to 40 gallons, inside the rock heat storage bin. The water in the tank will be warmed by the rocks before passing to the water heater. More information on domestic water heating is provided in the next chapter.

If cooling is needed in addition to heating, a liquid system is a more likely choice than an air system. Although some research has been done with air, most solar powered cooling systems use liquids. The same thermodynamic and physical properties which favor liquids for use in conventional cooling units also favor their use in solar cooling systems. Air systems, however, can be used successfully for cooling in some simple applications. In arid parts of the country, for example, cool night air can be blown through a rock bed and the coolness stored for daytime use.

The method of distributing heat or coolness to the rooms can help you determine which fluids to use. Forced-air circulation is most compatible with air systems but can also be used with liquid systems. Warm or cool water from the storage tank is passed through fan-coil units or heat exchangers, where the air blown across them is heated or cooled and delivered to the house. In the Solaris System, built by Dr. Harry Thomason, heat from the water tank warms the rocks packed around it. The rocks heat the house air circulating through them. Because of the hot or cold drafts that occur, forced-air heating and cooling systems *can* be uncomfortable to the people using them. But they do have the advantage of greater simplicity.

Most radiant heating systems use water to transport the heat, but some use hot air circulated through wall, ceiling, and floor panels. Hot water radiant systems, such as baseboard radiators, work well with liquid systems. Hot water from the collectors can circulate directly through the heating system or be sent to the heat storage tank. The main disadvantage of most liquid space heating systems is their need for high (140°F to 190°F) water temperatures. The higher the water temperature used, the lower the overall efficiency of the solar heating system. Steam heating systems are generally incompatible with solar collection because of the poor operating efficiencies of collectors at those high temperatures.

The amount of space allotted to heat storage is often a critical factor in the choice of fluids. Until phase-changing salts are cheap and reliable, the main choices for heat storage are (1) large tanks of water and (2) rock or gravel beds. Water tanks occupy from one third to one half the volume that rock beds need for the same amount of heat storage. This fact alone may dictate the choice of a liquid-type system. The options available for collection, storage, and distribution of heat are summarized in the accompanying chart.

A choice of heat transfer fluids is available for residences—but not for larger buildings. The larger the solar heated building, the greater the amount of heat that must travel long distances. If the fluid temperature is kept low to increase the collection efficiency, more fluid must be circulated to provide enough heat to the building. Liquid heat transport is therefore more likely in large buildings because piping occupies less valuable space than ductwork. For air to do a comparable job, large ducts or rapid air velocities are necessary. Both alternatives are usually expensive.

COMPONENT OPTIONS FOR INDIRECT HEATING SYSTEMS		
Collector Fluid	**Heat Storage**	**Heat Distribution**
air	rocks or gravel small containers of water small containers of phase-changing salts	gravity convection forced convection air-fed radiant panels or concrete slabs
water water-antifreeze solutions oil and other liquids	large tanks of water or other liquids large tanks of water embedded in rocks or gravel	baseboard radiators or fan-coil units water-fed radiant panels or concrete slabs forced convection past water-to-air heat exchangers

Climate may dictate the choice of transport fluid. In cold climates, where a house may require only heating, air systems would be the most likely choice. When a liquid system is subject to freezing conditions, an anti-freeze and water solution may be necessary. Two alternatives are to drain the water from the collector when the sun stops shining or to use piping that will not burst under repeated freezing and thawing.

Water and air are far less expensive than oil and anti-freeze. In cases where water is scarce, air is the least expensive alternative. But in cities or dusty regions, a sizable cash outlay may be required to filter, clean, and purify the air. You can usually build a collector for an air system more cheaply

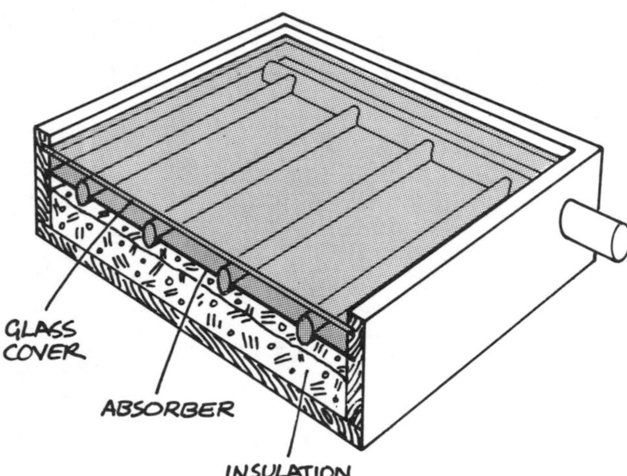

A liquid-type flat-plate collector.

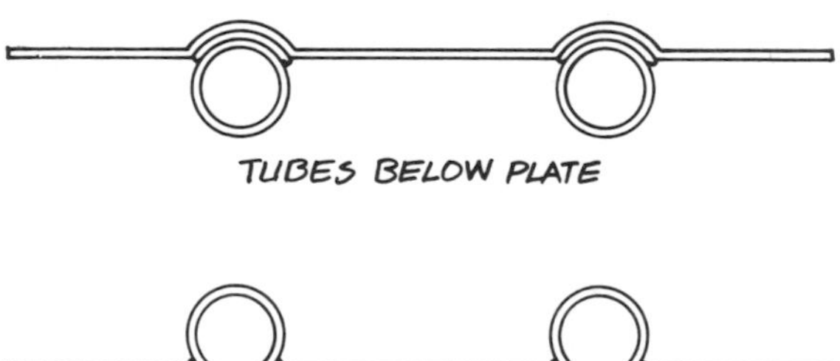

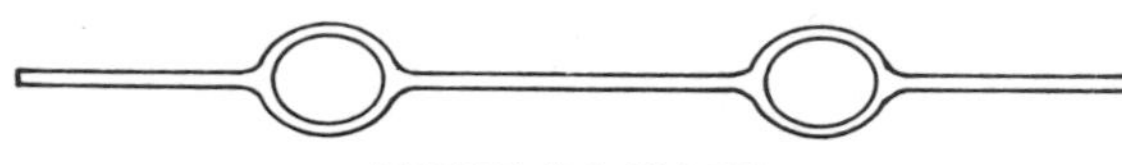

Tube-type absorbers—three possibilities.

than for a liquid system. Other components, including storage and heat exchangers (or the lack of them), also cost less for air systems. Local labor economics often favor the installation of air ducts over water pipes. But don't underestimate the cost of fans and automatic dampers.

Air systems can be cheaper to maintain because air leaks are nowhere near as destructive as water leaks. Anti-freeze solutions in liquid systems deteriorate and must be changed every 2 to 4 years. True enough, the cost of changing the anti-freeze in cars and trucks is minimal, but a residential liquid-type solar heating system requires up to *50 times* the anti-freeze a car does! Air systems, on the other hand, can be more costly to operate than liquid systems because more electrical power is required to move heat with air than with water.

In all fluid transport systems, the network of ducts and piping should be kept simple. Pipes or ducts should be well-insulated and as short as possible.

LIQUID-TYPE FLAT-PLATE COLLECTORS

The primary component of an indirect system is the solar collector. This device converts the sun's radiant energy into useful heat energy that is carried into the house by a fluid. The solar collectors already discussed, including concrete wall and roof pond collectors, have usually been an integral part of the house itself. But a flat-plate collector is an altogether separate piece of equipment—not unlike an appliance. It may be attached directly to the roof or walls of a house or propped up on its own support structure. Its distinguishing feature is just what the name suggests—the sun's energy is absorbed on a flat surface. Flat-plate collectors fall into two categories—*liquid-types* and *air-types*—according to the type of fluid which circulates through them to carry off the solar heat.

The basic components of a liquid-type flat plate collector are outlined in the first diagram on this page. The absorber stops the sunlight, converts it to heat, and transfers this heat to the passing liquid. Usually the absorber surface is painted black to improve the collection efficiency. To minimize heat loss out the front of the collector, transparent cover plates are placed above the absorber. Heat loss out the back is reduced by insulation. All of these components are enclosed in a wood or metal box for protection from wind and moisture.

Absorber Design

There are three categories of absorber designs—each characterized by the method used to bring liquids in contact with the absorber plate. The first category includes open-faced sheets with the liquid flowing over the front surface. The Thomason absorber, with water flowing in the valleys of corrugated sheet metal, is the premier example of these "trickle-type" collectors. A second category includes the tube-in-plate absorbers, which have liquid passages inside the metal plate. Tube-in-plate absorbers were first used in the Bridgers and Paxton Office Building. The third category of absorber design includes all flat metal absorber plates with metal tubes fastened to either side. Such tube-type absorbers were used in the first, third, and fourth MIT solar houses.

The open-face Thomason absorber shown in the diagram has the advantage of simplicity. Cool water from storage is pumped to a header pipe at the top of the collector and flows out into the corrugations through 1/32-inch diameter holes drilled into this header at each valley. A gutter at the base of the collector gathers the warm water and returns it to the storage tank. Although the sun and the trickling water eventually erode the black paint, Thomason reports little decrease in overall efficiency. He also says that condensation of water on the inside of the cover plate has little adverse effect. For low temperature applications (under 110°F) this open-face absorber appears to perform as well as a tube-type absorber, but the collection efficiency drops off sharply at higher temperatures. Its clearest advantage is that it is self-draining and needs no protection against corrosion or freezing. One disadvantage is that particles may be dislodged from the absorber and carried off by the water.

The best known tube-in-plate absorber is the aluminum product ROLL-BOND, made by Olin Brass for the refrigeration industry. The water passages are built into the plate itself during the process of fusing two sheets of aluminum together. These absorber plates have the advantages of low cost per square foot and very efficient heat transfer to the circulating liquid. Unfortunately, aluminum is extremely susceptible to to corrosion by water or water-antifreeze solutions. Similar tube-in-plate absorbers made with copper sheet have less corrosion problems but are more expensive.

Other absorbers with integral water passages can be fabricated with two sheets of corrugated or sheet metal spot-welded together. The water flows throughout the space between the sheets, making good contact with almost the entire surface. The absorbers now used in Ouroboros are of this type. Such "sandwich-type" absorbers can be easily fabricated by the home handyman or woman who is skilled at welding.

In most of the early experimental work with flat-plate collectors, the absorber plates consisted of flat metal sheets with copper tubes soldered, welded, wired, or clamped to them. Thousands of experimenters all over the world have struggled to develop cheap, effective methods of bonding tubes to plates. Good thermal bonds are of paramount importance. The conduction of heat into a single foot of tube can be as high as 1000 Btu/hr/°F for a securely soldered tube or as low as 3 Btu/hr/°F for a poorly clamped,

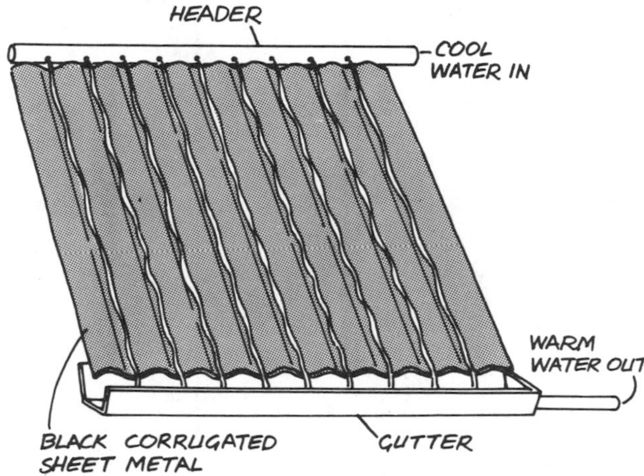

Thomason's trickle-type absorber.

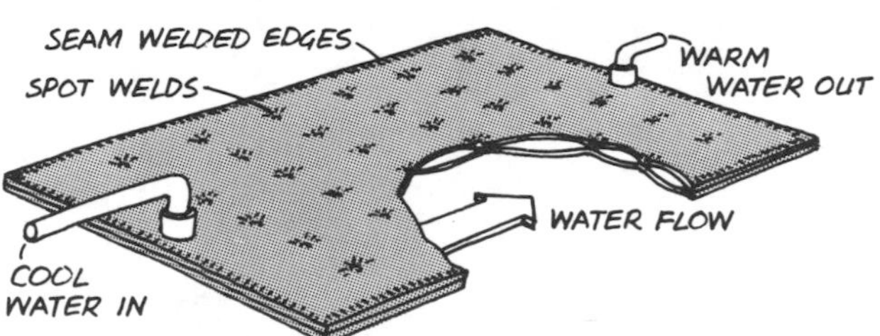

A typical sandwich-type absorber.

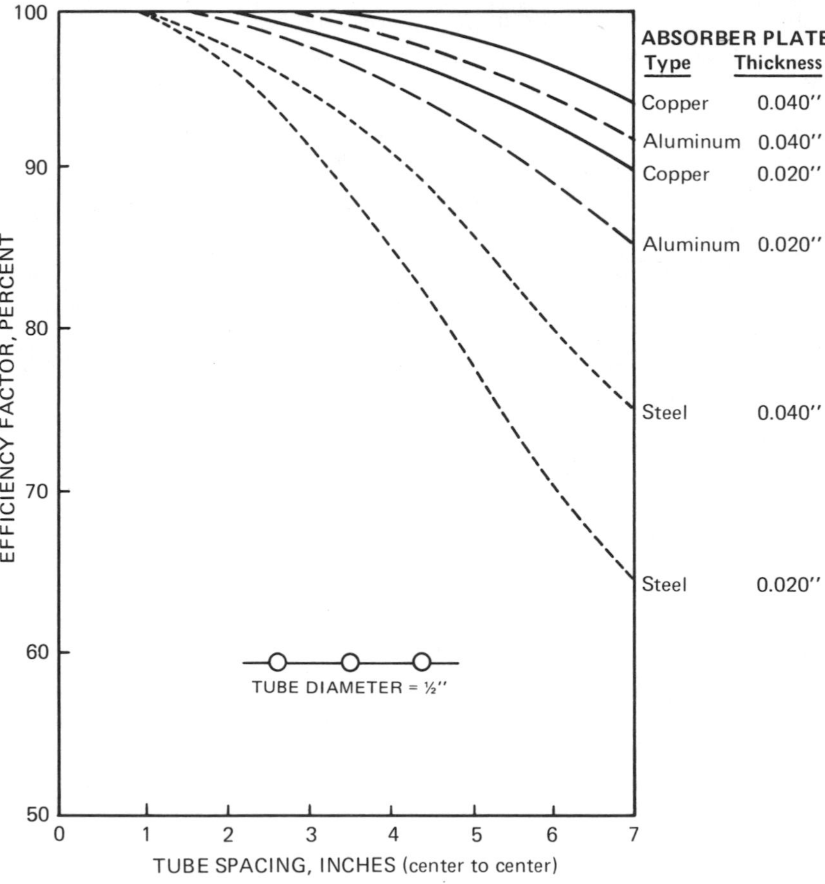

The variation in collector efficiency with tube spacing and absorber type.

wired, or soldered tube.

The major design choices for tube-type absorbers include the type and thickness of the metal, and the size and spacing of the tubes. An absorber with tubes spaced 6 inches apart and in good thermal contact with a blackened copper sheet 0.02 inches thick is 97 percent as good as a completely water-cooled black sheet. The accompanying graph by Ray Bliss illustrates the variation in absorber efficiency (the "efficiency factor" gauges the deviation from optimum) with tube spacing for various types and thicknesses of metals. In his 1959 study, he noted that cost rose *faster* than efficiency for increasing thicknesses of copper. Optimum cost and efficiency was achieved with a 0.010-inch thick copper sheet with tubes spaced at intervals of 4 to 6 inches. A tube spacing of 4 to 5 inches on 0.040-inch steel sheets was comparable. But prices of copper and steel have changed since 1959, and these studies could bear some re-examination.

Tube Sizing and Flow Patterns

The choice of tube size for an absorber involves tradeoffs between fluid flow rate, pressure drop, and cost. If cost were the only factor, the tube diameter would be as small as possible. But the smaller a tube, the faster a liquid must travel through it to carry off the same amount of heat. Corrosion increases with fluid velocity. And the faster the fluid flows, the higher the pumping costs. A good rule of thumb is to keep flow rates below 4 feet per second.

Typically, the *risers* (the tubes soldered directly onto the absorber plates) are 3/8 to 5/8 inches in diameter. The *headers* (those running along top and bottom of the plate) are 3/4 to 1 inch. Capillary effects in tubing smaller than 3/8-inch diameter can prevent proper emptying in self-draining collectors—and lead to freezing. Also, the collectors in a thermosiphoning system may need tube diameters twice those of pumped systems.

The pattern of the tubes (or channels) on (or in) the absorber plate is also important to the overall performance of the collector. Designers should strive to attain uniform fluid flow, low pressure drops, ease of fabrication, and low cost. Uniform fluid flow is the most important of these, as "hot spots" on the absorber plate will lose more heat than the other areas—lowering the overall efficiency.

Sample tube patterns are shown in the diagram. Patterns A and B are better than patterns C and D because flow rates in the risers are more nearly equal. But even in pattern A, the flow rates are higher in the end risers than in the middle ones, and the surface temperatures are higher toward the center of the absorber. *Serpentine* patterns F and G eliminate the problems of uniform fluid flow but have larger pressure drops. They are also much easier to fabricate because they have few plumbing connections. For self-draining collectors, the tubes and flow patterns must be arranged to allow for complete draining. In some past projects, air has become trapped in the tubing upon refilling, blocking fluid flow and creating hot spots in collectors. Serpentine patterns have few troubles of this nature.

In order to get all the solar collection surface needed for most applications, you will have to connect a number of independent

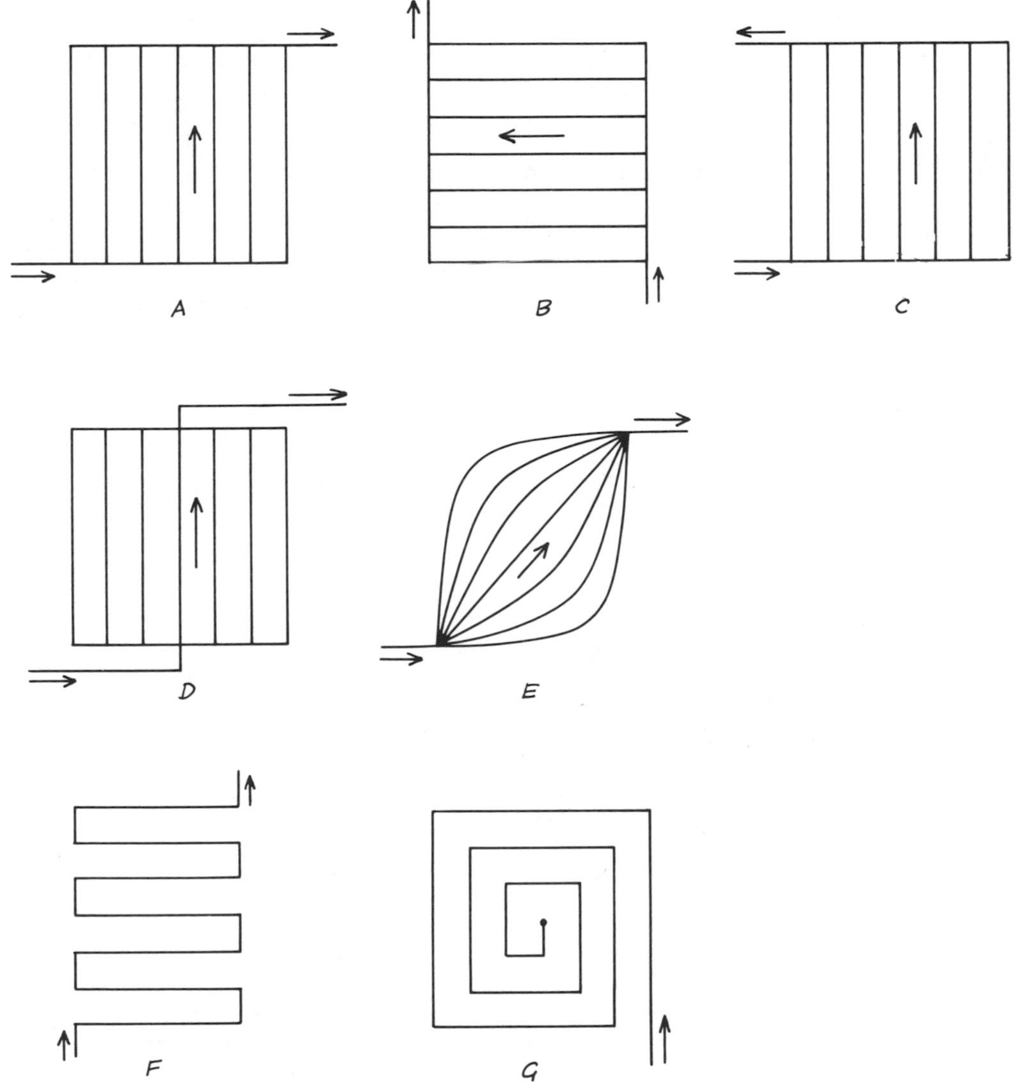

Sample flow patterns for liquid-type absorbers.

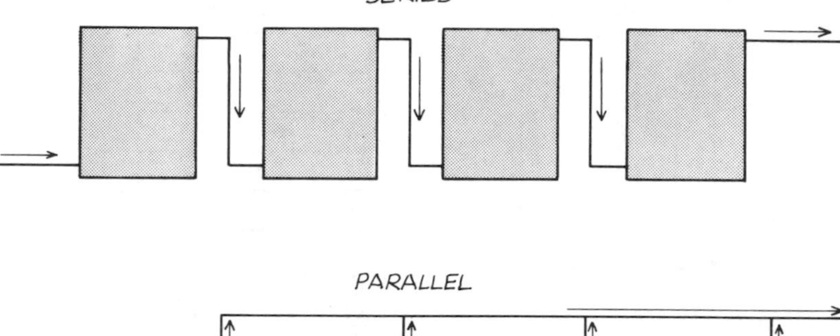

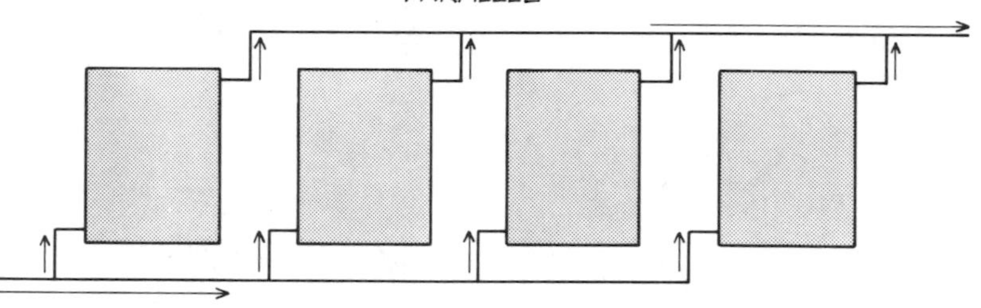

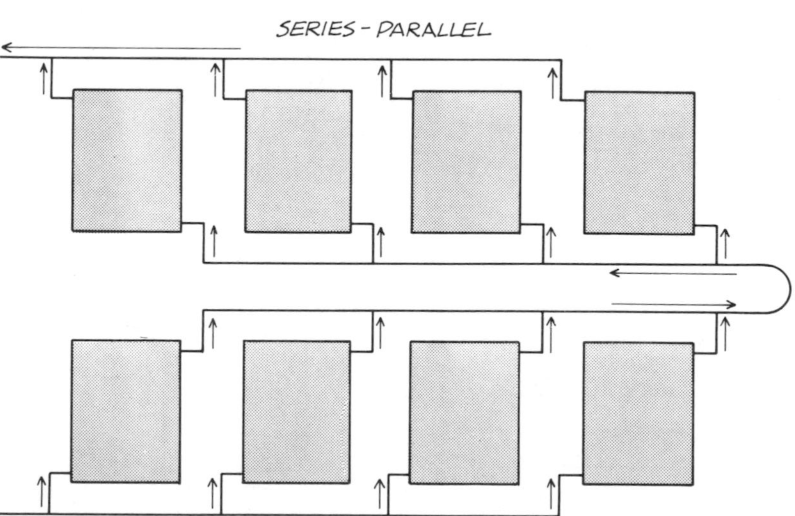

Series and parallel networks are the best ways to connect a number of flat-plate collectors.

collector panels with a network of piping. Series or parallel networks (see diagram) are the simplest. Again, the important criteria are uniform fluid flow and low pressure drop. A network of collectors piped in series has uniform flow but a high pressure drop, while a parallel hookup has just the opposite. For a large number of independent collector panels, a series—parallel network is your best bet. In any network, the exterior piping should be at least 1 inch in diameter and well-insulated.

Absorber Plates

The most difficult choice in designing and building a liquid-type flat-plate collector is the selection of an absorber plate. Traditionally, absorber plates have been made of copper, aluminum, or steel. But now plastics are making some inroads in a few limited, low-temperature applications. The criteria to watch in the choice of an absorber plate are its conductivity, durability, ease of handling, availability, and cost.

A metal need not be used for the absorber plate if the liquid comes in direct contact with every surface struck by sunlight. With almost all liquid systems now in use, however, the liquid is channelled through or past the plate. Heat must be conducted to these channels from those parts of the absorber that are not touching the fluid. If the conductivity of the plate is not high enough, the temperatures of those parts will rise, and more heat will escape from the collector—

lowering its efficiency. To reduce this heat loss, the absorber plate will have to be thicker or the channels more closely spaced. With a metal of high conductivity such as copper, the plate can be thinner and the channels spaced further apart. To obtain similar performance, an aluminum plate would have to be twice as thick and a steel sheet nine times as thick as a copper sheet. If costs were the only criterion, the choice would be simple. But this, unfortunately, is not the case.

Another very important criterion is the resistance of these metals to corrosion by the liquids. Copper resists corrosion far better than aluminum or steel, but under certain conditions water and anti-freeze can corrode all three. There are methods of reducing corrosion, but their cost and effectiveness must be compared. Steel must be either galvanized or stainless. A zinc lining for aluminum is expensive and not always available. You can also reduce corrosion by adding chromate-based inhibitors to the water or anti-freeze, but only in water of low mineral content.

These treatments only *reduce* corrosion in aluminum and steel. And there is no absolute guarantee of success. Until corrosion can be prevented, there are few alternatives to copper for liquid absorber plates. A notable exception is the open-face corrugated aluminum or steel sheets (painted black) used in Thomason's collector. It's also possible (but difficult!) to bond copper tubes to aluminum or steel sheets in tube-type absorbers.

Copper can be difficult to work with because it hardens as it is bent and it is singularly difficult to paint. Soldering copper tubes to copper plate is easy but expensive. Aluminum cannot be soldered or welded to any metal without extreme difficulty, although special mechanical bonds (see diagram) are proving successful. Steel requires special tools. And all metals require careful cleaning before application of the black surface.

The weight of the absorber, though not crucial to the design of a solar collector, ought to be considered if you don't want to huff and puff to get the collector on the roof. Typical collectors weigh less than 5 pounds per square foot, with the absorber accounting for about 1/5 of this weight.

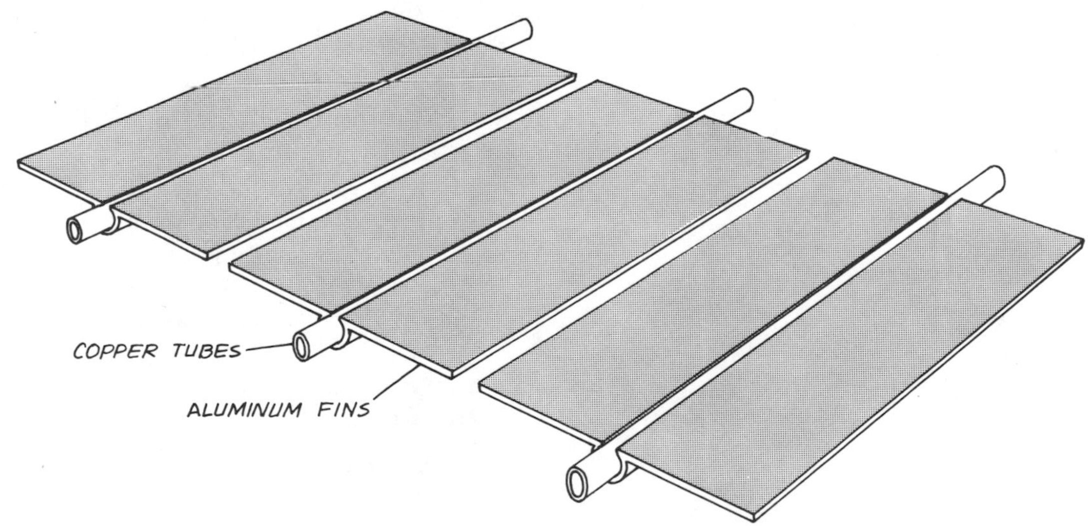

Mechanical bonding technique. Aluminum fins are clamped to copper tubes by tensile pressure.

Tips on Corrosion Prevention

Because oxygen can be very corrosive under certain conditions, air should be prevented from entering the heat transfer liquid. This can be very difficult in self-draining systems.

The pH of the transfer liquid (a measure of its acidity) is the most critical determinant of corrosion. Liquids coming in contact with aluminum must be neutral–with a pH around 6 or 7. Any deviation, whether lower (more acidic) or higher (more basic) severely increases the rate of corrosion. The pH must be measured frequently to prevent deviations from the norm. Anti-freeze should be replaced at 12 month intervals.

Systems in which the transfer liquid flows in contact with a number of different metals are susceptible to galvanic corrosion. If possible, you should avoid using several different metals. In particular, aluminum should be isolated from components made from other metals.

Avoid the use of aluminum tube-in-plate absorbers with swimming pool water–it can cause them to leak in a matter of months.

Costs of the various metals must be weighed against their relative thermal conductivity and resistance to corrosion. If thermal performance were the only criterion, the best buy would be aluminum. Unfortunately, its proclivity for corrosion precludes its widespread use. Copper is the next choice, but it is a nearly depleted resource. Stainless steel may be the most logical long-term choice. The mechanical bonding technique, in which copper contains the liquid flow and aluminum constitutes the absorber plate, is another promising alternative.

The availability of these metals is the final criterion to examine. Copper is in short supply and its price is soaring. Aluminum and steel are relatively easy to obtain, but even aluminum is becoming scarce. Steel absorber plates, particularly those treated for corrosion, are not as available commercially as aluminum. There is simply not enough copper on earth to provide every home and office building with an all-copper solar collector, even if the absorber lasts 50 years and the copper is recycled. The same limitation may be true for aluminum. Metals should be used only as a last resort. Air-type collectors that do not require metal absorber plates should be used whenever possible.

Absorber Coating

The primary function of the absorber surface or coating is to maximize the percentage of sunlight retained by the absorber plate. Any surface reflects *and* absorbs varying proportions of the sunlight striking it. The percentage it absorbs is called its *absorptance* α. This property was discussed in Chapter 4, along with the *emittance* ϵ, the tendency of a surface to emit longwave thermal radiation. An ideal absorber coating would have $\alpha = 1$ and $\epsilon = 0$, so that it would absorb all sunlight striking it and emit no thermal radiation. But there is no such substance, and we usually settle for flat black paints, with both α and ϵ close to 1.

Hardware store flat black paint is suitable for general use. But remember to clean the absorber thoroughly before applying the coating. Use an acid bath to insure maximum adhesion. You can compare the effectiveness of various paints by measuring the temperatures on a sunlit surface coated with samples of each. Asphaltum paints have been used successfully and so have high temperature black paints for wood stoves. A coating sold by Sears and Roebuck, the "Tar Emulsion Driveway Coating and Sealer," will also do the trick.

There are a few substances called *selective surfaces* which have high absorptance and low emittance. Selective surfaces absorb most of the incident sunlight but emit much less thermal radiation than ordinary black surfaces at the same temperature. Collectors with selective absorber surfaces attain higher collection efficiencies at higher temperatures than normal collectors. But they are unnecessary for systems which operate at temperatures below 100° F.

Selective surfaces must be evaluated on

the basis of their compatibility with a chosen metal plate, their durability, their availability, and their cost. Each selective surface can be applied to only a few materials. Those that can be used with copper will not necessarily work with aluminum. The higher cost of a selective surface should be offset by a decrease in the cost of the rest of the collector. For example, a second cover plate is superfluous when a selective surface is used. The improved performance of the collector may justify the added expense. Present costs of selective surfaces range from 25 cents to $2.50 per square foot.

Availability of selective surfaces is another problem. The process of applying the surface to a metal sheet is difficult because of the demands of quality control. Layers about ½ micron thick must be uniformly deposited upon the metal sheet by chemical baths, electroplating, or vapor deposition methods. Durability of the selective surface is also important. Destructive forces include moisture, high temperatures, and sunlight. You can find much more detailed information about selective surfaces in *Solar Energy Thermal Processes,* by Duffie and Beckman, from which the accompanying table was adapted.

The coating of an absorber should be chosen together with the collector cover plate. They have similar functions—keeping the solar heat in—and complement each other in a well-designed collector. For example, a selective surface with a single cover plate is usually more efficient than flat black paint with two cover plates. The accompanying graph compares the performance of flat blacks and selective surfaces for one and two cover plates. For collector temperatures below 150°F, a second cover plate is superfluous, but for temperatures above 180°F (needed, for example, in absorption cooling devices) a second cover plate or a selective surface may be necessary. For temperatures below 100°F, a selective surface performs no better than flat black paint. To date (1976), the added expense of a selective surface has only occasionally been justified by the improved performance. A few commercially available collectors use selective surfaces, including Miromit and Beasley solar water heaters and collectors marketed by Everett Barber's firm Sunworks.

Cover Plates

Cover plates are transparent sheets that sit about an inch above the absorber. Shortwave sunlight penetrates the cover plates and is converted to heat in the absorber. The escape of heat is retarded by the cover plates. They absorb the thermal radiation from the hot absorber, returning some of it to the collector, and prevent the escape of warm air. Commonly used transparent materials include glass, plexiglass, fiberglass-reinforced polyester, and thin plastic films. They vary in their ability to transmit sunlight and trap thermal radiation. They also vary in weight, east of handling, durability, and cost.

Glass is a favorite choice. It has very good

PROPERTIES OF SELECTIVE SURFACES FOR SOLAR ENERGY APPLICATION
α = Absorptance for solar energy
ϵ = Emmittance for long wave radiation

Surface	α	ϵ
"Nickel Black" on polished Nickel	0.92	0.11
"Nickel Black" on galvanized Iron*	0.89	0.12
CuO on Nickel	0.81	0.17
Co_3O_4 on Silver	0.90	0.27
CuO on Aluminum	0.93	0.11
Ebanol C on Copper*	0.90	0.16
CuO on anodized Aluminum	0.85	0.11
PbS crystals on Aluminum	0.89	0.20

*Commercial processes.
SOURCE: Duffie and Beckman, *Solar Energy Thermal Processes*, 1974.

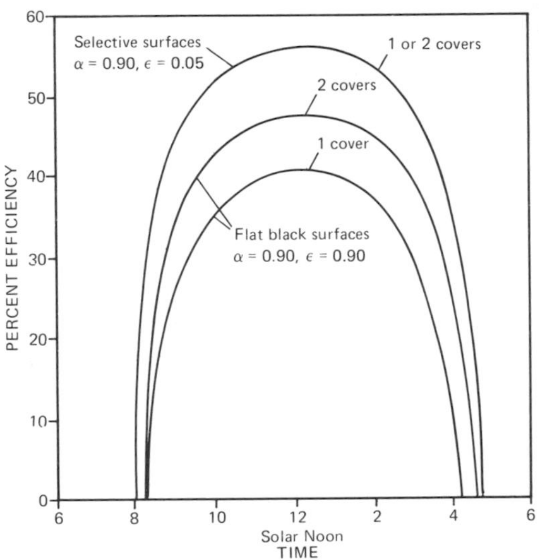

Flat-plate collector performance of selective and flat black absorber coatings.

solar transmittance and is almost totally opaque to thermal radiation. Depending upon the iron content of the glass, between 85 and 92 percent of the sunlight striking the surface of a 1/8 inch sheet of glass (at vertical incidence) is transmitted. It is stable at high temperatures and is relatively scratch- and weather resistant. Glass is readily available and installation techniques are familiar to most contractors.

But glass is heavy and fragile; it must be handled in small pieces and with great care. And it's better to set the glass in the collector frame at the building site rather than ship a pre-fit collector long distances. Size limitations require much intricate and expensive framework (glazing details). Glass is also expensive. Depending upon quality, hardness, and thickness, prices range from 50 cents to $2.00 per square foot.

You should try to obtain glass with a low iron content. Viewed on edge, the greener the glass the higher the iron content and the lower the transmittance. High quality glass absorbs 3 to 4 percent of the sunlight as compared to the 7 percent common for poorer grades. A 1/8-inch thickness is sufficient for most purposes. But in areas smitten by proverbial hailstones as large as golf balls, you might consider 1/4-inch glass or tempered glass. Anti-reflection coatings have been developed to increase transmittance, but as yet the improved performance (2 to 4 percent) isn't worth the premium price.

Glass cover plates must be able to expand and contract while maintaining a tight seal against moisture and air leakage. The glass should never rest directly on metal because of the potentially severe heat there. In collectors with multiple covers, the space between the outer layers should be vented to allow the escape of any moisture buildup. If insulating glass (thermopane, for example) is used, it must be properly tempered. In the late 1950's the MIT research team discovered that the inner layer of ordinary thermopane will break under intense thermal stresses.

Alternatives to glass include plastics such as plexiglass, Mylar, Tedlar, Teflon, and Lexan, and fiberglass-reinforced polyesters such as Sun-Lite. Plastics, many of them lighter and stronger than glass, have slightly higher solar transmittance because most of them are very thin films. Usually they are cheaper and available in large sheets—reducing the need for intricate glazing details. And their malleability works well with recent innovations in covering techniques.

But there's a fly in the polymers. Most plastics transmit some of the longwave radiation from the absorber plate. Longwave transmittance as high as 80 percent has been measured for some very thin films. The increased *solar* transmittance mitigates this effect somewhat—as does the use of a selective surface. But good thermal traps become very important at high collector temperatures, and many plastics just don't pass muster under these conditions.

Almost all plastics deteriorate after continued exposure to the ultraviolet rays of the sun. Thin films are particularly vulnerable to both sun and wind fatigue. They are consequently unsuitable for the outer cover plate of a multiply-covered collector. But they can still be used for inner covers, because glass screens out much of the ultraviolet radiation and wind is not a problem. Thin films like Tedlar can be used in this way. Some of the thicker plastics yellow and so decline in solar transmittance, even though they remain structurally sound. Other plastics like plexiglass and acrylics soften at high temperatures and remain permanently deformed.

In dirty or dusty regions, the low scratch resistance of many plastics make them a poor choice. Hard, scratch-resistant coatings are available—but only at increased cost. Teflon has very high solar transmittance (better than 95 percent) and good weatherability, but it is more expensive than some of the other plastics. Tedlar has good solar transmittance and is fairly opaque to thermal radiation, but its lifetime can be as short as 2 years when exposed to high collector temperatures. It also fatigues quickly if allowed to flap in the wind. Mylar is strong at first but weakens rapidly when exposed to ultraviolet radiation.

Sun-Lite, a fiberglass-reinforced polyester made by the Kalwall Corporation, is emerging as a good alternative to plastic films and glass. Its solar transmittance, initially above 90 percent, decreases under exposure to sunlight and high temperature, but eventually levels off above 80 percent. Its ease of handling, light weight, strength, and durability make it a very attractive choice. Sun-Lite should not be used as the inner cover of a multiply-covered collector, but it works fine as a single cover.

A critical choice in the design of a solar collector is the number of cover plates to use. Additional cover plates provide extra barriers to retard the outward flow of heat and insure higher collector temperatures. But the more cover plates, the greater the fraction of sunlight absorbed and reflected by them—and the smaller the percentage of solar energy reaching the absorber surface. At sharp angles of incidence, very little sunlight reaches the absorber of a collector having several cover plates (see the accompanying graph).

In general, the lower the temperature required from the collector, the fewer the cover plates. For example, solar collectors that heat swimming pools may not require any cover plate at all. For cooler climates, more cover plates may be needed. To obtain the same collector performance, for example, two covers may be necessary in New England and only one in Florida.

Although it increases the collector cost, an additional cover plate may reduce the costs of other parts of the system. Two covers will result in higher collection and storage temperatures—increasing the effective heat-storage capacity of a tank. Use of only one cover may force you to use a larger storage tank or a larger heat distribution system to make use of lower storage temperatures. Tybout and Löf found that the use of two covers is the most economical choice

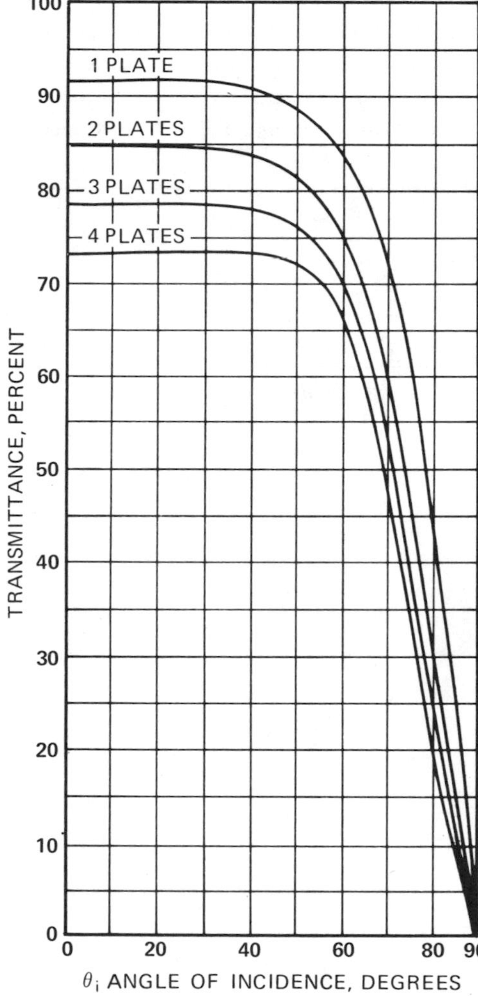

The percentage of sunlight transmitted through very clear glass. The angle of incidence is the angle between a ray of sunlight and a line drawn perpendicular to the plates.

OPTIMUM NUMBER OF COVER PLATES	
Collector Temperature minus Outdoor Temperature (°F)	Number of Covers
-10 to 10	0
10 to 60	1
60 to 100	2
100 to 150	3

SOURCE: A. Whillier, "Principles of Solar House Design," 1955.

for space heating throughout most of the United States. Spacings of ½ to 1 inch between cover plates are the most effective in reducing outward heat losses.

For uses other than space heating, the accompanying chart should be helpful. These are only rough guidelines. Your actual decision must be based on cost, collector configuration and materials, and the overall design of the heating and cooling system. Remember that a second cover can be expensive—and so can a selective surface. For low temperature applications, your best bet is usually a single cover and an absorber coated with flat black paint.

If you dislike washing windows, you may be pleased to learn that dirt on the cover plate can lower transmittance by at most 4 percent. But glazing supports and mullions can throw shade on the absorber. So be sure to design for a maximum of unshaded collector area. Gaskets and sealants should have good resistance to ultraviolet radiation and high temperatures. The glazing details should provide for drainage and keep out snow, ice, water, and wind.

Insulation

Insulation is added behind the absorber in order to cut heat losses out the back. If the collector is applied to the outside surface of a house, such as a wall or roof, heat lost out the back is transferred into the house. This can be an advantage during winter but not in the summer. Except in areas with cool summer temperatures, the back of the absorber should be insulated to minimize this heat loss and thereby raise the collector efficiency. Six inches of fiberglass batt insulation or its equivalent is adequate for roof top collectors, and as little as 4 inches is sufficient for vertical wall collectors. Where the collector sits on its own support structure separate from the house, 6 to 8 inches of fiberglass or its equivalent is the standard.

Because of their stability at high temperatures, fiberglass or mineral wool are better than styrofoam or the urethanes. Some urethanes will deform at high temperatures while others give off toxic gases and may even burst into flame. Whenever possible, the insulation should be separated from the absorber plate by at least a 3/4 inch air gap, and a reflective foil should cover the insulation. This foil reflects thermal radiation back to the absorber—thereby lowering the temperature of the insulation and increasing collector efficiency. These safety precautions are a *must* whenever using urethanes.

The perimeter of the absorber must also be insulated to reduce heat losses at the edges. Temperatures along the perimeter of the absorber are generally lower than those at the middle because of these losses, which can be reduced by massive insulation around the edges. But extra insulation should not be added if it seriously reduces the absorber surface exposed to the sun. Metal, plastic, or wood pans are often used as a protective casing around the outside of the insulation—

particularly in collectors available from manufacturers. Galvanized steel is commonly used, but pans with even better corrosion resistance are needed in wet climates.

Other Design Factors

The total collector weight, which ranges from 1 to 6 pounds per square foot, is well below the design loads for the roofs of most houses. Wind loads on collectors integrated into the walls or the roof are no problem either, since these surfaces must withstand wind conditions anyway. However, wind loads are a principle consideration in the design of independent support structures for separated collectors. Surface areas of individual collector panels should be kept small to reduce sail effects. You can maintain a low profile by placing a series of long, low collectors one behind the other in sawtooth fashion—as was done by Dr. George Löf in his Denver home.

Snow loads have not been a problem. The steep collector tilt angles needed at higher latitudes (where most of the snow falls!) are usually adequate to maintain natural snow run-off. And warm water from storage can be circulated through the collector in the morning to melt any remaining snow and induce it to slide off.

Be sure you have easy access to the collector surfaces *after* installation. Large collector surfaces can be difficult to maneuver on because of their relatively slick and fragile cover plates. Long, low collectors will reduce these access problems. Where local conditions require frequent window washing, this accessibility is even more important. In any case, the cover plate details must be designed to withstand some abuse from maintenance people.

A final design factor—the *thermal capacity* of the collector—is still a matter of some debate. This thermal capacity is equal to the amount of heat needed to warm a collector up to its operating temperature each morning. A collector with low thermal capacity responds more quickly to changes in sunlight and outside temperatures than one with high capacity. The comparison is similar to that between a light, wood-frame house and a heavy, masonry-block house. In the morning, the collector with low thermal capacity will reach its operating temperature (at which the pump turns on) more quickly than the collector with high capacity. Low thermal capacity also allows fuller use of intermittent bursts of sunshine on a partly cloudy day.

The thermal capacity of a collector is traceable mostly to the glass, metal, and water in it. More than half the thermal capacity of the collector used in the first MIT solar house was attributed to the glass, and another 20 percent to the copper absorber. When self-draining collectors are full, they have as much as twice their empty thermal capacity.

Several Liquid-Type Collectors

The basic collectors used by Hottel and

Woertz in their vanguard research had copper tubes soldered directly to the absorber plate. Construction details of one of these early collectors are shown in the diagram. The 1/2-inch copper tubes were soldered in a serpentine pattern to the backside of a 0.021-inch copper plate painted flat black on the side facing the sun. Two clear layers of glass, 3/16-inch and 1/8-inch thick, covered the assembly. Wood spacers insured a 3/4-inch air gap between the cover plates and above the absorber.

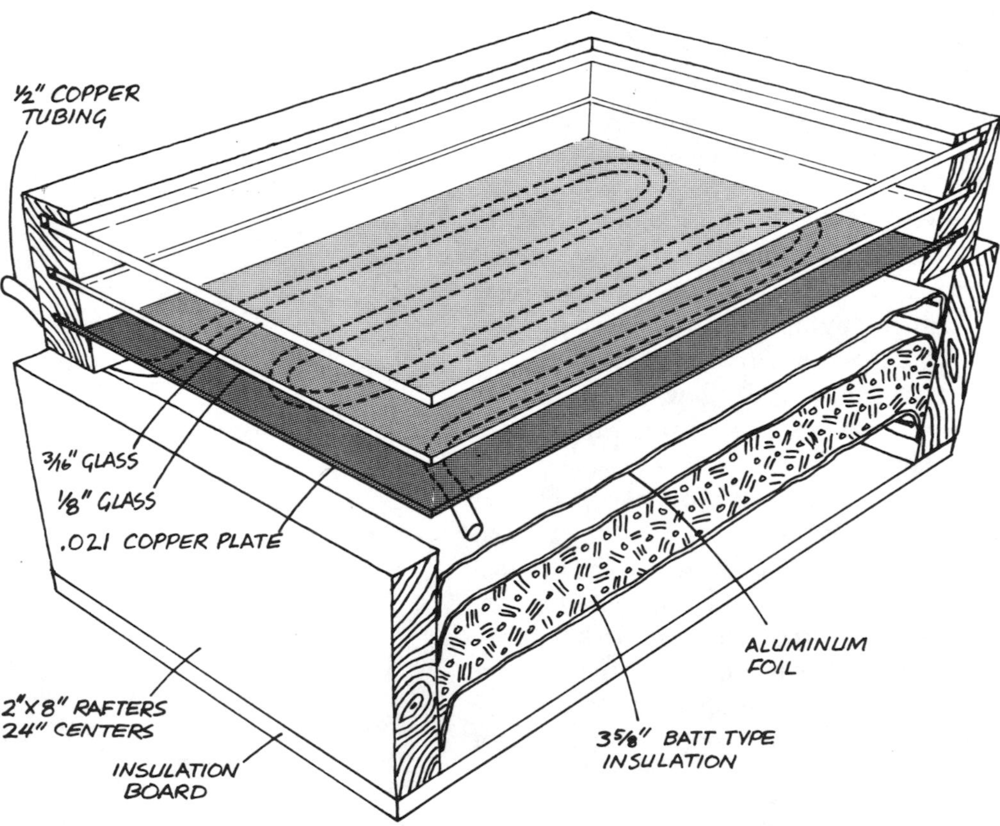

Construction details for one of the early MIT collectors.

A number of these collectors were mounted directly on the roof rafters. To keep out moisture, an aluminum cap (not shown in diagram) was set in the spaces between the individual panels. Underneath, the space between the rafters was filled with 3½-inch batts of mineral wool insulation—which were separated from the absorber by an aluminum foil barrier. The space between the insulation and the interior sheathing was vented in summer to keep the interior cool.

Anyone handy with tools can easily make this collector from readily available materials. Particular care must be taken to insure good thermal conduction through the solder joints. The inner glass must also have ample room to expand when the absorber gets hot. And beware—this design can be fairly expensive because of its heavy reliance on copper.

Another liquid-type flat-plate collector has been developed by William B. Edmondson, editor and publisher of the monthly newsletter, *Solar Energy Digest.* The modular version of his SolarSan™ collector was originally designed for use in solar water heating, but it can be readily adapted to space heating. This collector can be built for only \$3 to \$4 per square foot (cost of materials only—1976 prices). And his integrated roof-type collector costs only \$2 to \$3 per square foot more than the costs of materials normally used in a roof.

To use this patented design, you should

first contact Mr. Edmondson for a license to build one of these collectors on a single home. He sells this license together with a detailed handbook for $25.95; they can be obtained from Solar Energy Digest (P.O. Box 17776, San Diego, CA 92117).

Following the handbook, you lay down a layer of foil scrim paper on the roof decking. Over this comes a 1-inch layer of rigid fiberglass board insulation followed by a layer of dead soft aluminum foil (8 to 12 mils thick). On top of the foil sits the copper tubing. For simplicity, the tube pattern can be serpentine as shown in the diagram, but Mr. Edmondson recommends a grid pattern (with top and bottom headers) for ease in draining. The tubes are compressed down into the soft foil and the underlying insulation using long staples or several other means explained in the handbook. You then paint the tubes and foil a flat black.

There are a number of possibilities for the collector cover. You can place a ½ to 1 inch layer of honeycomb filter medium over the absorber surface. This layer allows sunlight to penetrate but blocks the convection heat flow back to the cover sheet. Nailing strips are then placed 16 to 24 inches apart on the top of this medium and secured to the roof decking with nails driven through the sandwich. A cover sheet of Kalwall Premium Sunlite or a Tedlar-coated fiberglass such as Filon sits ½ to ¼ inch above the filter. However, Edmondson has had difficulty finding a filter that did not yellow with age and now suggests removing it. Or you can include a 4-mil sheet of Tedlar between the cover and the absorber if double glazing is needed.

If the roof is uninsulated, you will need at least another 2 inches of fiberglass board insulation on top of the roof decking. But this insulation should be considered part of the house cost, not part of the collector cost. And the cover sheet replaces the shingles or other roofing material, so the only collector materials you pay extra for are the aluminum foil, copper tubing, black paint, nailing strips, and the filter medium or bottom glazing, if used. All told, these materials can cost as little as $2 per square foot of collector.

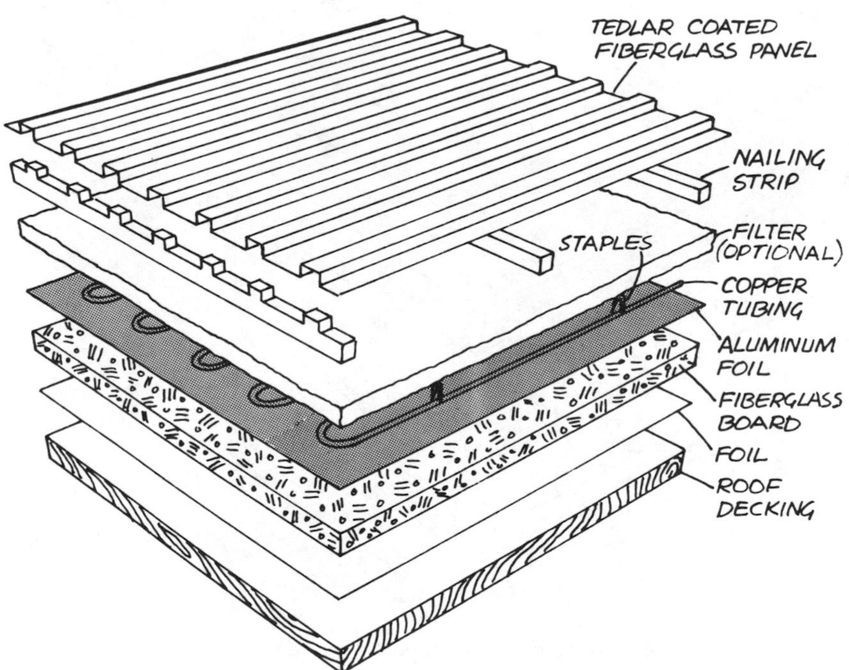

The integrated roof-type SolarSan collector—exploded view.

Everett Barber, whose Guilford house was discussed in Chapter 2, designed one of the first flush-mounted solar collectors available. This collector was designed to fit between roof rafters and replace the insulation, sheathing, and roofing materials.

Barber believes that standard construction materials produce the best results and the best economic payback in the long run. Both the fluid tubes and the absorber plates were made of copper, which has the longest lifetime of any material presently available for this type of absorber surface. The copper tubes were soldered to the absorber plate in a grid pattern (see diagram). A selective surface (with an absorptance of 90 percent and an emittance of 20 percent) was applied to the top face of the copper sheet and the whole assembly covered with a single layer of 3/16-inch glass. Galvanized sheet metal houses the collector.

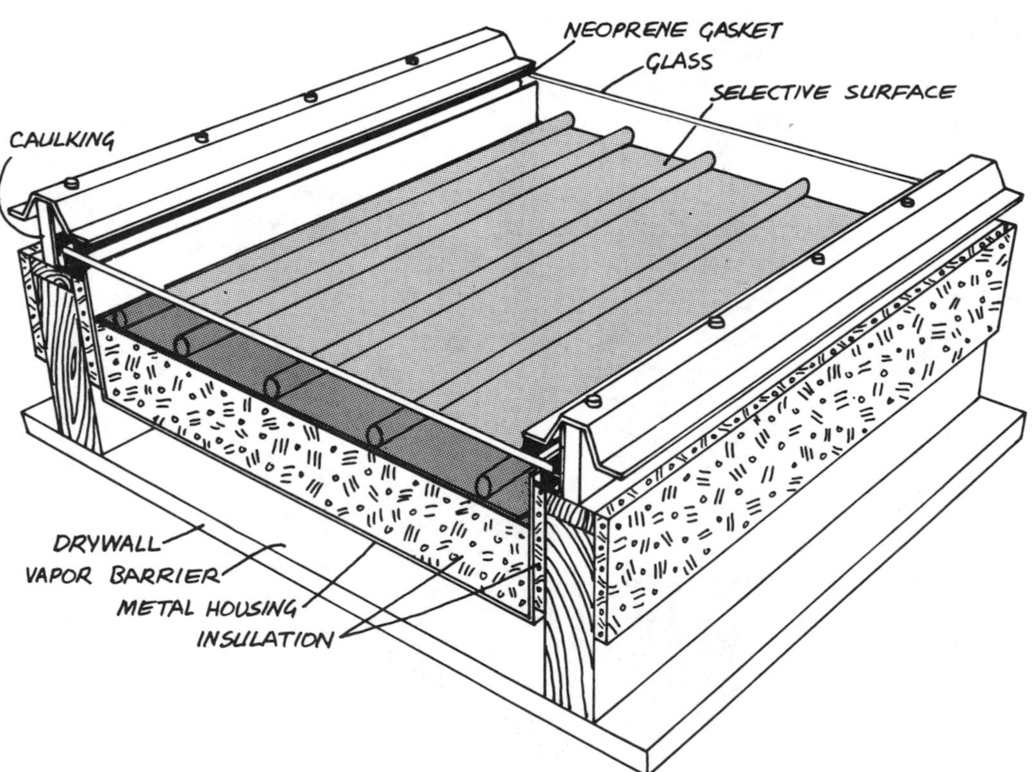

The flush-mounted collector originally marketed by Sunworks, Inc.

This "flush-mounted module" is no longer available from Barber's firm Sunworks, Inc. They are presently offering a "surface-mounted module," of similar design, that sits directly on top of a conventional roof. And they also produce an air-type collector. All in all, these are very good collectors. Independent tests by NASA, Desert Sunshine Exposure Tests, Inc., and the University of New Mexico have shown these collectors to have very good performance over a wide range of collector and outdoor air temperatures.

AIR-TYPE FLAT-PLATE COLLECTORS

Solar heating systems which use air as the heat transport medium should be considered for all space heating applications—particularly when absorption cooling or domestic water heating are not important. Air systems *do* lend themselves to nocturnal cooling methods such as blowing cool night air through the storage bin. And they don't have the complications and the plumbing costs inherent in liquid-type systems. Nor are they plagued by freezing or corrosion problems.

The relative simplicity of air-type solar heating systems makes them very attractive to people wishing to build their own. But precise design of an air-type system *is* difficult. All but the simplest systems should be designed by someone skilled in mechanics and heat transfer calculations. Once built, however, air-type systems are extremely easy to maintain or repair. Fans, damper motors, and controls may fail occasionally, but the collectors, heat storage, and ducting should last indefinitely.

The construction of an air-type collector is simple compared to the difficulty of plumbing a liquid-type collector and finding an absorber plate compatible with the heat transport liquid. Except for Thomason's collector, the channels in a liquid-type absorber must be leakproof and pressure-tight and must be faultlessly connected into a larger plumbing system at the building site. But the absorber plate for an air-type collector is usually just a sheet of metal or other material with a rough surface. Air-type collectors need not be built with terrific precision because air leakage and thermal expansion and contractions are only minor problems.

Absorbers

The absorber in an air type collector doesn't even have to be metal. In most collector designs, the circulating air flows over virtually every surface heated by the sun. The solar heat doesn't have to be conducted from one part of the absorber to the flow channels—as in liquid-type collectors. Almost any surface heated by the sun will surrender its heat to the air blown over it.

This straightforward heat transfer mechanism opens up a wide variety of possible absorber surfaces. Ray Bliss and Mary Donovan used four layers of black cotton screening for the absorber surface in the Desert Grassland Station collector (Chapter 2). George Löf used sheets of glass painted black in the overlapped-plate collectors he used in the Boulder and Denver Houses. A variation tried at the University of Florida had blackened aluminum plates instead of the glass plates of Löf's collector.

Other possible absorber materials include metal scraps fixed to plywood, metal lath, fiberglass mesh, crushed glass or rock, cloth, paper, and even that old black veil that granny wore to church on Sunday! Many of these can be obtained very cheaply—as recycled or reused materials. If you have any old metal scraps lying around after installing your ductwork, for example, just nail, glue, solder, or weld them onto the existing absorber surface. But the entire absorber surface must be black, must be heated directly by the sun, and must come in contact with the air flowing through the collector.

A sheet metal absorber plate, the old standby for liquid collectors, is still a good choice. Metal is preferable for collectors in

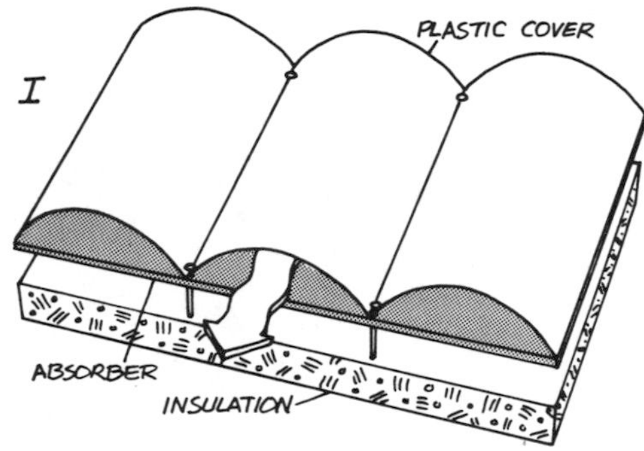

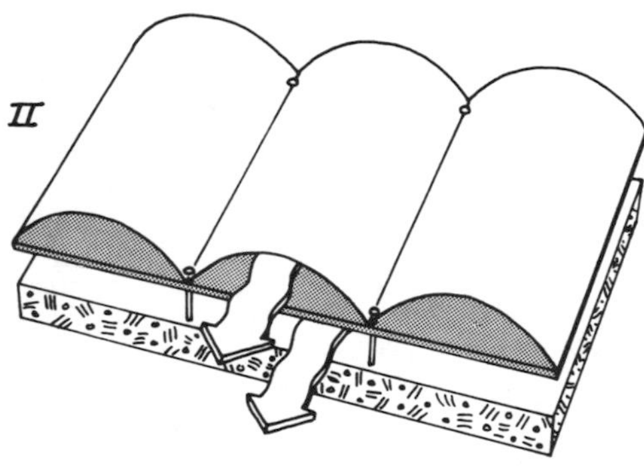

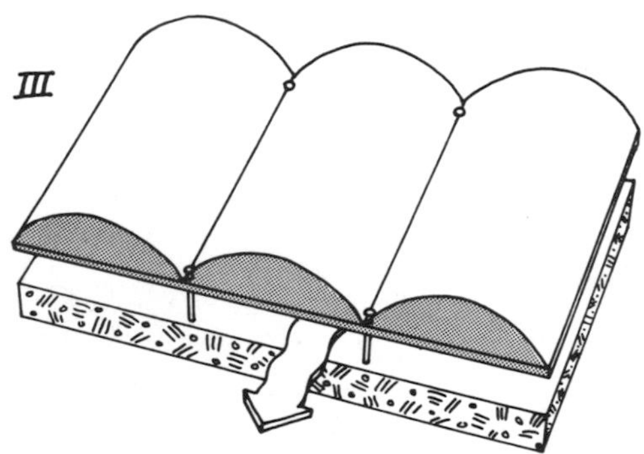

The three types of warm air solar collectors.

which the sun cannot reach every last surface in contact with the moving air. Because of its high conductivity, metal can also alleviate the "hot spots" caused by an uneven air flow. Excess heat merely flows to other areas where the air is making better contact.

Air Flow and Heat Transfer

Just *where* to put the air ducts relative to a blackened metal absorber is a question that merits some attention. The three basic configurations shown in the diagram were studied by J. D. Close (see references). In type I, air flows between a transparent cover and the absorber; in type II, another air duct is located behind the absorber; and in type III, only the duct behind the absorber is used. The type II collector has the highest efficiency when the collector air temperature is only slightly above that outdoors. But as the collector temperature increases, type III is dramatically better.

The rate of heat flow from the absorber to the passing air stream is also crucial. The *heat transfer* coefficient *h* is one measure of this flow. It is similar to the U-value, which is a measure of the heat flow through a wall or roof. The higher the value of *h,* the better the heat transfer to the air stream and the better the collector performance. Good values of *h* fall in the range of 6-12 Btu/hr/ft^2 /°F. At a temperature 25°F above that of the air stream, one square foot of good absorber surface will transfer 150-300 Btu per hour to the passing air—or about as much solar radiation as is hitting it. The value of *h* can be increased by increasing the rate of air flow, by increasing the effective surface area of the absorber, or by making the air flow more turbulent. As long as pumping costs do not get out of hand, higher values of *h* are definitely preferred.

In his study, Close concluded that a higher value of *h* is more important at higher operating temperatures. For summer crop drying, in which the collector temperature may be only 30-40°F above that outdoors, a single sheet of metal is about as effective as a finned absorber or one with "vee" corrugations (see diagrams). But when the temperature difference approaches 100°F, a collector with a finned absorber is 5 to 10 percent more efficient than one with a flat metal plate. And a "vee" corrugated absorber is another 10 to 15 percent more efficient than a finned absorber.

Paint the back side of an absorber black if air flows in contact with it. And the surface that separates the back air duct from the insulation should be lined with aluminum foil to reflect thermal radiation back to the absorber. But if a perforated absorber is used (see diagram), this separating surface can be painted black to serve as yet another heat transfer surface. The use of a blackened screen that allows half of the solar radiation to penetrate to such a lower surface doubles the value of *h* (per square foot of collector area) and increases the solar heat collection by 10 to 15 percent.

Whether the absorber surfaces are metal

or not, *turbulent* flow of the air stream is very important. Poor heat transfer occurs if the air flows over the absorber surface in smooth, undisturbed layers. The air next to the surface is almost still and becomes quite hot, while layers of air flowing above it do not touch the absorber surface. Two levels of turbulent flow will help ameliorate this situation. Turbulence on the macroscopic level can be observed with the naked eye when smoke blown through the air tumbles over itself. Turbulence on the microscopic level involves this tumbling right up next to the absorber surface.

To create turbulent flow on either level, the absorber surface should be irregular—not smooth. The finned plate and "vee" corrugations create macroscopic turbulence by breaking up the air flow—forcing the air to move in and out, back and forth, up and down. To create microscopic turbulence, the surface should be rough or coarse, with as many fine, sharp edges as possible. Surfaces such as gravel, air filters, cloth, mesh, and pierced metal plates do the trick.

But increased air turbulence means a greater pressure drop across the collector. Too many surfaces and too much restriction of air flow will require that a larger fan be used to push the air. The added electrical energy required to drive the fan may cancel out the solar heat savings.

Absorber Coatings and Cover Plates

While considerations for absorber coatings,

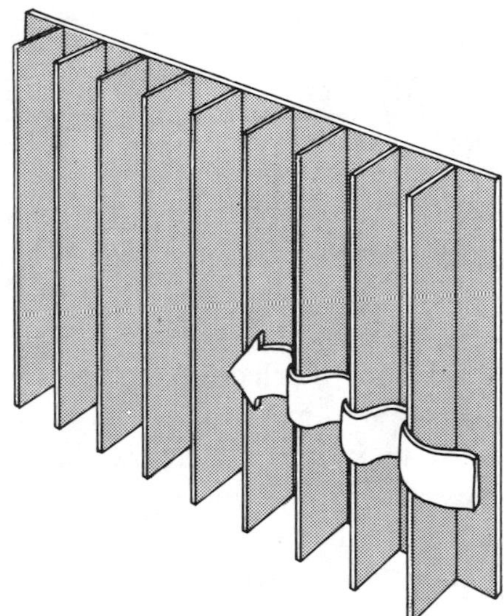

Finned absorber plate

Vee-corrugated absorber

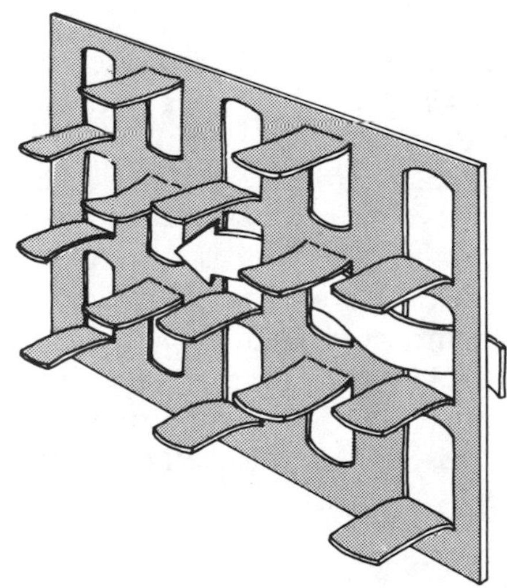

Perforated absorber plate

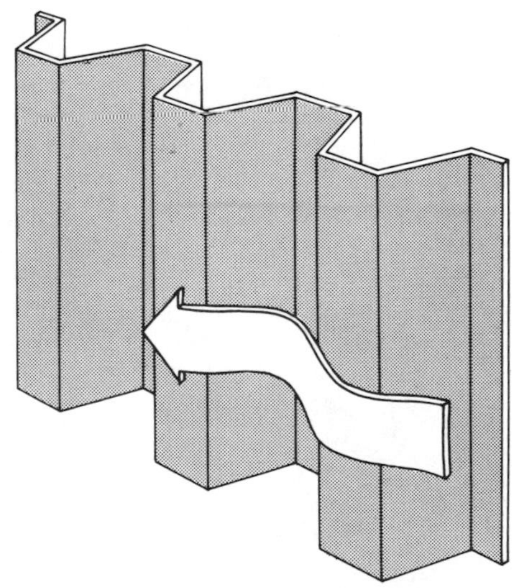

Steel-flooring absorber

selective surfaces, and cover plates are similar for air-type and liquid-type collectors, there are a few differences. One of the primary drawbacks of a non-metallic absorber is the extreme difficulty of applying a selective surface to it. Until this technology improves, metal absorbers will be preferred in applications where a selective surface is desirable. Low-cost, efficient air-type collectors will be readily available if selective surfaces can ever be applied to non-metal absorbers with ease.

As with liquid-type collectors, the use of a selective surface is about equivalent to the addition of a second cover plate. For type I and II collectors, in which air flows between the absorber and the glazing, the addition of a second cover plate may be preferred because it creates a dead air space in front of the absorber.

The use of a "vee" corrugated absorber plate is somewhat analogous to the use of a selective surface. This plate increases the overall solar absorption (and hence the "effective" absorptance) because direct radiation striking the vees is reflected several times, with more absorption occurring at each bounce. Oriented properly, its absorptance is higher than that of a flat metal sheet coated with the same substance. But the increase in the emitted thermal radiation is small by comparison.

Other Design Factors

Air leakage, though not as critical a problem as water leakage in a liquid-type collector, should still be kept to a minimum. Because the solar heated air is under some pressure, it will escape through the tiniest crack. Prevention of air leakage helps to raise the collector efficiency. Take special care to prevent leakage through the glazing frames. By using large plastic sheets instead of many small glass panes you can reduce the number of glazing joints and cut the possibility of leakage. And just as storm windows cut the air infiltration into your home, second and third cover plates reduce air leakage from a collector.

For type I and II collectors, the turbulent flow through the air space in front of the absorber results in somewhat larger convection heat flow to the glass than is the norm with water-type collectors. Thermal radiation losses from the absorber are a relatively smaller part of the overall heat loss. Plastic cover plates such as Tedlar, with a longwave transmittance as high as 40 percent, can still be fairly effective in these collectors. The absorber in a type I collector becomes relatively hot and loses a lot of heat out the back. But in type II and III collectors, a turbulent air flow cools the back side somewhat and less insulation is required.

One drawback of air as a heat transfer fluid is its low heat capacity. The specific heat of air is 0.24 and its density is about 0.075 pounds per cubic foot under normal conditions. By comparison, water has a specific heat of 1.0 and a density of 62.5 pounds per cubic foot. For the same temperature rise, a cubic foot of water can store about 3500 times more heat than a cubic foot of air. It takes 260 pounds, or about 3500 cubic feet of air, to transport the same amount of heat as a cubic foot of water.

Because of this low heat capacity, large spaces through which the air can move are needed—even in the collector itself. Air ducts in collectors range from ½ to 6 inches thick. The larger the air space, the lower the pressure drop, but the poorer the heat transfer from absorber to air stream. And larger ducts mean more unusable space and higher costs for materials. For flat, sheet-metal absorber surfaces, the duct size usually is ½ to 1 inch. Duct sizes ranging from 1.5 to 2.5 inches are standard for large collectors using natural convection or having unusually long (more than 15 feet) ducts.

COLLECTOR PERFORMANCE AND SIZE

The performance of flat-plate collectors has been studied extensively since the landmark work of Hottel and Woertz 35 years ago. Most researchers try to predict the collector efficiency—the percentage of solar radiation hitting the collector that can be extracted as useful heat energy. A knowledge of this efficiency is very important in sizing a collector. If you know the available solar energy at your site, the average collector efficiency, *and* your heating needs, you're well on your way to determining the size of your collector.

The collector efficiency depends upon a number of variables—the temperature of the collector and the outside air, the rates of insolation and fluid flow through the collector, and its construction and orientation. By manipulating the variables he or she can control, a designer can improve overall collector performance. Unfortunately, few gains in efficiency are made without paying some penalty in extra costs. Beyond the obvious requirements of good collector location and orientation, many improvements in efficiency just aren't worth the added expense. Keep a wary eye turned toward the expenses involved in any schemes you devise to improve the efficiency.

A completely rigorous and detailed analysis of collector performance is beyond the scope of this book. But a survey of the various alternatives is included along with an assessment of their impact upon collector performance. The reader who is interested in looking further is encouraged to consult some of the many available works on this subject. Among these, we recommend the classic "Performance of Flat-Plate Solar-Heat Collectors," by Hottel and Woertz, "The Derivation of Several 'Plate-Efficiency Factors' Useful in the Design of Flat-Plate Solar-Heat Collectors," by Raymond Bliss, and the excellent "Design Factors Influencing Solar Collector Performance," by Austin Whillier (see references). The book, *Solar Energy Thermal Processes,* by John Duffie and William Beckman, is also helpful in this regard.

Collector Heat Losses

A portion of the sunlight striking the collector glazing never makes it to the absorber. Even when sunlight strikes a single sheet of glass at right angles, about 10 percent is reflected or absorbed. The maximum possible efficiency of a glazed collector is therefore about 90 percent. Even more sunlight is reflected and absorbed when it strikes at sharper angles—and the collector efficiency is further reduced. Over a full day, less than 80 percent of the sunlight will actually reach the absorber and be converted to heat.

Further decreases in efficiency can be traced to heat escaping from the collector. The heat transfer from absorber to outside air is very complex—involving radiation, convection, and conduction heat flows. While we cannot hope to analyze all these processes independently, we *can* describe some important factors, including:

- average absorber temperature
- outdoor air temperature
- wind speed
- number of cover plates
- amount of insulation

Perhaps you've already noticed that very similar factors determine the rate of heat escape from a house!

More heat escapes from collectors having hot absorbers than from those with relatively cool ones. Similarly, more heat escapes when the outdoor air is cold than when it it is warm. The *difference* in temperature

Energy Flows in a Collector

Because energy never disappears, the total solar energy received by the absorber equals the sum of the heat energy escaping the collector and the useful heat energy extracted from it. If H_a represents the rate of solar heat gain (expressed in Btu/ft^2/hr) by the absorber, and H_e is the rate of heat escape, then the rate of useful heat collection, H_c, is given by

$$H_c = H_a - H_e$$

Usually H_c and H_a are the easiest quantities to calculate, and H_e is expressed as the difference between them. The rate of solar heat collection is easily determined by measuring the fluid flow rate R (in lb/ft^2/hr) and the inlet and outlet temperatures T_{in} and T_{out} (in °F). The solar heat extracted, in Btu per per square foot of collector per hour, is then

$$H_c = R \times C_p \times (T_{out} - T_{in})$$

where C_p is the specific heat of the fluid—1.0 Btu/lb for water and 0.24 Btu/lb for air. Knowing H_c and the rate of insolation I, you can immediately calculate the collector efficiency E (in percent):

$$E = 100 \times \frac{H_c}{I}$$

The instantaneous efficiency can be calculated by taking this ratio at any selected moment. Or an average efficiency may be determined by dividing the total heat collected over a certain time period (say an hour) by the total insolation during that period.

Of the total insolation I, the amount actually converted to heat in the absorber, H_a, is reduced by the transmittance τ of the cover plates and by the absorptance α of the absorber. The value of H_a is further reduced (by 3 to 5 percent) by dirt on the cover plates and by shading from the glazing supports. Therefore, the rate of solar heat gain in the absorber is about

$$H_a = 0.96 \times \alpha \times \tau \times I$$

Both α and τ depend upon the angle at which the sunlight is striking the collector. Glass and plastic transmit more than 90 percent of the sunlight striking perpendicularly. But during a single day, the average transmittance can be as low as 80 percent for single glass, and lower for double glass. The absorptances of materials commonly used for collector coatings are usually better than 90 percent. If no heat escapes from the collector and all this solar radiation is converted to heat that is absorbed in the collector fluid, then

$$H_c = H_a = 0.96 \times \alpha \times \tau \times I$$

and the average collector efficiency (for a whole day) would still be less than 80 percent. Unfortunately, there are large heat losses from a flat-plate collector, and efficiencies rarely get above 70 percent.

between the absorber and the outdoor air, $\Delta T = T_{abs} - T_{out}$, is what drives the overall heat flow in that direction. The heat loss from a collector is roughly proportional to this difference.

As the absorber gets hotter, a point is eventually reached where the heat loss from a collector equals its solar heat gain. At this equilibrium temperature, the collector efficiency is zero—no useful heat is being collected. Fluids are usually circulated through a collector to prevent this occurrence. They carry away the accumulated heat and keep the absorber relatively cool. The higher the fluid flow rate, the lower the absorber temperature and the higher the collector efficiency.

Some fluids cool an absorber better than others. Although it has the disadvantages of freezing and corrosion, water is unmatched as a heat transfer fluid. It has the advantages of low viscosity and an extremely high heat capacity. Solutions of ethylene glycol in water solve the freezing problems but they also have lower heat capacity. For the same flow rates, a 25 percent solution of glycol in water will result in a 5 percent drop in collector efficiency.

As a heat transport fluid, air rates a poor third. While optimum water flow rates are 4 to 10 pounds per hour for each square foot of collector, 15 to 40 pounds of air are usually needed. And the rate of heat transfer from absorber to fluid must also be considered. Rough corrugated surfaces work best in air-type collectors. Good thermal bonds and highly conductive metal absorbers are needed with water-type collectors. The quicker the heat transfer to the passing fluid, the cooler the absorber and the higher the collector efficiency.

As with houses, the collector heat losses can be lowered by adding insulation or extra glass. But extra glass also cuts down the sunlight reaching the absorber. The relation of these two factors to the collector efficiency is illustrated in the next two graphs. In the first, the equivalent of 2 inches of fiberglass insulation is placed behind the absorber. The back of the second absorber is very heavily insulated—so that virtually no heat escapes. In all cases, the daily average collector efficiency falls with increasing differences between the absorber and outside air temperatures. For absorber temperatures less than 40°F above the outside air, a single pane of glass is best, and the added insulation is only a marginal improvement. For temperature differences beyond 60°F, the extra cover plates and insulation are obviously helpful. These remarks apply to a specific collector, but are generally true for most others.

Insolation

The collector efficiency also depends upon the amount of sunlight beating down. Under cloudy morning conditions, for example, the absorber will be much too cool to have fluid circulating through it, and no useful heat can be extracted. But at noon on a sunny day, the collector will be operating

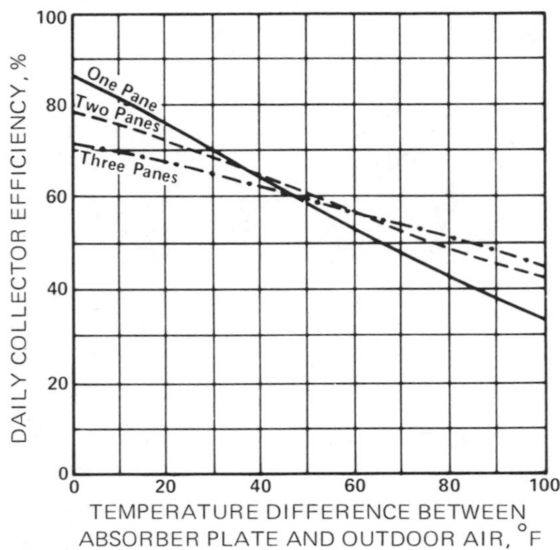

Performance of a moderately-insulated collector.

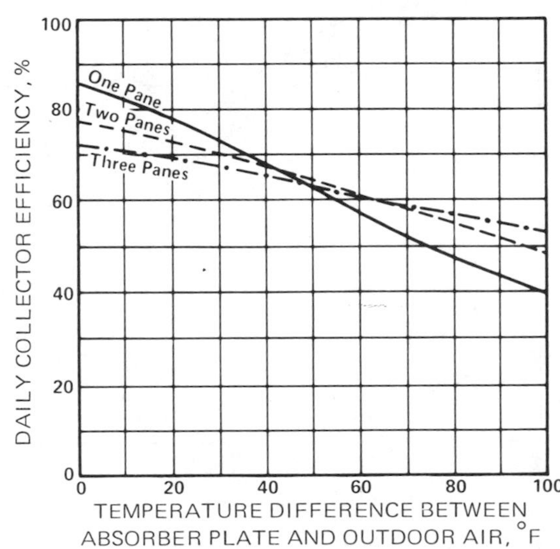

Performance of a well-insulated collector.

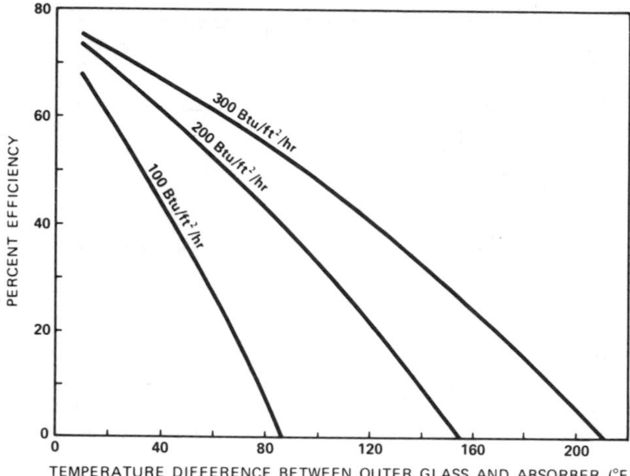

The effect of different insolation rates on collector efficiency. The outdoor air temperature is assumed constant.

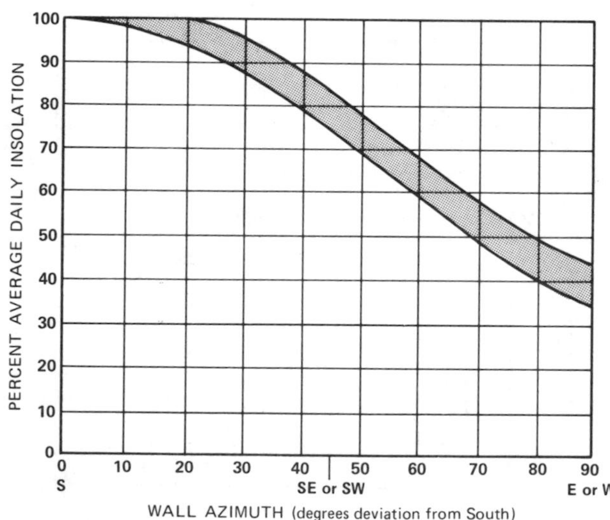

The percentage of insolation on vertical walls for orientations away from true south.

at full tilt, delivering 60 percent of the solar energy to storage. If the fluid flow can be increased to keep a constant absorber temperature, the collector efficiency will *increase* as the insolation rate increases.

The actual value of the insolation at a particular spot is very difficult to predict. Weather conditions vary by the hour, day, month, and year, and a collector designed for average conditions may perform quite differently at other times. For example, the collector described above may have an average daily efficiency of 40 percent, but its efficiency at any one moment can be anywhere in the range from 0 to 60 percent. Usually, we have to resign ourselves to using the *average* collector efficiency in our calculations. But that's not as bad as it may sound. The average daily insolation multiplied by the average collector efficiency gives us the solar heat collected per square foot on a typical day. With sufficient heat storage to tide us over times of shortage, why worry?

The Clear Day Insolation Data and the Percentage of Possible Sunshine Maps (given in Appendix 1) provide a suitable method for calculating the insolation at most sites. They are limited to south-facing surfaces, unless you want to do a little trigonometry. But as outlined in Chapter 3, they are capable of providing good estimates of the average insolation for any day and time. And fortunately a precise knowledge of the insolation isn't critical. Variations of 10 percent in the insolation will change collector efficiency by only 3-4 percent out of a total efficiency of 40 percent.

Collector Orientation and Tilt

Two other factors which determine a collector's performance are its orientation and tilt angle. A collector facing directly into the sun will receive the most insolation. But flat-plate collectors are usually mounted in a fixed position and cannot pivot to follow the sun as it sweeps across the sky each day or moves north and south with the seasons. So the question naturally arises, "What is the best orientation and tilt angle for my collector?" In addition, designers need to know how much they can deviate from optimum—without getting sued for negligence should a collector-dependent family suffer unexpected frostbite.

Although true south is the most frequent choice for the collector orientation, slightly west of south may be a better choice. Because of early morning haze which reduces the insolation and because of higher outdoor air temperatures in the afternoon, such an orientation can give slightly higher collector efficiencies. Harry Thomason oriented his first house 10° west of south for just these reasons. On the other hand, afternoon cloudiness in some localities may dictate an orientation slightly east of south. Fortunately, deviations of up to 25° from true south cause relatively small reductions in collector efficiencies. The designer has a fair amount of flexibility in his choice of collector orientation.

A useful diagram on the previous page shows the approximate decrease in the insolation on a vertical wall collector facing away from true south. The graph is valid for latitudes between 30° N and 45° N—almost the whole United States, if we forget Alaska and Hawaii. And it applies to the coldest part of the year—from November 21 to January 21. You can use this graph together with the Clear Day Insolation Data to get a rough estimate of the clear day insolation on surfaces that do not face true south. Simply multiply the data by the percentage appropriate to the orientation you have selected.

The effect orientation has upon the required size of a collector is illustrated by two examples in the next diagram. The vertical wall collectors in all cases are sized to provide 50 percent of the heating needs of a 1000 square foot house in either Boston or Charleston, South Carolina. Note that southwest (or, for that matter, southeast) orientations require an extra collector area of only 10 percent in Boston and 30 percent in Charleston. Tilted surfaces are affected even less by such variations.

The collector tilt angle depends upon its intended use. A steeper tilt is needed for winter heating than for summer cooling. If a collector will be used the year round, the angle chosen will be a compromise. If heating and cooling needs do not have equal weight, your selection of a tilt angle should be biased toward the most important need.

The general consensus holds that a tilt angle of 15° greater than the local latitude is

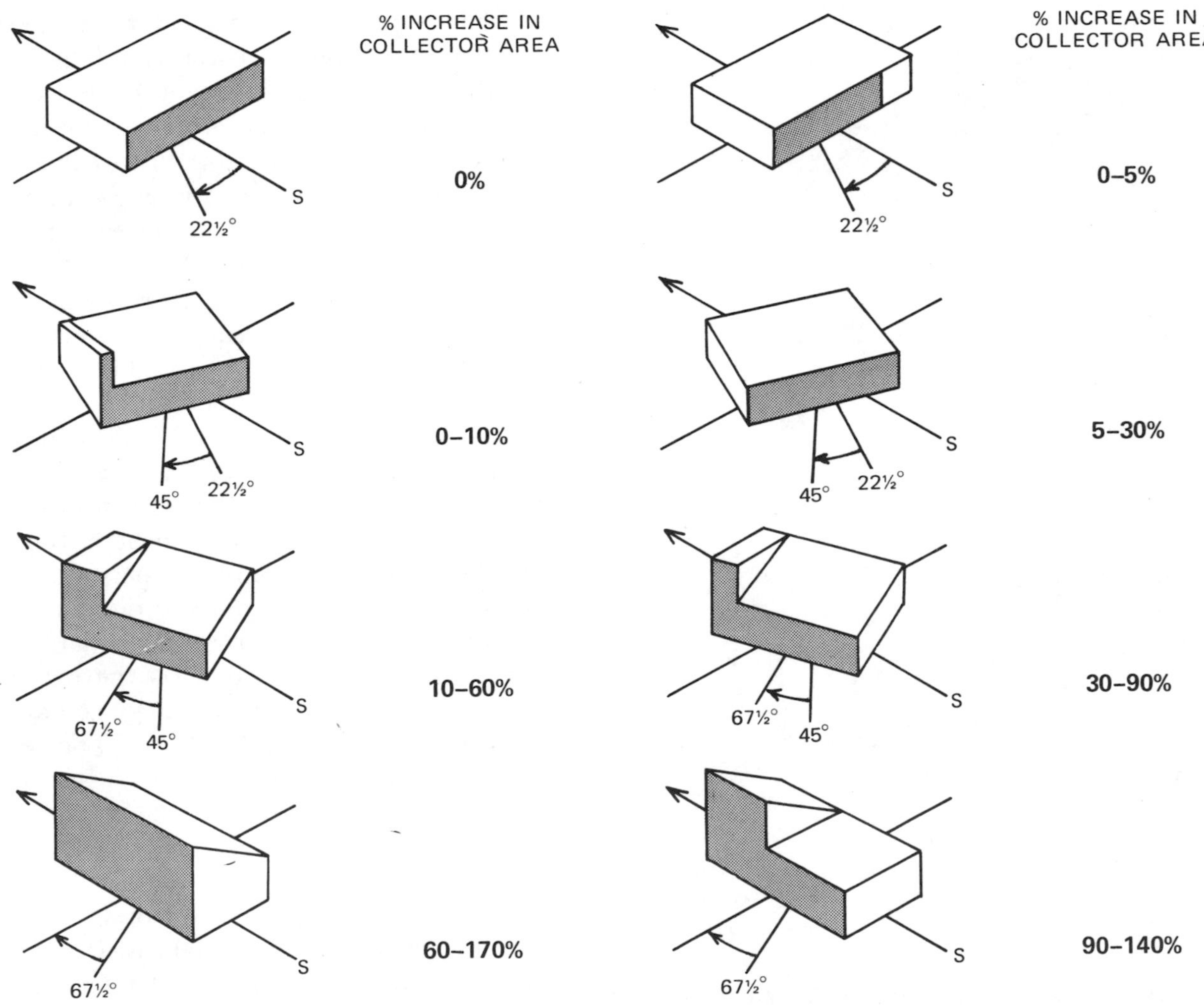

The change in area of a vertical wall collector with orientations away from true south. The collectors (shaded areas) have been sized to provide 50 percent of the winter heating needs of well-insulated homes in Boston and Charleston.

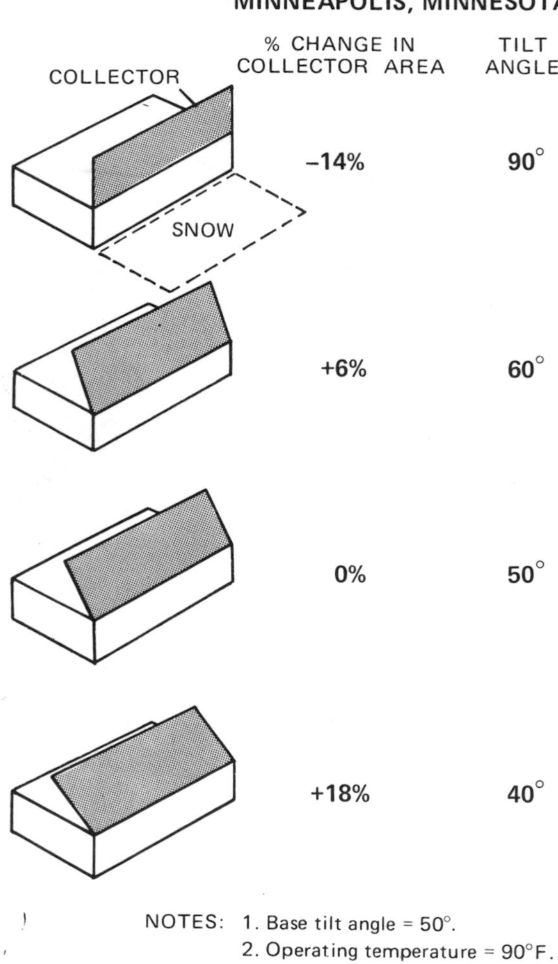

Small changes in collector size are required when the tilt angle differs from the optimum.

the optimum for winter heating. For year round uses, the collector tilt angle should equal the local latitude. And for maximum summer collection, the best choice is 15° less than the latitude. But your house won't freeze up or boil over if you don't have exactly the right tilt angle. The collector can be tilted 15-20° away from the optimum and still get more than 90 percent of the maximum possible insolation.

For areas with severe winter cloudiness, steeper tilts may be required. Sometimes vertical wall collectors are a good idea. Such a collector receives its peak insolation in the winter months, but gets very little in the summer, so don't expect to heat a swimming pool with it. When reflections from the ground are added to the sunlight hitting a vertical collector directly, its performance can surpass that of a collector tilted at the "optimum" winter heating angle (latitude + 15°). Clean, fresh snow has the highest reflectance–87 percent–of any common surface. It can add another 15 to 30 percent to the solar heat output of a vertical collector. Other surfaces such as asphalt, gravel, concrete, and grass have reflectances ranging from 10 to 33 percent.

The accompanying diagram shows the relationship between tilt angle and collector size. The collectors in all four cases are sized to provide 50 percent of the January heating for a typical Minneapolis home oriented facing true south. Relatively small changes in collector size can compensate for the reductions in solar collection that result from deviations from the January "optimum" (50°) tilt angle. With additional sunlight reflected from fallen snow, a vertical wall collector can be 14 percent *smaller* than the rooftop collector tilted at 50°.

Sizing the Collector

Accurate performance predictions are needed for sizing a collector. It is important to know whether 35 or 40 percent collector efficiencies (on the average) can be expected in any given month. With this knowledge and some predictions of the average daily insolation for that month, you'll have a good idea of how much solar heat you can expect from each square foot of collector. The size then follows from the average monthly heat demands.

But the performance of a collector is even more difficult to predict when it is tied into an entire heating system. Although some rules of thumb are beginning to emerge, the subject is still clouded in mystery as far as the average homeowner or builder is concerned. Site conditions, the heating demands of the house, and specific design choices (such as the collector tilt, the operating temperature, and the heat storage capacity) all affect the average collector performance.

A well-designed collector might be able to collect 1200 Btu per square foot on a sunny winter day. But not all of this heat will reach the rooms. There will be heat losses from the heat storage container, which may already be too hot to accept additional heat. Solar energy will be rejected during extended periods of sunny weather—even if the outdoor temperatures are quite cold. Only when the proper sequence of sunny and cloudy days occurs can all the available solar energy be used.

Consider a 1000 square foot house with a heating demand of 12,000 Btu per degree day, or 84 million Btu in a 7000 degree day climate. This demand is distributed over the heating season (late September to early May) as shown in the diagram. Little heat is needed in October or April and the bulk of the heat is needed from December through February—just when sunlight is at a premium.

Such a house needs about 600,000 Btu on a day when the outside temperature averages 15°F. On a sunny day, this amount of solar heat can be supplied by 500 square feet of a good collector. And with a 35°F temperature rise (from 85°F to 120°F, for example), about 2000 gallons of water will absorb all of the solar heat. But under *average* operating conditions, a square foot of this collector gains only 350 Btu of usable solar heat per day—or 84,000 Btu for the entire seven-month heating season. The 500 square feet of collector will supply only half of the seasonal heating load of 84 million Btu.

Contrary to what you might expect, doubling the collector size from 500 to 1000 square feet does *not* provide 100 percent of the heating load. Instead, it provides about 75 percent because the larger collector does not work to full capacity as often as the smaller one. In this particular system, the usable heat per square foot of collector drops from 84,000 Btu to 69,000 Btu because the house just cannot use the added heat in the fall and spring. And the system must reject more heat during four or five successive days of January sunshine. Even if the storage size were doubled, the system would have to reject excess heat during the autumn and spring. As collector size increases for a fixed heat demand, the amount of solar energy provided by each square foot drops because of the decreased *load factor* on each additional square foot.

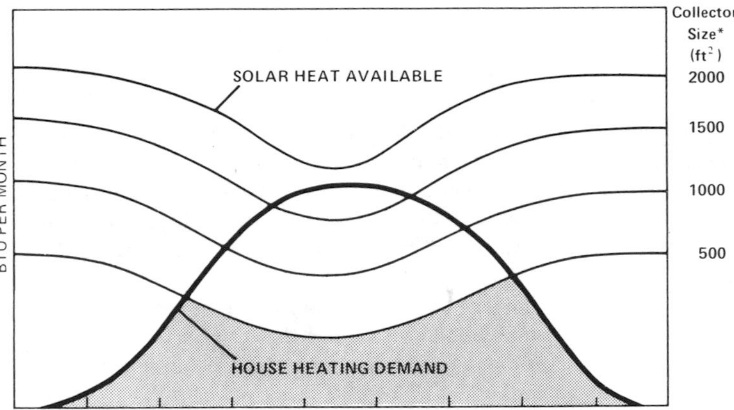

The solar heat delivered per square foot of collector decreases as the collector size increases. The shaded area represents the portion of the heating demand supplied by a 500 square foot collector. Doubling the collector size may double the solar heat supplied in January, but it doesn't help much in October or April.

Collector Size* (ft²)	Solar Heat Used per Square Foot of Collector	Total Solar Heat Supplied per Season	
		MBtu†	% of Demand
2000	42,000	84	100%
1500	56,000	75	89%
1000	69,000	63	75%
500	84,000	42	50%

*Storage size remains fixed.
†1 MBtu = 1 million Btu.

Estimating Collector Performance

The following method will help you to estimate the average efficiency and monthly solar heat output of a well-built solar collector. This method is accurate to about 20 percent and results in conservative estimates of performance. A running example is included for illustration.

1. *Find the total number of hours of sunshine for the month from the maps of Mean Percentage of possible Sunshine in Appendix 1; for example, 148 hours for Boston in January.*
2. *Find the average day length for the month from almanacs or the weather bureau; 10 hours for Boston in January.*
3. *From the Clear Day Insolation tables in Appendix 1, calculate the number of "collection hours" per day. For the selected collector tilt angle, this is the number of hours that the insolation is greater than 150 Btu/ft². Count ½ hour for insolation rates between 100 and 150 Btu/ft²; 7 hours for 40°N latitude and 60° tilt.*
4. *Determine the total collection hours per month by multiplying the sunshine hours per month (#1) by the collection hours per day (#3) and dividing by the average day length (#2); 148 hours × 7 hours ÷ 10 hours = 104 hours.*
5. *Determine the total daily useful insolation, defined as the total insolation during collection hours, by adding the hourly insolation rates (from the Clear Day Insolation tables) for those collection hours described above; 187 + 254 + 293 + 306 + 293 + 254 + 187 = 1774 Btu/ft²/day.*
6. *Determine the average hourly insolation rate I during the collection period by dividing the total daily useful insolation (#5) by the number of collection hours per day (#3); 1774 Btu/ft²/day ÷ 7 hours/day = 253 Btu/ft²/hr.*
7. *Determine the average outdoor temperature during the collection period from ½ the sum of the normal daily maximum temperature and the normal daily average temperature for the month and locale. These are available from the local weather bureau and from the* Climatic Atlas of the United States, *pages 2-24;* T_{out} *= ½(38°F + 30°F) = 34°F.*
8. *Select the average operating temperature* T_{abs} *of the collector and find the difference* $\Delta T = T_{abs} - T_{out}$. *In general, you should examine a range of possible values for* T_{abs}; ΔT *= 120°F − 34°F = 86°F.*
9. *Refer to a performance curve for the collector to determine the average collector efficiency from a knowledge of* ΔT *(#8) and I (#6). The sample curves provided on the opposite page apply to a tube-on-plate liquid-type collector manufactured by Revere Copper and Brass Co; but they should be fairly accurate for most collectors of moderate to good construction; average collector efficiency = 38% for a double-glazed collector.*
10. *Determine the average hourly collector output by multiplying the average hourly insolation rate I (#6) by the average collector efficiency (#9); 0.38 × 253 Btu/ft²/hr = 95 Btu/ft²/hr.*
11. *The useful solar heat collected during the month is then the average hourly collector output (#10) multiplied by the number of collection hours (#4) for that month: 95 Btu/ft²/hr × 104 hours/month = 9880 Btu/ft²/month for January in Boston.*

This procedure should be repeated for a number of other collector operating temperatures and tilt angles.

A simplified method for collector sizing, developed by Total Environmental Action, Inc., lies somewhere between educated guesswork and detailed analytical calculations. It is accurate to within 20 percent and is biased toward conservative results. Architects, designers, and owner-builders often need such a "first-cut" estimate of collector size to proceed with their designs. Purists in need of better estimates should consult the work of Liu and Jordan in *Low Temperature Engineering Applications of Solar Energy.*

First the monthly output of a flat-plate collector is calculated as the product of the *useful* sunshine hours in the month times the average hourly solar heat output of the collector during that month. The number of useful sunshine hours is less than the total number of sunshine hours because the insolation rate is not high enough in early morning or late afternoon to justify collector operation. The two most influential factors that determine the hourly solar output—the hourly insolation rate and the temperature difference between the absorber and the outside air—change rapidly during a single day. Therefore, we only try to determine an *average* hourly solar output. The method outlined in "Estimating Collector Performance" is an attempt to determine the reasonable mean monthly output from average insolation rates and temperature differences.

A sample calculation of mean monthly solar heat output is provided for illustration. The hypothetical collector is sited in Boston and oriented south at a tilt angle of 60°. At an average operating temperature of 120°F, this collector has an efficiency of 38 percent and gathers 9880 Btu per square foot during the month of January.

Monthly solar output for the rest of the heating season has been calculated with the same method and listed in the accompanying table. The output of a vertical collector (including 20 percent ground reflectance) is included in the table, as are the monthly outputs when 90°F and 140°F operating temperatures are allowed. The seasonal output is the sum of all these monthly figures. In your design work, it's extremely useful to consider a number of alternative collector tilts and operating temperatures—instead of proceeding single-mindedly with a preconceived design. Almost every collector operates over a range of temperature and its efficiency varies in a corresponding fashion. It's instructive to determine the solar heat collection for a few of these conditions.

In general, the larger the percentage of house heating you want your collector to supply, the more difficult it is to estimate its size using these simplified methods. The actual sequence of sunny and cloudy days becomes more important as the percentage of solar heating increases. If a full week of cold, cloudy days happens to occur in January, your collector (or storage) would have to be enormous to insure 90 percent solar heating. But good approximations of collector size can be made for systems that are designed to supply 60 percent or less of the seasonal heating needs.

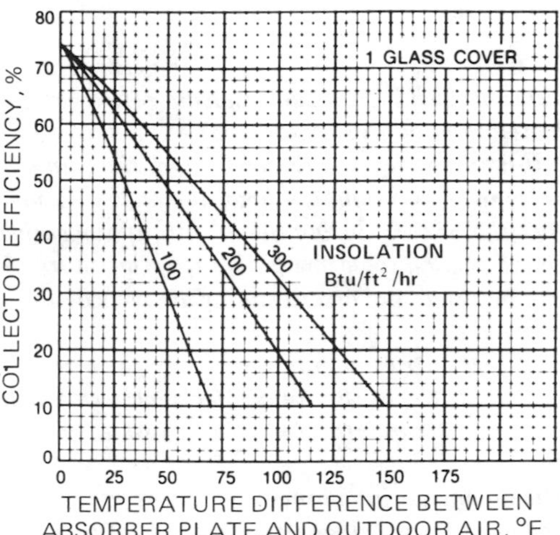

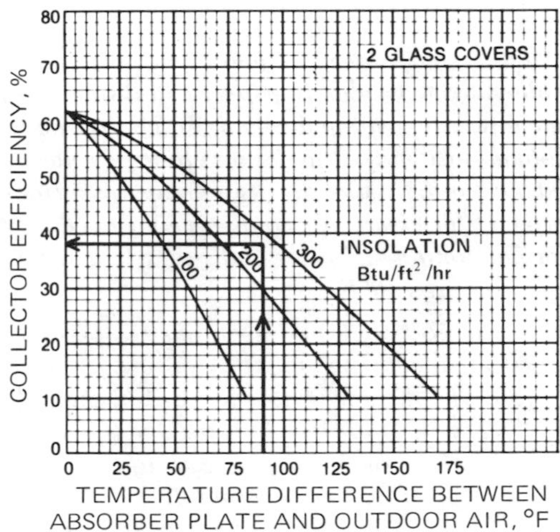

Performance curves for single and double-glazed collectors. Use these graphs to estimate collector efficiency from the insolation rate and temperatures of the absorber and outdoor air.

EXAMPLE: Find the efficiency of a double-glazed collector operating at 120°F if the outdoor air temperature is 34°F and the insolation rate is 253 Btu/ft²/hr.

SOLUTION: The value of $\Delta T = T_{abs} - T_{out}$ is 120°F − 34°F = 86°F. Start from this value at bottom of right hand graph and move vertically to intersect (an imaginary) curve for 253 Btu/ft²/hr. Then move horizontally left to find estimated collector efficiency, 38%.

ESTIMATES OF MONTHLY COLLECTOR OUTPUT*

Collector:		Average Solar Heat Collected (Btu/ft²)									
Temp. (°F)	Tilt	Sep	Oct	Nov	Dec	Jan	Feb	Mar	Apr	May	TOTALS
90°	60°	21,700	19,630	13,780	13,080	12,480	14,640	18,000	14,720	15,225	143,255
90°	90°†	14,615	19,781	14,310	14,170	13,000	15,250	10,925	5,040	2,520	109,611
120°	60°	18,600	15,855	11,130	9,810	9,880	11,590	14,250	12,160	12,325	115,600
120°	90°†	12,700	16,006	11,660	11,445	10,400	12,200	8,625	3,240	1,800	88,076
140°	60°	16,275	12,835	9,010	8,175	7,800	9,150	11,250	10,240	9,425	94,160
140°	90°†	10,160	12,986	9,540	9,265	8,320	9,760	5,750	2,160	1,080	69,021

*Location: Boston.
†With 20% ground reflection.

Estimating the Collector Size

The following procedure helps you to estimate the collector size needed to supply a desired percentage of the yearly heating demand. To use it, you need the monthly output per square foot of collector, as calculated in "Estimating the Collector Performance." The Boston example is continued here for illustration—with the collector tilted at 60° and operating at 120°F.

1) *For the tilt angle and operating temperature selected, enter the monthly output per square foot of collector in column A. Add them to get the heating season output for one square foot.*
2) *Enter the monthly degree days of the location (see Appendix 1) in column B.*
3) *Enter the monthly heat loss of the house in column C. This is the product of the (monthly degree days) times the (heat loss per degree day—or 9500 Btu per degree day for our Boston home).*
4) *Add the entries in column C to determine the seasonal heat loss. Divide this total by the total of column A (step 1) and take 60% of the result as a first guess at the collector area needed to supply 50% of the seasonal heat demand:*

$$0.60 \times (53.46 \text{ MBtu} \div 115{,}600 \text{ Btu/ft}^2) = 277.5 \text{ ft}^2 .$$

5) *Multiply this collector area by the entries in column A and enter the resulting solar heat collected in column D.*
6) *Subtract entries in column D from those in column C to obtain the heat demand NOT met by solar energy during the month. If a negative result occurs, solar energy is supplying more than can be used, and a zero should be recorded in column E.*
7) *Subtract entries in column E from those in column C to get the total solar heat used by the house in the month. Enter these results in column F.*
8) *Divide entries in column F by those in column C to get the percentage of monthly heat losses provided by solar (column G).*
9) *Divide the seasonal total of column F by that of column C to get the percentage of the seasonal heat loss provided by solar, or 47% in the Boston home. If this result is too low (or high) the collector area can be revised in step 4 and steps 5 to 9 repeated until satisfaction is achieved.*

The total of column F is the "useful" solar energy output of the collector. It can be used to predict the economic return on the initial expenses of the system.

House Heat Loss: 9500 Btu/degree day
Collector Area: 277.5 ft^2

Month	A Collector Output (Btu/ft^2)	B Degree Days (°F-days)	C Heat Loss (MBtu*)	D Solar Heat Collected (MBtu*)	E Auxiliary Heat (MBtu*)	F Solar Heat Used (MBtu*)	G Percent Solar Heated
September	18,600	98	0.93	5.16	0	0.93	100%
October	15,855	316	3.00	4.40	0	3.00	100%
November	11,130	603	5.73	3.09	2.64	3.09	54%
December	9,810	983	9.34	2.72	6.62	2.72	29%
January	9,880	1,088	10.34	2.74	7.60	2.74	26%
February	11,590	972	9.23	3.22	6.01	3.22	35%
March	14,250	846	8.04	3.95	4.09	3.95	51%
April	12,160	513	4.87	3.37	1.50	3.37	69%
May	12,325	208	1.98	3.42	0	1.98	100%
Heating Season Totals	**115,600**	**5,627**	**53.46**	**32.07**	**28.46**	**25.00**	**47%**

*Millions of Btu

A simplified method of calculating the collector size from monthly output and heating demand figures is outlined in "Estimating the Collector Size." The monthly output figures are those of our hypothetical collector, tilted at 60° and operating at an absorber temperature of 120°F. The heating demand figures are for a well-insulated Boston home of 1000 square feet that loses 9500 Btu per degree day. In this particular example, we strive to provide 50 percent of the seasonal heat demands of the house. If the initial guess at the appropriate collector size does not provide the desired percentage, it can be revised up or down and the calculations repeated until the desired results are achieved.

The final size of the collector should reflect other factors besides heating demand—for example, the size of the heat storage container, the solar heat gain through the windows, and cost. Much more information and detail on this simplified method of collector sizing can be found in pages 42-61 of the TEA publication, *Solar Energy Home Design in Four Climates.*

CONCENTRATING COLLECTORS

Because flat-plate collectors are difficult to mass-produce and transport, *concentrating* and *focussing collectors* may soon emerge as favorite children of industrial giants. These collectors use one or more reflecting surfaces to concentrate sunlight onto a small absorber area. Use of expensive materials like copper or glass can be kept to a minimum and sometimes even eliminated. And collector performance is enhanced by the added sunlight hitting the absorber.

Owner-builders have been tinkering with concentrating collectors for years. Henry Mathew's rooftop collector on his Coos Bay house (see Chapter 2) and Steve Baer's Drumwall both use aluminized surfaces to reflect additional sunlight onto their absorbers. In the Winters House, the inner surfaces of the insulating lids reflect sunlight down onto the roof pond collectors. Depending upon their total area and orientation, flat reflectors such as these can direct 50 to 100 percent more sunlight at the absorber. And they work well in cloudy or hazy weather—when diffuse sunlight is coming from the entire sky.

The reflecting surface can be *curved* to reflect incoming sunlight onto an even smaller area. A parabolic surface (a fly ball hit to the outfield traces out a parabolic path) can focus sunlight on an area as small as a blackened pipe with fluid running through it. Such a focussing collector will perform extremely well in direct sunlight but will not work at all under cloudy or hazy skies because only a few of the rays coming from the entire bowl of the sky can be caught and reflected onto the blackened pipe. And even in sunny weather, the reflecting surface must pivot to follow the sun so that the absorber remains at the focus. The mechanical devices needed to accomplish this tracking can be expensive and failure-prone. But the higher temperatures and efficiencies possible with a focussing collector are sometimes worth this added cost and complexity.

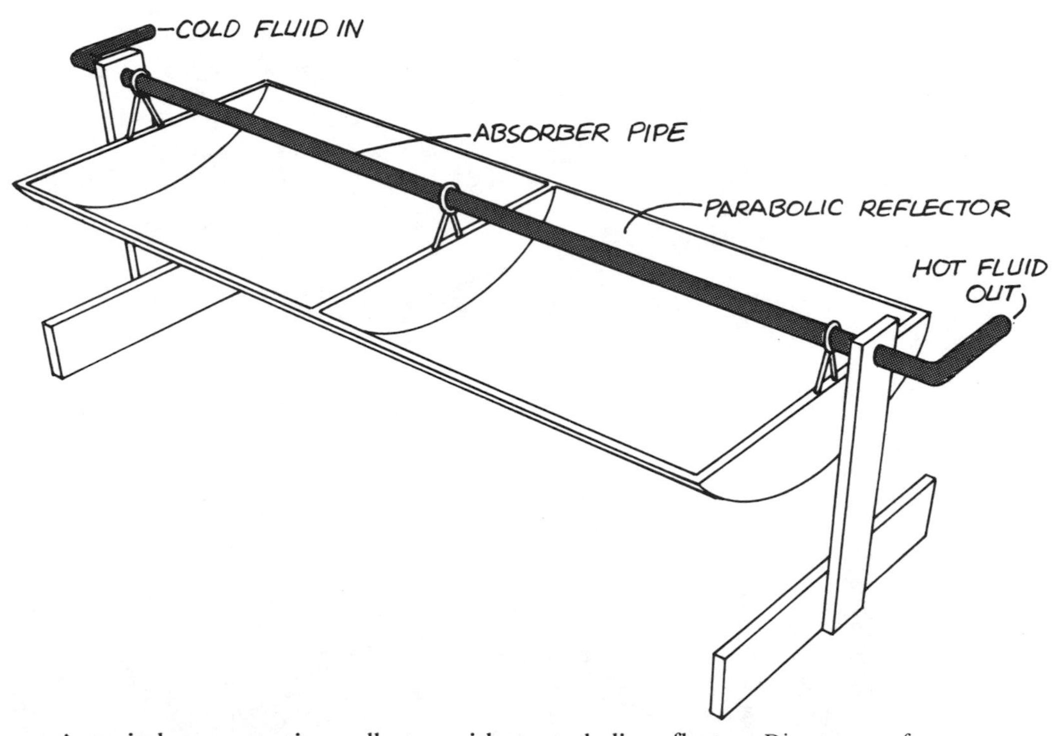

A typical concentrating collector with a parabolic reflector. Direct rays from the sun are focussed on the black pipe, absorbed, and converted to heat.

Pyramidal Optics Collector

Inventor Gerald Falbel designed a concentrating collector that can be built into

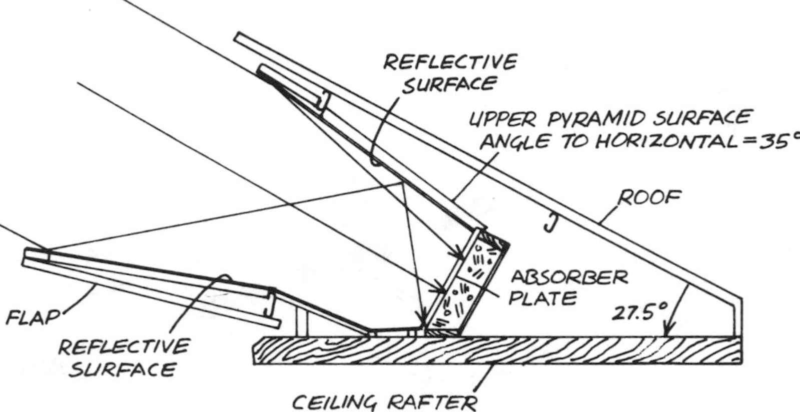

Cross-section of the pyramidal optics collector.

A pyramidal optics collector installed at the home of Gerald Falbel.

an attic. Using what he calls "reflective pyramidal optics," the collector concentrates sunlight onto a flat plate absorber located at the back of the attic. This system allows a reduction in the size and cost of the absorber plate and glass.

A prototype system was installed in 1974 at Falbel's home in Stamford, Connecticut. The inside surfaces of the attic are lined with aluminized mylar to reflect sunlight onto a 2' x 8' aluminum ROLL-BOND absorber plate. At the mouth of the collector is a hinged reflector door that can be closed at night or during cloudy weather. When the sun shines, the tilt of the door can be adjusted to throw maximum insolation on the absorber. A concentration ratio of 3 to 4 times the normal insolation rate can be obtained on sunny days and 1 to 2 when it is overcast.

One of the best features of this design is its potential for eliminating the freezing problems that plague water-type collectors in cold climates. With the door closed at night, temperatures in the attic stay above freezing. Simpler absorber designs and materials can be used—at correspondingly lower cost. This helps offset the added expenses of the attic structure, reflectors, and door. And because of the lower heat loss from a sheltered absorber, higher fluid temperature can be attained without sacrificing collector efficiency. Further information about this system can be obtained from the Wormser Scientific Corporation (88 Foxwood Road, Stamford, Connecticut 06903).

Compound Parabolic Collector

Another concentrating collector has been developed at the Argonne National Laboratory by Dr. Roland Winston, a physicist. His collector uses an array of parallel reflecting troughs to concentrate *both* direct and diffuse solar radiation onto a very small absorber—usually blackened copper tubes running along the base of each trough. The two sides of each trough are sections of a parabolic cylinder—hence the name "compound parabolic collector." Depending upon the sky condition and collector orientation, a three to eightfold concentration of solar energy is possible. The collector performs at 50 percent efficiency while generating temperatures 150°F above that of the outside air.

The real beauty of Winston's collector is its ability to collect diffuse sunlight on cloudy or hazy days. Virtually all the rays entering a trough are funneled to the absorber at the bottom. With the troughs oriented east-to-west, the collector need not track the sun. Winston merely adjusts its tilt angle every month or so. After publishing his initial designs, he discovered that the same optical principles have been used by horseshoe crabs for thousands of centuries. These antediluvian creatures have a similar structure in their eyes to concentrate the dim light that strikes them as they "scuttle across the floors of silent seas."

Several prototypes of this "crab-eye" collector have been built by licensed manufacturers. If costs can be kept down, this collector has great promise for such applications of solar energy as absorption cooling and power for steam turbines that require high temperatures and efficiencies. Conceivably, one could even put photovoltaic cells at the base of the troughs and generate electricity directly.

Many other concentrating collectors have been developed recently, but the designs are too varied and numerous to discuss here. Most of these collectors can produce temperatures above 300°F with efficiencies between 50 and 75 percent. But because of their high cost and the complexity they cause in a total system, these collectors are not yet (1976) economically feasible for solar space heating. They may soon find applications in solar absorption cooling, which requires temperatures of 200°F or higher.

HEAT STORAGE

Some capacity for storing solar heat is almost always necessary because the need for heat continues when the sun doesn't shine. And more heat than a house can use is generally available when the sun *is* shining. By storing this excess, an indirect system can provide energy as needed—not according to the whims of the weather.

If costs were not a factor, you would probably design a heat storage unit large enough to carry a house through the longest

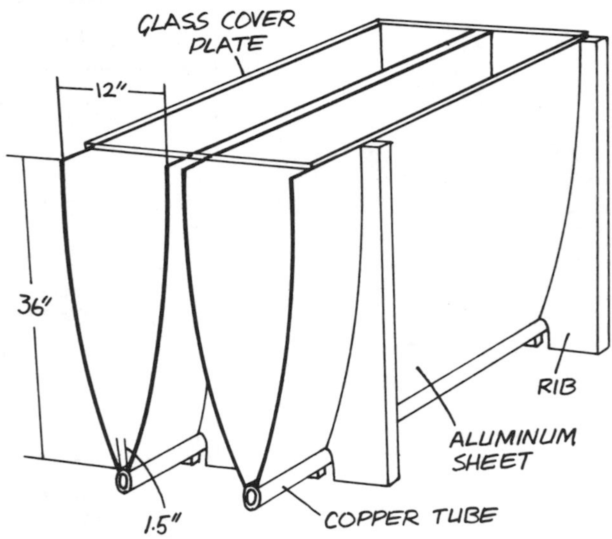

Compound parabolic collector invented by Dr. Roland Winston.

PROPERTIES OF HEAT STORAGE MATERIALS				
Material	Specific Heat (Btu/lb/°F)	Density (lb/ft³)	Heat Capacity (Btu/ft³/°F)	
			No voids	30% voids
Water	1.00	62	62	43
Scrap Iron	0.12	490	59	41
Scrap Aluminum	0.23	170	39	27
Concrete	0.23	140	32	22
Stone	0.21	170	36	25
Brick	0.20	140	28	20

SOURCE: Hottel and Howard, *New Energy Technology, Some Facts and Assessments*, 1971.

periods of sunless weather. The well-insulated 17,000 gallon storage tank in the basement of the first MIT house could store heat from the *summer* for use in *winter!* But most of us do not have the money to spend on an enormous storage tank, and our designs are limited by what we can afford. Some of the major factors influencing heat storage costs are:

- choice of storage medium
- amount or size of the storage medium
- type and size of a container
- location of the heat storage
- use of heat exchangers, if required
- choice of pumps or fans to move the heat transfer fluid

In designing solar heat storage, you must weigh these cost considerations against the performance of the system. All of the above factors influence performance to some extent. Other factors include the average operating temperature of the entire system, the pressure drop of the heat transfer fluid as it passes through or by the storage medium, and the overall heat loss from the container to its surroundings.

In general, the heat storage capacity of different materials varies according to their specific heat—the number of Btu required to raise the temperature of 1 pound of a material 1°F. The specific heats of a few common heat storage materials are listed in the table along with their densities and heat capacities—the amount of heat you can store in a cubic foot of the material for a 1°F temperature rise. Heat energy stored in this way (often called *sensible heat*) is reclaimed as the temperature of the storage medium falls. It takes high temperatures or large volumes of material to store enough sensible heat (say 500,000 Btu) for a few cold, sunless days. Rocks and water are by far the most common storage media because they are cheap and plentiful.

Some materials absorb a lot of heat as they melt, and surrender it as they solidify. A pound of Glaubers salt absorbs 104 Btu and a pound of paraffin 65 Btu when they melt at temperatures not far above normal room temperatures. The heat absorbed by such phase-changing materials is often called *latent heat.* The storage of heat in phase changing materials can reduce heat storage volumes drastically. But problems of cost, containment, and imperfect re-solidification have kept their use very limited thus far.

A solar collector and heat storage medium should be chosen together. Liquid-type collectors almost always require a liquid storage medium. Most air-type collectors require a storage medium consisting of small rocks, or small containers of water or phase-changing salts. These allow the solar heated air to travel around and between—transferring its heat to the medium. Within these two basic categories of heat storage, there are many possible variations.

Tanks of Water

Water is cheap and has a high heat capacity. Relatively small containers of water will store large amounts of heat at low temperatures. From 1 to 10 gallons are needed per square foot of collector—or 500 to 5000 gallons for a 500 square foot collector. Another advantage of water heat storage is its compatibility with solar cooling. But there are several problems with water storage, such as the high cost of tanks and the threat of corrosion.

Water containment has been simplified in recent years by the emergence of good waterproofing products and large plastic sheets. Previously, the only available containers were leak-prone galvanized steel tanks. Their basement or underground locations made replacement very difficult and expensive. Glass linings and fiberglass tanks helped alleviate corrosion problems but increased initial costs. Until recently, the use of poured concrete tanks has been hampered by the difficulty of keeping them water tight—concrete is permeable and develops cracks. But large plastic sheets or bags now make impermeable liners having long lifetimes. And with lightweight wood or metal frames supporting the plastic, the need for concrete is eliminated.

The most straightforward heat storage system (see diagram) is a water filled container in direct contact with both the collector and the house heating system. The container shown is made of concrete or cinder blocks with a waterproof liner, but it might well be a galvanized or glass-lined steel tank. The coolest water from the bottom of the tank is circulated to the collector for solar heating and then returned to the top of the tank. Depending upon the time of day, the temperature difference between the bottom and top of a 3 to 4 foot high tank can be 15 to 25°F. The warmest water from the top of the tank is circulated directly through baseboard radiators or radiant heating panels inside the rooms. This heat storage and distribution scheme was used in the third MIT solar house (see Chapter 2).

A variation of this basic scheme has a heat exchanger—a copper coil or finned tube—immersed in the tank of solar heated water. Water or another liquid circulates through the heat exchanger, picks up heat, and carries it to the house. Or warm water in the

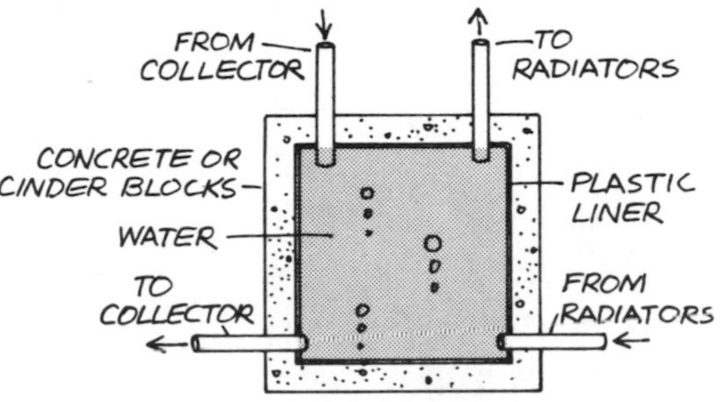

Heat storage tank tied directly to both the collector and the house heating system.

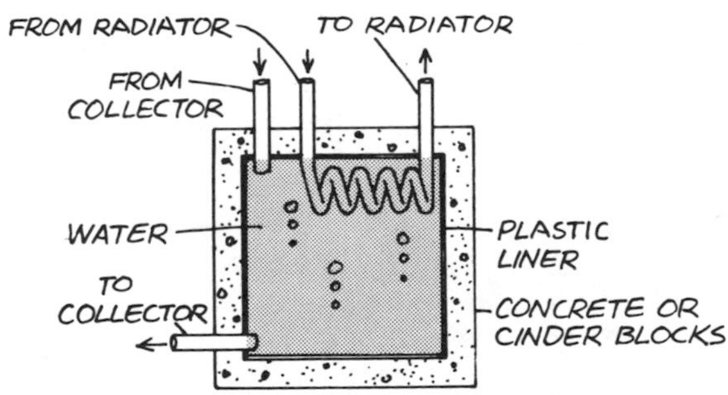

Use of a heat exchanger to extract solar heat from a storage tank.

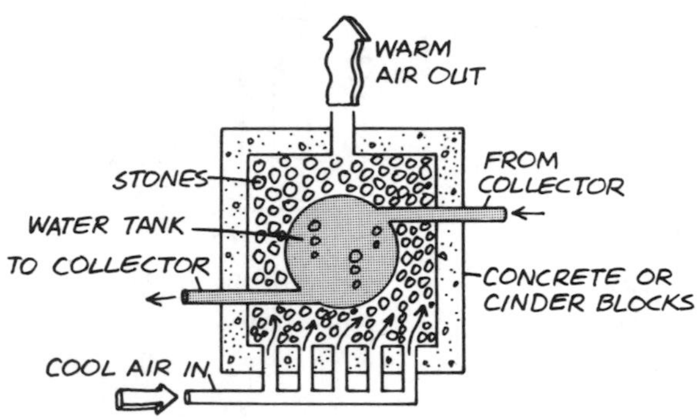

Harry Thomason's approach uses both water and stone as storage media.

tank can be pumped through heat transfer coils located in an air duct. Cool room air is blown past the coils and heated. This latter method was used to deliver solar heat to the rooms of the fourth MIT Solar House.

An ingenious approach to heat storage was developed by Dr. Harry Thomason. He surrounded a tank of solar heated water with tons of fist-sized stones. Not only do the stones provide extra heat storage but they also act as a cheap, effective heat exchanger. Solar heat from the tank travels slowly but continuously to the surrounding stones. Large volumes of cool room air are circulated slowly through the warm stones and returned to the house.

Heat exchangers are necessary when the water in the tank cannot be used for purposes other than heat storage. For example, an antifreeze solution used in a solar collector is often routed through a heat exchanger to prevent mixing with water in the tank. And heating engineers often insist that water in the tank *not* be used in the room radiators—particularly when the tank water is circulated through the collector. Because of their large size, some of these heat exchangers can be expensive. For a typical metal heat exchanger submerged in the water tank, the total metal surface can be as much as 1/3 the surface area of the solar collector. The beauty of Thomason's storage scheme is that it uses an incredibly cheap and plentiful material for a heat exchanger—rocks! And because of the large total surface area of these rocks, storage temperatures as low as 75°F can be used to heat the house.

For the designer who wishes to include heat storage as an integral part of a total design, the placement of a large unwieldly tank can be a problem. Self-draining systems require a tank located below the bottom of the collector, and thermosiphoning systems need it above the collector top. If the storage tank is linked to other equipment such as furnace, pumps, or the domestic water heater, it will probably have to be located near them.

Rock Beds

Rocks are the best known and most widely used heat storage medium for air-type systems. Depending upon the dimensions of the storage bin, rock diameters of 1 to 4 inches will be required. But through much of New England, for example, the only available rock is 1 to 1½ inch gravel. Even if the proper size is available, a supplier may be unable or unwilling to deliver it. Collecting rocks by hand sounds romantic to the uninitiated but becomes drudgery after the first thousand pounds. And many thousands of pounds—from 80 to 400 pounds per square foot of collector—are required because of the low specific heat of rock.

A large storage bin must be built to contain the huge quantities of rock needed. With 30% void space between the rocks, the bin requires about 2½ times as much volume as a tank of water to store the same amount of heat over the same temperature rise.

And the large surface area of a rock storage bin leads to greater heat losses. But the slow natural circulation of heat through rocks (as compared to water) offsets this loss somewhat.

Contrary to the belief of many engineers, rock storage bins can be used in systems which combine cooling and domestic water heating with space heating. Cool night air is blown over the rocks, and the coolness stored for daytime use. And conventional air conditioners can be used to cool the rocks at night. Domestic water can be preheated in a small tank buried in the rocks. Or cold water from city mains can pass through a heat exchanger located in the air duct returning from the collector to the storage bin.

The location of a rock storage bin must take into account its great volume and weight. It can be located in a crawl space under the house at small additional cost. Putting it inside the basement or other living space is usually more difficult and expensive. Dr. George Löf developed an ingenious method of containing the rocks in vertical fiberboard cylinders placed within the house. He used this simple method in his own Denver home. Because of the great weight, however, strong foundations under such containers are a must.

To distribute heat from a rock storage system, air is either blown past the hot rocks or allowed to circulate through them by gravity convection. From there, the air carries solar heat to the rooms. In general, a fan or a blower is needed to augment the natural circulation and give the inhabitants better control of the indoor temperature.

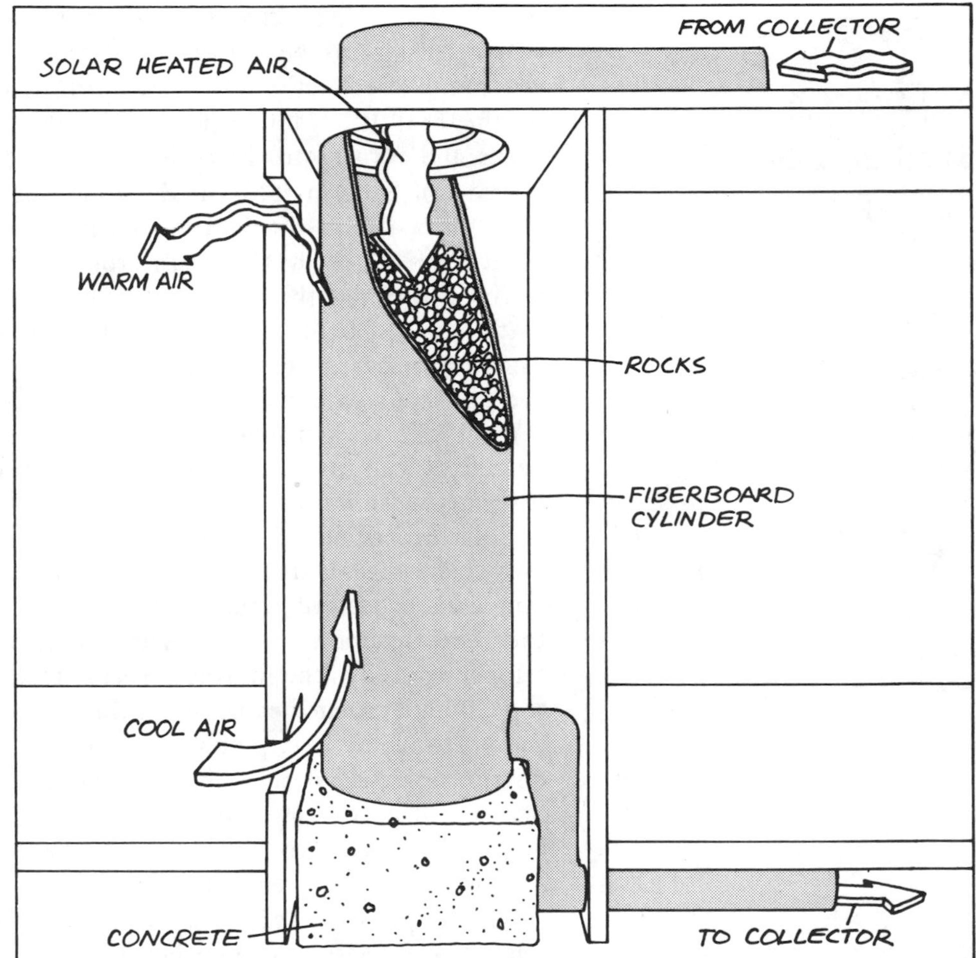

A vertical fiberboard cylinder used to contain a rock heat storage bed. Heat can flow to the rooms by natural convection as shown, or another fan can be used to circulate cool air through the rocks and back to the rooms.

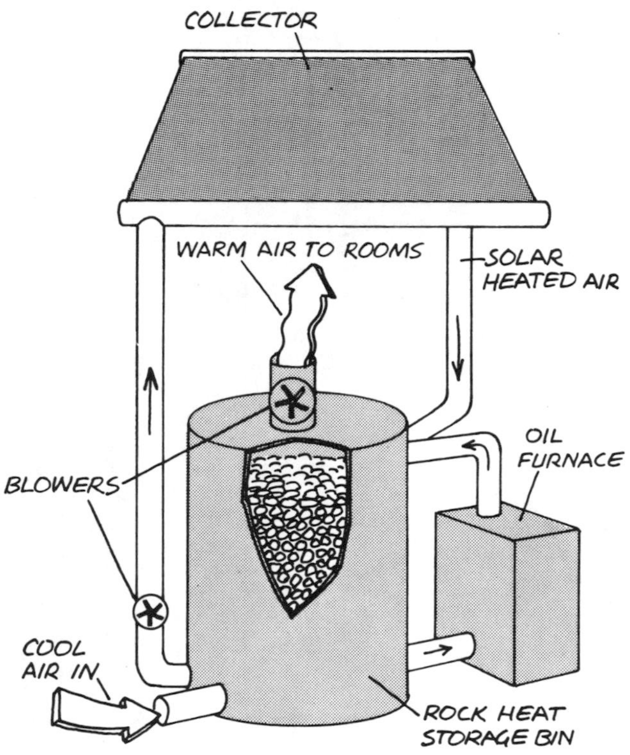

Schematic diagram of an air system with rock heat storage. The auxiliary oil furnace heats only the room air—not the rocks.

A basic method of transferring heat to and from a heat storage bed is shown schematically in the diagram. Solar heated air from the collector is delivered to the *top* of the bin. It is drawn down through the rocks and returns to the collector from the bottom of the bin. To heat the house, cool air is drawn in at the bottom and is heated as it rises through the warm rocks. The warmest rocks at the top transfer their heat to the air just before it is sent to the rooms. The furnace heating cycle (also shown) allows the house air to bypass the rock storage. In general, the rock storage bed should not be heated by the furnace except when the bed is located within the living space.

Solar heated air is brought in at the top of a rock storage bed in order to encourage temperature stratification. House air can then be heated to the highest possible temperature by the warmest rocks at the top. But if solar heated air comes in at the bottom, the heat percolates upward and distributes itself evenly through the entire bed—resulting in low temperatures throughout. Bringing cool room air in at a warm top also promotes this even distribution.

The shape of a rock storage bin is closely related to rock size. The farther the air must move among the rocks, the larger the rock diameter required to keep the pressure drop and fan size small. If a storage bin is more than 8 feet tall, the rocks should be at least 2 inches in diameter—and larger for taller containers. For squat, horizontal bins such as you might put in a crawl space, 1 to 2 inch gravel can be used.

The optimum rock diameter depends a lot on the velocity of the air moving through the rocks. The slower the air speed, the smaller the rock diameter or the deeper the bed of rocks can be. And the smaller the rock diameter, the greater the rock surface area exposed to the passing hot air. A cubic foot of 1 inch rock has about 40 square feet of surface area while the same volume of 3 inch rock has about 1/3 as much. In general, the rocks, stones, gravel, or pebbles should be large enough to maintain a low pressure drop but small enough to insure good heat transfer. Papers by Löf, Close, and Bird (see the references at the end of this chapter) will help you determine the relationship between rock diameter, air velocity, and pressure drop.

Small Containers of Water

One-gallon or smaller containers of water can also be used as the heat storage medium in air-type systems. They can be arranged in racks, on shelves, or in any fashion that allows an unobstructed air flow around them. Possible containers include plastic, glass or aluminum jars, bottles, or cans. Perhaps the best arrangement is to sandwich the containers between large trays and blow the air past them horizontally. Placed inside wall cavities, they serve the dual purpose of heat storage and room partitions.

A preliminary house design that uses containerized water for heat storage was developed by TEA, Inc. for a location in Phoenix. In this design, air circulates up through a vertical south-wall collector and down through a vertical partition filled with small containers of water. Such a wall configuration is not very feasible with rocks. One distinct advantage of containerized water is that it can be readily incorporated into a total house design. Leakage is not a serious problem because no more than one gallon of water can escape from any single leak.

Another advantage of containerized water is that a smaller total volume can be used to contain the same amount of heat as in rock beds. With 50 percent void space between the containers, water stores 32 Btu per cubic foot for a 1°F temperature rise. With 30 percent voids, rock stores 25 Btu. If the water containers can be arranged with only 30 percent voids, they can store 43 Btu per cubic foot.

Should this type of heat storage find widespread applicability, special containers of rigid molded plastic will probably become available. Such containers would stack one above the other and be self-supporting to some extent. Similar containers of soft vinyl could be shipped easily and filled on location. Proper designs would incorporate integral air passages to allow for circulation and rapid heat transfer.

Eutectic Salts

Eutectic, or low melting point, salts are

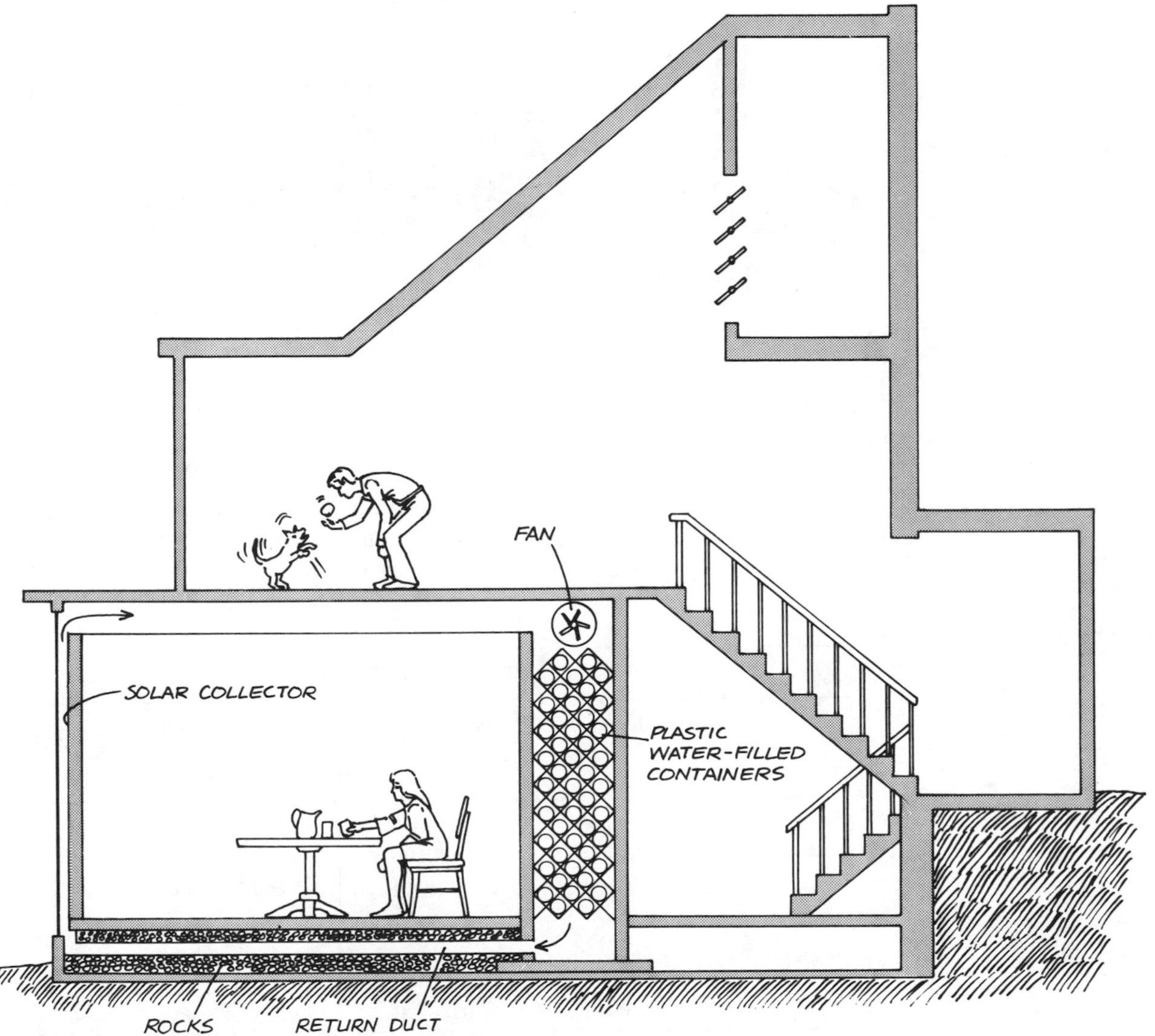

Solar air-heating system using many small containers of water as the storage medium.

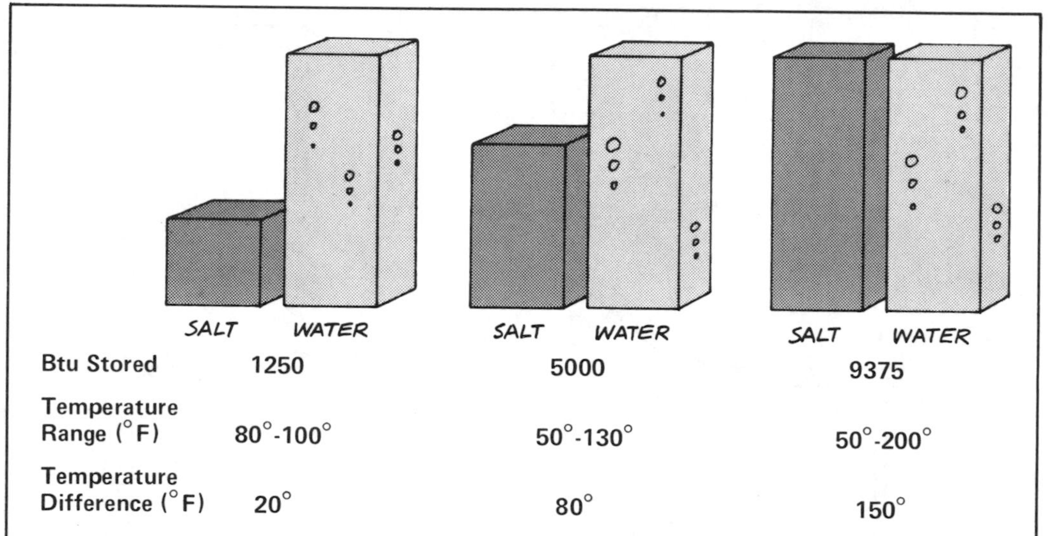

The volume of Glaubers salt needed to store the same amount of heat as a cubic foot of water. The salt volume indicated includes 50% voids between the containers of salt.

PROPERTIES OF EUTECTIC SALTS				
Eutectic Salt	Chemical Formula	Melting Point (°F)	Latent Heat (Btu/lb)	Density (lb/ft^3)
Calcium chloride hexahydrate	$CaCl_2 \cdot 6H_2O$	84-102	75	102
Sodium carbonate decahydrate	$Na_2CO_3 \cdot 10H_2O$	90-97	106	90
Disodium phosphate dodecahydrate	$Na_2HPO_4 \cdot 12H_2O$	97	114	95
Calcium nitrate tetrahydrate	$Ca(NO_3)_2 \cdot 4H_2O$	102-108	60	114
Sodium sulfate decahydrate (Glaubers Salt)	$Na_2SO_4 \cdot 10H_2O$	88-90	104	97
Sodium thiosulfate pentahydrate	$Na_2S_2O_3 \cdot 5H_2O$	118-120	90	104

SOURCE: Maria Telkes, "Storage of Heating and Cooling," 1974.

the only real alternative to rocks and containers of water as the heat storage for an air-type system. A eutectic salt absorbs a large amount of heat as it melts at a low temperature and releases that heat as it solidifies. A pound of Glaubers salt, the most widely studied and used, absorbs 104 Btu as it melts at 90°F and about 21 Btu as its temperature rises another 30°F. To store the same 125 Btu in the same temperature range requires about 4 pounds of water or 20 pounds of rocks.

Much smaller storage volumes are possible with eutectic salts. Consequently, they offer unusual versatility in storage location. Closets, thin partitions, structural voids, and other small spaces within a house become potential heat storage bins.

But this advantage is less pronounced when you increase the temperature range over which the salt cycles. The diagram illustrates the volume of Glaubers salt needed to store the same amount of heat as a cubic foot of water over three different temperature ranges. With 50 percent voids between the containers of salt, twice as much total volume is needed. Clearly, the advantages of Glaubers salt decline as the storage temperature range increases.

A number of similar eutectic salts have been studied over the past 30 years by Dr. Maria Telkes. Some of them are listed here in one of her tables. The various melting temperatures available make it possible to select a salt that provides a system with its most effective storage temperature—

as was done in the University of Delaware solar house.

But the costs of these salts can often demolish the bestlaid plans of enthusiastic designers. Off the shelf, Glaubers salt costs little more than a penny a pound. But preparing and putting it in a container can run the costs up. It is unlikely that Glaubers salt can ever be installed in a solar heat storage system for less than 20 cents a pound. The other salts can cost significantly more. Glaubers salt is available in most parts of the United States, but this is not necessarily true for the other salts.

Until recently, a recurring problem with eutectic salts has been their imperfect resolidification. During the liquid state the various chemicals would stratify in their containers and only partial solidification would occur as the temperature dropped. The salts used in the Dover House became practically useless as a heat storage medium after a single heating season. Dr. Telkes seems to have solved this problem by using thickening and nucleating agents in the salt solutions and by storing them in shallow pans or long, thin tubes. More than a thousand melting cycles have been tallied and further experimentation is increasing the count. For more information on eutectic salts, consult the work by Dr. Telkes that is listed in the references.

Paraffin lies somewhere between water and eutectic salts in its ability to store heat. Its specific heat is only 0.7, but it is almost as dense as water. And when it melts, it absorbs 65 Btu per pound. It resolidifies easily and is much cheaper than eutectic salts—particularly if its shipping container can be used in the solar heat storage system.

But paraffin has a few peculiar problems of its own. It corrodes plastics and some metals such as copper. And unless tightly sealed, paraffin will slowly oxidize and disappear into the far reaches of the atmosphere. It also expands by 20 percent when it melts—causing problems of containment. When it solidifies and releases the stored heat, the paraffin shrinks and pulls away from the container walls—drastically slowing the transfer of heat out of the container. But its greatest disadvantage is its combustibility. Few building codes will permit its use inside a house, and heat storage units with paraffin as the medium have to be located outside.

Insulation

Every storage system—whether of water, rock, or eutectic salt—requires a massive amount of insulation. The higher its average temperature and the colder the surroundings, the more insulation required. For low temperature (below 120°F) storage units inside the house, at least 6 inches of fiberglass insulation (or its equivalent) is the norm. The same unit in the basement needs 8 inches of fiberglass or more. And if located outside, it must be shielded from the wind and insulated even more heavily. The ground can provide insulation if the water table is low. But be careful—even a small amount of moisture movement through the soil will ruin its insulating value.

All ducts or pipes should be just as well insulated as the storage unit. Heat loss from the ducts or pipes can be further reduced by putting the collector close to the storage. The shorter the ducting or piping, the lower the total heat loss. And you'll save on construction and operating costs too.

Storage Size

The higher the temperature a storage medium can attain, the smaller the storage bin or tank needs to be. For example, 1000 pounds of water (about 135 gallons or 16 cubic feet) can store 20,000 Btu as its temperature increases from 80°F to 100°F, and 40,000 Btu from 80°F to 120°F. It takes almost 5000 pounds of rock (or 40 cubic feet, assuming 30 percent voids) to store the same amounts of heat over the same temperature rises.

Offhand, you might be tempted to design for the highest storage temperatures possible in order to keep the storage size down. But the storage temperature is linked to those of the collector and heat distribution system. If the average storage temperature is 120°F, for instance, the heat transfer fluid will not begin to circulate until the collector reaches 125°F. And collector efficiency plummets as the temperature of its absorber rises. A collector operating at 90°F may collect *twice* as much heat per square foot as one operating at 140°F. On the other hand, the storage must be hot enough to feed your baseboard radiators, fan coil units, or radiant panels. For example, a fan coil unit that delivers 120°F air cannot use storage tank temperatures of 100°F very well. In general, the upper limit on the storage temperature is determined by the collector performance and the lower limit by the method of heat distribution. You can increase the possible range of storage temperatures and keep the storage size at a minimum by using collectors that are efficient at high temperatures and heat distribution systems that operate on low temperatures.

It's a good idea to allow some flexibility in your initial designs so that you can alter the heat storage capacity after some experience under real operating conditions. For example, an over-sized concrete water tank can be filled to various levels until the best overall system performance is attained. The same can be done with a rock storage bin. If you're not too sure of your calculations, the storage should be oversized rather than undersized—to keep its average temperature low.

The capacity of a heat storage unit is often described as the number of sunless days it can keep the house warm. But this approach can be misleading. A system that provides heat for two sunless days in April is much smaller than one that can do so in January. It's better to describe the heat storage capacity as the number of degree days of heating demand that a system can provide in the absence of sunlight. For example, a well-insulated 1200 square foot house in Minneapolis loses about 10,000 Btu per degree day. In the basement, a tank with 15,000 pounds of water (about 2000 gallons) stores 600,000 Btu as its temperature rises from 80°F to 120°F. Assuming the heating system can use water at 80°F, this is enough heat to carry the house through 60 degree days (or through one full sunless day when the average outdoor temperature is 5°F).

Generally, the storage should be large enough to supply a home with enough heat for at least one average January day. In Minneapolis, there are about 1600 degree days each January, so the storage unit in our Minneapolis example should be designed to supply at least 52 degree days of heating demand—or 520,000 Btu. Depending upon available funds, the storage can be even larger. Or you can sink your money into better collectors that are efficient at higher temperatures. Solar heat can then be stored at higher temperatures—increasing the effective storage capacity of a tank or bin.

At the very least, the storage should be large enough to absorb all the solar heat coming from the collectors in a single day. If you can be satisfied with 60 percent solar heating or less, the simplified method described in "Estimating the Storage Size" will be useful. First you size the collector according to the method provided earlier under "Estimating the Collector Size." Then determine the volume of storage medium required per square foot of collector. Multiplying this

Estimating the Storage Size

The following procedure helps you to calculate the volume of water or rocks needed to store all the solar heat coming from a collector on an average sunny winter day. It assumes that the collector performance and size have already been determined according to procedures described on pages 178 to 180.

First you need to determine the maximum storage temperature to be expected. This is 5°F less than the maximum collector operating temperature—considered earlier in "Estimating the Collector Performance" on page 178. The temperature range of the storage medium is the (maximum storage temperature) minus the (lowest temperature that the heat distribution system can use). For example, if the collector can operate at 140°F, the maximum storage temperature is 135°F; and if the heating system can use 85°F, the temperature range is (135 – 85) = 50°F.

Next, determine the amount of heat you can store in a cubic foot of the storage medium over this temperature range. This amount is the (specific heat of the medium) times the (density of the medium) times the (temperature range). For example, a cubic foot of water can store

$$1.0 \text{ Btu/lb/°F} \times 62.4 \text{ lb/ft}^3 \times 50\text{°F} = 3120 \text{ Btu/ft}^3$$

over a 50°F temperature range, or 417 Btu per gallon. If the collector gathers 1000 Btu/ft^2 on an average sunny day in winter, you need (1000 ÷ 417) = 2.4 gallons of water heat storage per square foot of collector. For the collector on the Boston home, with an area of 277.5 ft^2 (see page 180), that's 666 gallons of water at the very minimum.

We recall from the Boston example that there are 1088 degree days in January, or 35 per day. The house loses 9500 Btu per degree day, or (35 × 9500) = 332,500 Btu on an average January day. But the 666 gallons of water can store only 278,055 Btu over a 50°F temperature rise. Therefore, the storage volume must be increased to 796 gallons, or 2.9 gallons per square foot of collector, to satisfy the storage needs of a single January day.

If rock were the storage medium, even more volume would be necessary. A cubic foot of solid rock weighs about 170 pounds and has a specific heat of 0.21 Btu/lb/°F, so it can store

$$0.21 \text{ Btu/lb/°F} \times 170 \text{ lb/ft}^3 \times 50\text{°F} = 1785 \text{ Btu/ft}^3$$

over the same 50°F temperature range. To store the 1000 Btu from a single square foot of collector, you need (1000 ÷ 1785) = 0.56 cubic feet or 95 pounds of rock. To store the 332,500 Btu required for an average January day, you need (332,500 ÷ 1785) = 186 cubic feet of solid rock.

The total volume occupied by the heat storage container must include void spaces in the storage medium. If there are 30% voids between the rocks, for example, this Boston home would need a 266 cubic foot storage bin for the rocks. Or if containerized water with 50% voids were used, the 796 gallons or 106 cubic feet of water would occupy 212 cubic feet of house volume.

volume by the total collector area gives a reasonable "first-cut" estimate of storage size. This estimate should be close enough for preliminary design work. If this storage volume fails to meet the heat demand for an average January day, revise your estimate upward until it does.

To get more than 60 percent solar heating, it helps to know the normal sequence of sunny and cloudy days in your area. If sunny days followed cloudy days one after the other, you would only have to size the collector and storage for one sunny day and the following cloudy day. Almost 100 percent of the heating demand could then be provided if the system were designed for the coldest two day period. If the normal sequence were one sunny day followed by two cloudy days, both collector and storage size would have to be doubled to achieve the same percentage. At the Blue Hills weather station near Boston, for example, about 80 percent of the sunless periods are two days long or less. A collector and storage system that could carry a house in Blue Hills through two cloudy days of the coldest weather will supply more than 80 percent of the house's heat needs. But the wide variation in weather patterns at a single location makes such a practice little more than educated guesswork. And this kind of weather data is hard to obtain.

HEAT DISTRIBUTION

An indirect solar heating system usually requires another sub-system to distribute the heat to the rooms. With integrated solar heating methods such as roof ponds, concrete walls and south windows, the solar heat is absorbed directly in the fabric of the house and heat distribution came naturally. But an indirect system usually needs more heat exchangers, piping, ducts, pumps, fans, and blowers to get the heat inside the house. And there must still be some provision for backup heating in the event of bad weather.

The heat distribution system should be designed to use temperatures as low as 75°F to 80°F. If low temperatures can be used, more solar energy can be stored and the collector efficiency increases dramatically. Thomason's warm air distribution system extracts heat at temperatures as low as 75°F. The water-to-air heat exchanger in the fourth MIT house could use water temperatures as low as 84°F. In general, warm air heating systems use temperatures from 80°F to 130°F, while hot water radiant heating systems require temperatures from 90°F to 160°F. At temperatures above 212°F, steam heating is virtually out of the question. So incorporating solar heating into an existing house equipped with steam heating will require a completely separate heat distribution system.

Most designers of new homes opt for forced warm air systems or radiant heating panels. By using larger volumes of air or oversized panels, these solar heating systems can operate at lower temperatures. Of the two, forced air heating makes the best use of low grade heat. Radiant heating panels take time before the room is comfy and usually require slightly higher storage temperatures. The oversized panels needed to provide adequate heating can be expensive.

Radiant heating systems with copper pipes embedded in concrete floor slabs are becoming more popular. They can use very low storage temperatures and provide comfortable, even heat. But these systems work slowly because it takes time to warm (or cool) the entire mass of concrete.

Auxiliary Heating

Even a system with a very large storage capacity will encounter times when the heat is used up. So the house must have an auxiliary heat source. This is a major reason why solar heating has not yet met with widespread acclaim.

The severe consequences of a single sustained period of very cold, cloudy weather are enough to justify a full-sized conventional heating system as backup. Small homes in rural areas can probably get by with wood stoves. If the climate is never too severe, as in Florida and most of California, a few small electric heaters may do the trick. But most houses will require a full-sized gas, oil or electric heating system. At this writing, solar energy is a means of decreasing our consumption of fossil fuels—not a complete substitute.

If the auxiliary system provides only a small percentage of the total heat demand, you might well consider electric heating. But remember that 10,000 to 13,000 Btu are burned at the power plant to produce 1 kilowatt-hour of electricity—the equivalent of only 3400 Btu in your house. At an efficiency of 65 percent, an oil furnace burns only 5400 Btu to achieve the same result. And electric heating can be very expensive, although electric heaters themselves are usually cheaper than a gas or oil furnace.

Usually, the auxiliary heater should not be used to heat the storage tank or bin because the collector will operate at a higher temperature and lower efficiency. And there will be costly heat losses from the storage container if an auxiliary system provides continuously higher storage temperatures. The heat lost from the storage container is usually 5 to 20 percent of the solar heat collected—and even more where the auxiliary keeps it warm.

Heat Pumps

The heat pump has served as a combination backup and booster in a number of solar energy systems. It is basically a refrigeration device working in reverse. The heat pump takes heat from one location (the *heat source)* and delivers it to another (the *heat sink).* The heat source is cooled in the process and the heat sink is warmed. Heat pumps in the Bridgers and Paxton Office Building and in the Tucson Laboratory boosted storage temperatures for use in radi-

Heat Pump Principles

A heat pump is a mechanical device that transfers heat from one medium to another, thereby cooling the first and warming the second. It can be used to heat or cool a body of air or a tank of water (or even the earth). The cooled medium is called the "heat source" and the warmed medium is the "heat sink." A household refrigerator is a heat pump that takes heat from the food compartments (the heat source) and dumps it in the kitchen air (the heat sink).

The heat pump transfers heat against the grain—from cool areas to warm. This sleight-of-hand is accomplished by circulating a heat transfer fluid or "refrigerant" (such as the freon commonly used in household refrigerators) between the source and sink and inducing this fluid to evaporate and condense. Heat is absorbed from the source when the heat transfer fluid evaporates there. The vapor is then compressed and pumped through a heat exchanger in the sink, where it condenses—releasing its latent heat. The condensed liquid returns to the heat exchanger in the source through an expansion valve, which maintains the pressure difference created by the compressor.

The packaged, self-contained heat pump used in residential applications generally reverses the direction of the refrigerant flow to change from heating to cooling or vice-versa. A four-way valve reverses the direction of flow through the compressor so that high pressure vapor condenses inside the conditioned space when heating is needed and low pressure liquid evaporates inside when cooling is desired.

Heat pumps are classified according to the heat source and sink, the fluid used in each, and the operating cycle. The heat pump shown here is a water-to-water pump with reversible refrigerant flow. A household refrigerator is an air-to-air heat pump with a fixed refrigerant flow.

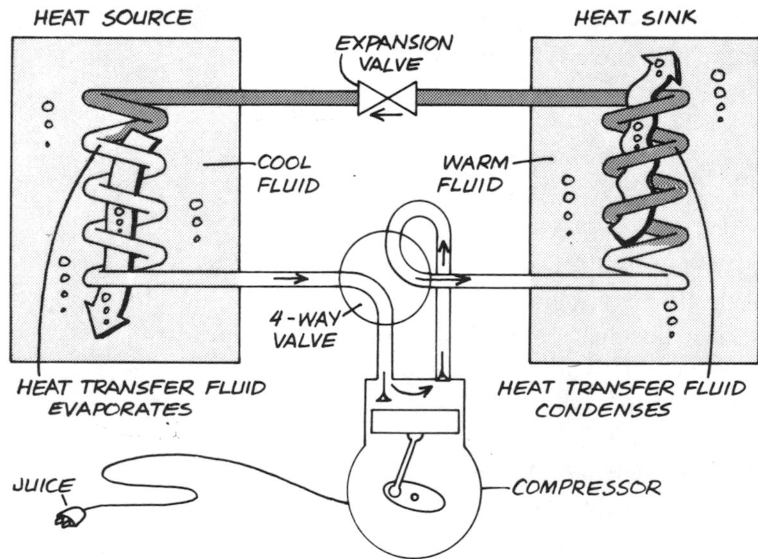

ant heating panels and provided backup heat. They also helped provide summer cooling.

A heat pump can deliver about 3 times the energy required for its operation. For every 2 Btu which a heat pump takes from a source, it needs the equivalent of 1 Btu of electricity for its operation. It delivers all 3 Btu to the heat sink. Thus, its *Coefficient of Performance* (COP), or the ratio of the heat energy delivered to the energy required for operation, is 3. Typically, this coefficient ranges from 2.5 to 3 for good heat pumps. By contrast, electric resistance heating has a coefficient of performance equal to 1, because it delivers 1 Btu of heat for every 1 Btu of electricity expended.

When heat pumps are used in conjunction with solar heating systems, the stored heat is useful over a wider temperature range. Without heat pumps, a forced warm air system would use storage temperatures from 80°F to 130°F and a hot water radiant system would use 90°F to 160°F. But with a heat pump, both systems can use 40°F storage temperatures! The heat pump takes low grade heat from storage and delivers it at higher temperature to the heat distribution system. This increased temperature range results in an increased heat storage capacity and markedly enhanced collection efficiency. The extra Btu's that a cool collector can gather each year often justify the added cost of installing a heat pump.

But heat pumps require electricity for operation. About 11,000 Btu are burned at the power plant when a heat pump uses

1 kilowatt-hour of electricity (or 3400 Btu) to deliver a total of 10,400 Btu. So, including losses at the power plant and in electrical transmission, the real Coefficient of Performance is closer to 1 than 3. And electricity *is* expensive. High electricity bills have been a major shortcoming of past solar heating systems that relied on heat pumps.

SOLAR COOLING

Indirect solar energy systems can also cool a house during the summer. And the sun is usually shining the brightest when cooling is most needed. The hottest months and times of day occur at times of nearly peak solar radiation. Systems that provide both heating *and* cooling can operate the year round—with additional fuel savings and a shorter payback period.

Passive Solar Cooling

Solar cooling seems paradoxical. How is it that a heat source can be used to *cool* a house? One answer is that solar energy is also a source of *power* that can move room air in ways that enhance comfort. The "chimney effect" cooling used in the 1974 Odeillo residences and the Tyrrell house (see Chapter 5) epitomize this approach. Solar heat absorbed in a glazed concrete south wall warms the air in front of it. In the summer this warm air escapes through vents at the top—drawing cooler air into the rooms through vents in the north wall. The "make-up" air can be cooled further by drawing it in through an underground duct—as in the Jackson house. Instead of a conventional fan, solar energy provides all the power needed to move this air.

Nocturnal Cooling

Substantial cooling can be obtained by using nocturnal radiation to cool the storage volume at night. Warm objects radiate their heat to the cooler night sky—particularly in arid climates. Warm air or water from the storage is cooled as it circulates past a surface exposed to the night sky. The cooled fluid returns to the storage container, which is cooled in the process. During the next day the storage releases this coolness to the house while it simultaneously absorbs the excess heat from the rooms.

One of the earliest solar houses to use nocturnal cooling was the Desert Grassland Station at Amado, Arizona (see Chapter 2). Ray Bliss and Mary Donovan stretched a black gauze cloth over an opening in the ground that led to the rock storage bin. The night air drawn into the storage was further cooled as it passed through the cloth—which radiated the absorbed heat to the night sky.

Harry Thomason used another nocturnal cooling method in his first solar house. Warm water from the storage tank was pumped to the roof ridge and allowed to trickle down the north-facing roof. This water was cooled

Coefficient of Performance

A heat pump uses electrical energy to manipulate heat transfer from source to sink. The heat deposited in the sink is a combination of the heat generated by compressing the refrigerant (which requires electrical power) and the latent heat released by the condensing vapor. The heat removed from the heat source is just the latent heat of evaporation. The effectiveness of a heat pump is indicated by its Coefficient of Performance, or COP:

$$COP = \frac{\text{heat energy deposited (or removed)}}{\text{electrical energy consumed}}$$

The electrical energy required to run the compressor (in kwh) can be converted to Btu by multiplying by 3413 Btu/kwh. Because the heat of compression is part of the heat deposited in the sink, the COP of a heat pump used for heating is usually greater than the COP of the same heat pump used for cooling.

by evaporation and radiation as it trickled over the black shingles. The cool water was subsequently trapped in a gutter and returned to the storage tank for use during hot days.

The performance of these systems was not exceptional. Bliss rated the comfort level attained with his system as equal to that of a conventional evaporative cooler but inferior to that of a high quality air conditioner. Thomason's cooling system did not work very well when the nights were humid or cloudy—a common occurrence in Maryland summers.

Auxiliary Cooling

The storage volume can also be cooled using a small refrigeration compressor. Most through-the-wall air conditioners use such compressors to cool the indoor air. This unit acts as the backup or auxiliary cooling system—analogous to the backup heating system. If operated only at night, its capacity can be as small as half that of an independently functioning unit and still meet peak cooling demands. Nighttime operation will be particularly wise if electric companies charge more for electricity during times of peak loads on hot summer afternoons. An even smaller compressor can be used if it operates continuously night and day—cooling the storage when not needed by the house.

Systems with heat pumps can also use them for summer cooling. The heat pump can be used to cool the storage—the same as it did in the winter. Or it can take heat from the house and discharge it into the storage. This stored heat is then released to the night air.

In systems with dual storage units, the heat pump transfers heat from one to the other—cooling the first and warming the second. The cool fluid in the first unit is circulated to the house while the concentrated heat in the second is discharged to the outdoors. In the Bridgers and Paxton Office Building, for example, this heat is discharged by an evaporative cooler. But in the Tucson Laboratory, the second storage unit was cooled by nocturnal radiation. The warm water in this tank was circulated through unglazed solar collectors exposed directly to the night sky. Both projects used large quantities of electricity to run the heat pumps and other cooling devices.

Absorption Cooling

Solar collectors can provide the heat needed by an *absorption cooling* device—making solar-powered air conditioners a distinct possibility. An absorption cooling unit uses two working fluids—an *absorbent* such as water, and a *refrigerant* such as ammonia. Solar heat from the collector boils the refrigerant out of the less volatile absorbent. The refrigerant then condenses and passes to cooling coils inside the room. Here it vaporizes once again and absorbs heat from the surrounding air. The refrigerant vapor is subsequently reabsorbed in the absorbent, releasing heat into cooling water or the atmosphere.

Unfortunately, most absorption cooling devices work best with fluid temperatures between 250° F and 300° F. The lowest possible working fluid temperature that can be used is about 180°F—where flat-plate collectors have sharply reduced efficiencies. And the collectors have to operate at temperatures about 15° F to 20° F above this lower limit.

Dr. George Löf has shown (see references) that if 210° F water is supplied by a collector, the working fluids will receive solar heat at 180° F and the water will return to the collector at 200° F. On a hot summer day, a square foot of collector might deliver 900 Btu—or about 40 percent of the solar radiation hitting it. About 450 Btu will be removed from the interior air, so that a 600 square foot collector can provide a day time heat removal capacity of about 270,000 Btu or 30,000 Btu per hour.

Solar collectors designed for absorption cooling systems are more expensive than those used only for winter heating. But substantial fuel savings are possible if the same collector can be used for both purposes. Concentrating collectors are particularly well suited to absorption cooling because they can supply high temperatures at relatively high efficiency. The Owens-Illinois collectors used in the third Colorado State University house generate liquid temperatures of 200° F with about 70% efficiency. Flat plate collectors with selective absorber surfaces are another possibility—especially if they have

Absorption Cooling Principles

Just like window air-conditioners and heat pumps, an absorption cooling device uses the evaporation of a fluid refrigerant to remove heat from the air or water being cooled. But window air-conditioners and heat pumps use large quantities of electricity (see page 196) to compress this evaporated fluid so that it condenses and releases this heat to the "outside." The condensed fluid then returns to the evaporating coils for another cycle.

In an absorption cooling cycle, the evaporated refrigerant is absorbed in a second fluid called the "absorbent." The resulting solution is pumped to the "re-generator" by a low-power pump. Here, a source of heat—which can be fuel heat or solar energy—distills the refrigerant from the solution. The less volatile absorbent remains a liquid and returns to the absorber. The refrigerant liquid returns to the evaporating coils—where it evaporates and cools the room air, completing the cycle.

Absorption cooling devices can use hot fluid from a solar collector to boil the refrigerant from the absorbent. Unfortunately, most absorption cooling devices work best with fluid temperatures between 250°F and 300°F. Flat-plate collectors are inefficient at such high temperatures, but concentrating collectors can produce these temperatures easily. If the costs and complexity of concentrating collectors can be brought down, they may find ready application in solar absorption cooling.

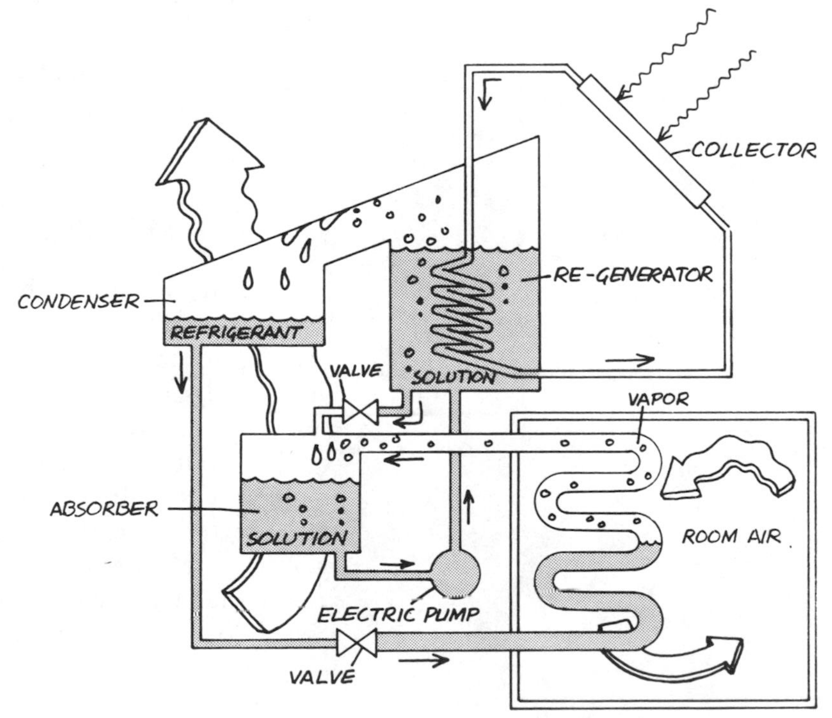

multiple transparent covers. Almost all absorption cooling equipment requires liquid-type collectors. It is unlikely that air-type collectors will be used for this purpose.

Experimental solar houses at the University of Florida and Colorado State University (see Chapter 2) have been used to test these approaches to solar cooling. The Florida house has achieved 100 percent solar cooling with an absorption cooling unit that uses water as the absorbent and ammonia as the refrigerant. The first and third Colorado State houses have Arkla-Servel units that use lithium bromide as the absorbent and water as the refrigerant. Solar powered absorption cooling is still in the demonstration stage. Because of its complexity and high initial cost, absorption cooling has limited applicability to single family homes.

SYSTEM DESIGN

Simplicity is the watchword in the overall design of an indirect solar energy system. It's tempting to design more and more complex systems—always trying to squeeze another ounce of performance or a little more comfort out of them. But this added complexity usually means higher initial costs and greater operating and maintenance expenses. It's better to design a simple system that may require the inhabitants to toss a log or two in a fire every now and then.

If you're installing the system in a new house, by all means design the house to collect and store solar heat in its walls and floors. On a sunny winter day, enough solar energy streams through a hundred square feet of south facing windows or skylights to keep a well-insulated house toasty well into the evening. And if the house has a concrete floor slab or masonry block walls insulated on the outside, any excess heat can be stored for use later at night. The solar heat gathered in the rooftop collectors can then be husbanded in the storage units until it is *really* needed rather than squandered heating the house during a cold, sunny day. Even if you're installing the system in an existing house, you might try adding a greenhouse onto the south side. The solar heat collected there can be ducted into the rooms to keep them warm by day.

Rudimentary Air Systems

The very simplest of indirect solar heating systems has collectors that function only when the sun is shining and the house needs heat. Air is ducted from the house to the collector, heated by the sun, and blown back into the rooms by a fan. The only heat storage container is the fabric of the house itself—and the heavier it is, the better. The fan operates only when the collector temperature is warmer than that inside the house. It shuts off when the collector cools in the late afternoon or when room temperatures become unbearably hot. The more massive the house, the more heat it can store before temperatures get out of hand and the longer it can go without backup heating.

Another simple indirect system delivers solar heated air to a rock storage bin just beneath the house. Heat flows up to the rooms by natural convection or percolates through the floor. Peter van Dresser's solar adobe in Santa Fe (see Chapter 2) uses this approach. After almost two decades of operation, his system still provides better than 60 percent of the heat needed by the house. The rest comes from a fireplace and a small gas heater.

Other Air Systems

Better control of the room temperature is possible with another fan added to blow solar heat from the storage bin to the rooms. The fan draws cool room air through the storage bin and blows warm air back to the rooms. The backup heater (which can be a wood stove, electric heater, or an oil or gas furnace) is completely independent of the solar heating system. Ideally, it isn't needed when the sun is shining because the solar heat gain through the windows keeps the house warm. When the sun isn't shining and the house needs heat, solar heat is drawn from the storage bin—if available. If not, the backup heater is put to use. There are 4 possible modes of operation for this system and they are detailed in the diagrams.

In larger houses, it's expensive to have separate delivery systems for solar and auxiliary heat. Integrating the two into a

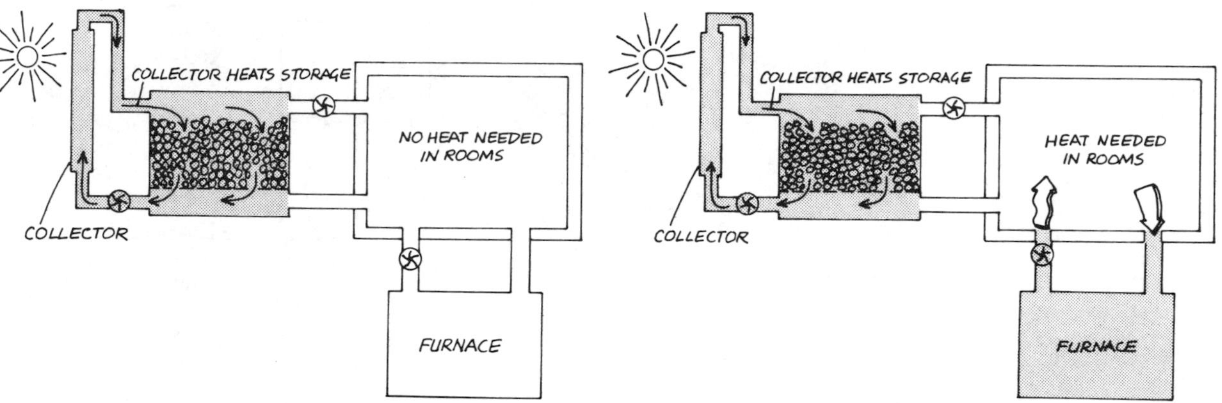

When the sun shines, the collector heats the storage. The furnace heats the rooms when necessary.

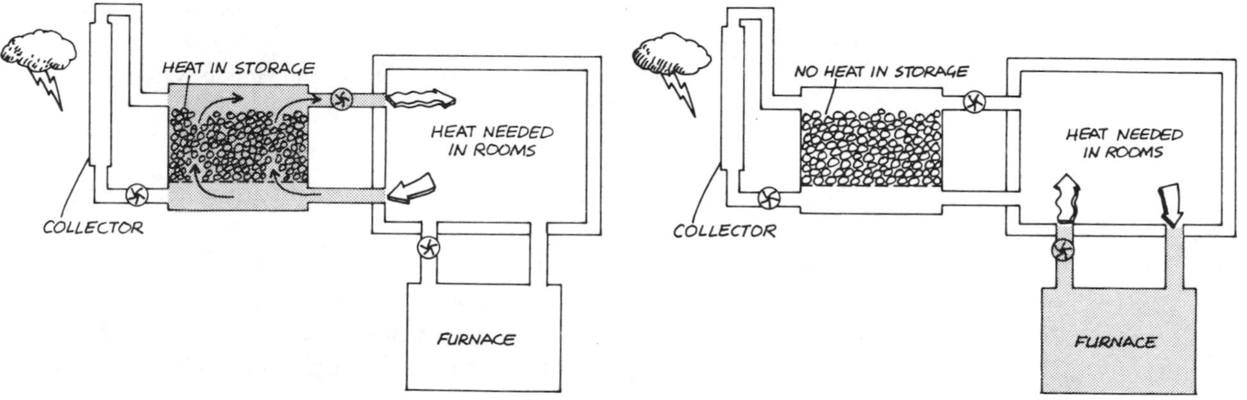

When the sun isn't shining, stored heat is delivered to the rooms as needed. If there is no heat in storage, the furnace comes on.

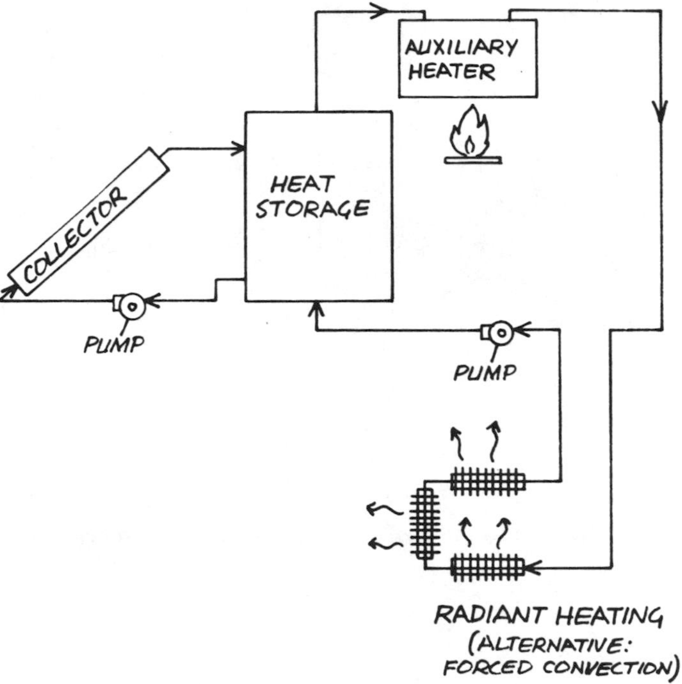

Piping system design for a simple liquid system (from Revere Copper and Brass Co.).

single delivery system requires extra dampers and controls but can be cheaper in the long run. In Dr. George Löf's Denver house, a gas furnace heats the air coming from the rock storage bins when it isn't warm enough. The warm air then passes to the rooms in its normal cycle. And solar heated air from the collectors goes directly to the rooms if they need heat on a sunny day.

In any air-type system, ducting and dampers should be kept to a minimum. Try to design the system so that fan pressure opens the dampers for maximum air flow and closes them tightly as needed—eliminating the need for expensive damper motors and controls.

Liquid System Designs

Usually a liquid-type solar energy system is not as economical as an air system for heating a single family dwelling. But with larger dwellings and increasing needs for domestic water heating and absorption cooling, a liquid system becomes more feasible. It doesn't have to be elaborate. Henry Mathew built one of the most basic liquid systems for his Coos Bay home (see Chapter 2). A single pump circulates water through the rooftop collectors and back to the 8000 gallon storage tank underneath the house. The collected solar heat rises into the rooms via natural convection of the warm air surrounding the tank.

Another very basic liquid system is illustrated in the first of the two accompanying diagrams. Water from the storage tank is heated by the auxiliary, if necessary, before delivery to the baseboard radiators or radiant heating panels. Only two pumps are needed to circulate the water through the two heat transport loops. With some small differences, this is the system used in the third MIT house (see Chapter 2 for a much more detailed diagram).

A somewhat more complex system is illustrated in the second diagram. Because of the threat of freezing, a water-glycol solution circulates through the collector and surrenders its solar heat upon passing through a heat exchanger immersed in the storage tank. Heat is distributed to the rooms by a warm air heating system that uses an additional fan to blow cool room air past a water-to-air heat exchanger. Cold inlet water from a city main or well pump passes through yet another heat exchanger immersed in the storage tank. This water is preheated before it travels to the conventional hot water heater.

The solar heating system in the fourth MIT house is basically the same as this one. But as the diagram of that system (in Chapter 2) illustrates, an indirect solar energy system can be much more involved than these simplified diagrams indicate. Additional pipes, controls, and valves are required for the various modes of operation. And the heat exchangers degrade the overall performance of the system. According to Duffie and Beckman (see references) the use of a heat exchanger substantially increases the collector operating temperature and lowers its efficiency. The greater the number of heat exchangers in a system, the lower the collector efficiency. Even more pumps, fans, and heat exchangers are needed if solar absorption cooling is desired. By now you can probably understand why upwards of $10,000 have been spent on some indirect solar energy systems.

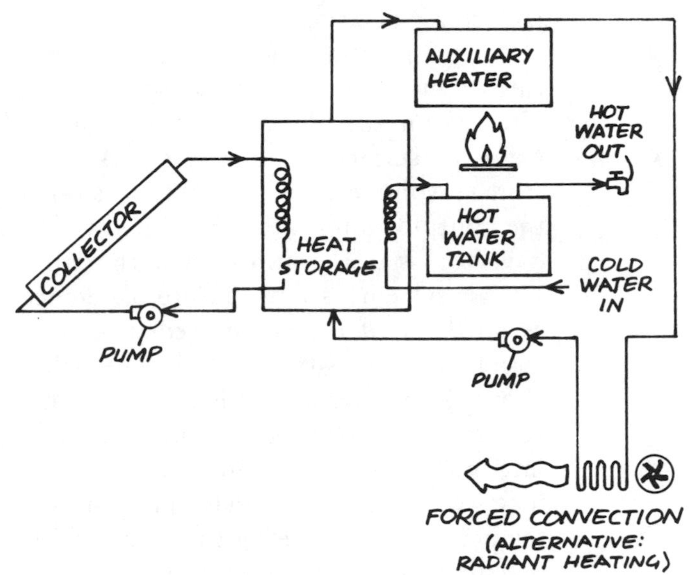

A liquid system designed for forced convection heating and preheating of domestic hot water.

Controls

One set of controls governs the delivery of heat (or coolness) to a house from the collector, heat storage, or backup heating (or cooling) system. Its operation is determined by the needs of the household and the limits of the entire system. In general, the thermostat governing the energy flow from storage can operate at a different temperature level than the thermostat on the backup heater. The first one might be set at 70°F and the second at 68°F. If the heat storage cannot maintain 70°F room temperatures, the backup system springs into action when the temperature falls below 68°F.

Controls to govern collector operation are relatively simple and are becoming readily available. Most of these controls determine collector operation by comparing the collector temperature and the storage temperature. One temperature sensor is placed directly on the absorber or on a similarly exposed surface. The other sits in the storage volume near its exit. Customarily the collector pump starts working when the collector is 5°F warmer than the storage. For air systems, a temperature difference of as much as 20°F may be needed before the circulation fan is triggered. A time delay of about 5 minutes is needed to prevent the system from turning on and off during intermittent sunshine. Some liquid systems may need controls that prevent liquid temperatures from rising to the point where pressures can burst the piping.

Performance and Cost

The tradeoff between performance and cost is crucial to the design of any solar energy system. The performance of a system is measured by the amount of energy it can save a household per year. The dollar value of the energy saved is then compared with the initial (and operating) costs of the system. The initial costs must not get so high that they can never be recouped over the life of the system. One doesn't have to be quite as careful in the design of conventional heating systems because the fuel costs are far and away the major heating expense. But the initial costs of an indirect solar heating system are usually so high that more than 10 years of trouble-free operation are needed before the energy savings make it a good investment. Some crucial questions include:

- How much energy will I save by using solar energy?
- What is the projected dollar value of that energy?
- How much am I willing to spend now in order to save that much energy each year?

A look at some earlier solar projects will help to answer the first question.

Good performance data are available for a number of the early solar buildings discussed in Chapter 2. In the Boulder house, a total collector area of 463 square feet supplied about 20 million Btu during the first winter—or about 43,000 Btu per square foot. Perhaps as much as 75,000 Btu per square foot would have been supplied to the house had the system worked smoothly. The 315 square feet of air-type collector in the Desert Grassland Station saved about 14 million Btu during the heating season—or about 44,000 Btu per square foot. During the 1959-60 heating season, the 640 square feet of liquid-type collectors on the fourth MIT house collected just over 40 million Btu. Of this amount, only 34.5 million Btu were used for space and water heating—or about 54,000 Btu per square foot. Almost 10 years earlier, the collector on the third MIT house delivered between 60,000 and 70,000 Btu per square foot in each heating season.

Clearly, a solar heating system is doing very well if it supplies 100,000 Btu per square foot of collector per year—the equivalent of about a gallon of oil burned at 75 percent efficiency. Present costs of heating oil are about 50 cents a gallon (1976 prices). If inflation stopped dead in its tracks, it would take at least 20 years to recoup the initial expenditures for a system that cost $10 per square foot of collector. But oil prices are sure to rise and the payback period may be only 10 to 15 years. If the only alternative is electric heating, solar energy seems a good bet. At 4 cents per kilowatt-hour, it costs $1.20 to produce 100,000 Btu, and you might recoup the initial expense of a solar heating system in less than 10 years. Of course the maintenance, operating, and interest costs have been ignored in this simple analysis. But so have tax incentives and the real costs of pollution.

So the real trick lies in building a good solar heating system for about $10 to $15 per square foot of collector. If it can supply 50,000 to 100,000 Btu per square foot and last 15 to 20 years, it's probably a good investment. In general, indirect systems with a *seasonal efficiency* (the ratio of solar energy collected and used to that hitting the collector) of 30 to 40 percent are attainable with present technology. They will deliver from 50,000 to 100,000 Btu and cost from $15 to $25 per square foot of collector—a bit more than we'd like. If you are good with your hands and do much of the labor yourself, you might get these costs down to $10 per square foot.

On the other hand, many of the integrated collector-storage systems described in Chapter 5 have seasonal efficiencies ranging from 25 to 50 percent, but cost only $5 to $10 more per square foot than the wall or roof they replace. South windows can save up to 100,000 Btu per square foot per year if covered with a good insulating shutter on winter nights. And the lifetimes of most of these simple systems are easily better than 20 years.

Where to Go from Here

This chapter has given you a broad survey of almost all the practical alternatives for indirect solar heating and cooling. But it stops short of providing detailed design and construction information. And only an outline of the economics has been presented. The interested reader will need further information before he or she can begin to build an indirect solar energy system.

The recent book *Other Homes and Garbage,* by Jim Leckie and others, is one of the first places to look. Chapter 4 of their book discusses solar heating and gives a detailed (if somewhat single-minded and academic) description of the design process for a liquid-type solar heating system. There are many useful tips on sizing the various components and suggestions about where to find materials cheaply. But the book all but ignores air-type systems, which are probably more suitable for small dwellings.

Solar Energy Home Design in Four Climates, a recent TEA publication, fills this void. The reader is led step-by-step through the design of solar heated houses in four representative U.S. cities (Boston, Minneapolis, Phoenix, and Charleston, S.C.). An air-type indirect solar heating system is used in each house. The book presents several useful design tools than have only been summarized in this work. Sizing of collectors and storage units (as well as south glass and concrete slabs) is treated extensively. But many of the necessary engineering details have been omitted.

Those interested in the economics of solar heating and cooling are advised to consult the papers by Löf and Tybout. Their work is universally acclaimed as the best on the subject.

FURTHER READING

ASHRAE. *Handbook and Product Directory: 1974 Applications.* New York: American Society of Heating, Refrigerating and Air Conditioning Engineers, 1974 (see Section 59 on solar energy).

Anderson, Bruce. *Solar Energy: Fundamentals in Building Design.* New York: McGraw-Hill Book Company, 1977.

Bird, R.B., W.E. Stewart, and E.N. Lightfoot. *Transport Phenomena.* New York: John Wiley & Sons, Inc., 1960.

Bliss, R.W. Jr. "The Derivations of Several 'Plate-Efficiency Factors' Useful in the Design of Flat-Plate Solar Heat Collectors." in *Solar Energy,* Vol. III, No. 4, December 1959.

Buelow, R.H. and J.S. Boyd, "Heating Air by Solar Energy." in *Agricultural Engineering,* January 1957.

Close, D.J. "Solar Air Heaters for Low and Moderate Temperature Applications." in *The Journal of Solar Energy Science and Engineering,* Vol. VII, No. 3, July 1963.

Davis, Albert J. and Robert P. Shubert. *Alternative Energy Sources in Building Design.* Blacksburg: Virginia Polytechnic Institute, 1974 (available for $7.00 plus $0.75 postage from International Compendium, 1762 Tucker Street, Beltsville, Maryland 20705).

Duffie, John A. and William A. Beckman. *Solar Energy Thermal Processes.* New York: John Wiley & Sons, Inc., 1974.

Hottel, Hoyt C. and J.B. Howard. *New Energy Technology, Some Facts and Assessments.* Cambridge, Massachusetts: The MIT Press, 1971.

Hottel, Hoyt C. and B.B. Woertz. "The Performance of Flat-Plate Solar-Heat Collectors." *ASME Transactions,* Vol. 64, 1942.

Jordan, Richard C., editor. *Low Temperature Engineering Applications of Solar Energy.* New York: American Society of Heating, Refrigerating and Air Conditioning Engineers, 1967.

Leckie, J., G. Masters, H. Whitehouse, and L. Young. *Other Homes and Garbage.* San Francisco: Sierra Club Books, 1975.

Löf, G.O.G. "Cooling with Solar Energy." in *World Symposium on Applied Solar Energy, Phoenix, Arizona 1955.* Menlo

Park, California: Stanford Research Institute, 1956.

Löf, G.O.G. and R.W. Hawley. "Unsteady-State Heat Transfer Between Air and Loose Solids," in *Industrial and Engineering Chemistry,* Vol. 40, No. 6, June 1948.

Löf, G.O.G. and R.A. Tybout, "The Design and Cost of Optimized Systems for Residential Heating and Cooling by Solar Energy." in *Solar Energy,* Vol. 16, 1974.

NSF/University of Maryland. *Proceedings of the Solar Heating and Cooling for Buildings Workshop, March 21-23, 1973.* Washington; National Science Foundation, 1973 (prepared by the Department of Mechanical Engineering, University of Maryland, under NSF-RANN Grant GI-32488; Vol. 1 available for $3.00 from the National Technical Information Service, Springfield VA 22161).

Telkes, Maria. "Storage of Solar Energy." in *ASHRAE Transactions,* Vol. 80B, 1974.

Total Environmental Action. *Solar Energy Home Design in Four Climates.* Harrisville, New Hampshire: Total Environmental Action, Inc., 1975 (available for $12.75 from the publisher, Church Hill, Harrisville, NH 03450).

Tybout, Richard A. and G.O.G. Löf. "Solar House Heating," in *Natural Resources Journal,* Vol. 10, No. 2, April 1970.

Whillier, Austin. "Principles of Solar House Design." in *Progressive Architecture,* May 1955.

Whillier, Austin. "Solar House Heating—A Panel." in *World Symposium on Applied Solar Energy, Phoenix, Arizona, 1955.* Menlo Park, California: Stanford Research Institute, 1956.

Whillier, Austin. "Black-Painted Solar Air Heaters of Conventional Design." in *Solar Energy,* Vol. VIII, No. 1, 1964.

7
Do It Yourself
Putting Solar Energy to Work in Existing Houses

A twelfth century English astronomer examines a twentieth century apparatus.

Much of the technology involved in harnessing the 'alternative sources of energy' is simple in principle, requires no exotic materials, and is suitable for construction in the home handyman's workshop—though still offering plenty of scope for ingenuity and resourcefulness.

Philip Steadman
Energy, Environment, and Building

If you are not buying or building a new house, there are still many ways to harness solar energy and slash skyrocketing fuel bills. A number of these solar heating projects are within range of the average home handyman or woman. They include solar water heaters, greenhouse additions, and other solar heating systems applicable to existing houses. The principles of solar home design learned in earlier chapters can then be applied to your own living situation. The experience you gain in such a project will prove invaluable should you later decide to pull up your roots, buy some land in the country, and build a solar home on your own.

This chapter is divided into three separate sections containing a number of specific projects. The first section is a detailed survey of solar water heating methods—with emphasis on simple, low-cost systems. The middle section discusses energy conservation in an existing house. The final section examines several ways of solar heating an existing home, with emphasis on passive systems—thermosiphoning window and wall collectors, greenhouses, and the use of south windows. Most of the projects discussed here require only moderate skills, cost under $1000, and should pay for themselves in reduced fuel bills within 10 years. Many of the systems can be dismantled and relocated so that the investment and labor is worthwhile even for renters and nomadic homeowners. .

SOLAR WATER HEATING

The use of solar energy to heat household water supplies has been technically feasible since the 1930's, when solar water heaters were commonly used in California and Florida. Before low-priced natural gas and electricity became available, they were often the cheapest source of hot water, especially in rural areas. They were either owner-built or bought from heating contractors. Many of the current designs were developed in those early days, and solar water heating is a well-proven technology undergoing a vigorous rebirth today.

The use of solar energy to heat domestic water supplies is quite similar to its use in home heating. However, the smaller scale and lower cost of solar water heaters put them within easier reach of the homeowner's skills and pocketbook. Furthermore, they are easily integrated with existing water heating systems.

In several aspects, solar water heating is more favorable than space heating. For one, hot water needs are fairly constant throughout the year. The collector and other parts of the system will be operating all year, and initial costs can be recouped in 3 to 7 years. By contrast, a solar space heater is fully operational only during the coldest months of the year, and the payback period is longer.

A solar water heater can also be sized more closely to the average demand. Space heating systems must be large enough to meet those extreme loads which occur only once or twice a year. A water heater, on the other hand, has roughly the same load day in and day out and does not have to accommodate wide fluctuations in demand.

A problem common to *all* types of solar heating is the fluctuation of available sunlight. But variable weather conditions are less problematic for household solar water heating because hot water requirements are more flexible. If the supply of hot water runs out during extended cloudy weather, the consequences are less severe than if the house were to lose its heat. It's the difference between letting the laundry wait a bit longer or having the pipes freeze and burst. If you can tolerate occasional shortages of hot water, your solar water heater can have a very straightforward design—free of the complications that provide for sunless periods.

When a more constant hot water supply is needed, a conventional hot water heater can take up the slack. Such heaters are readily available, reliable, and automatic. Controls are simple enough because this auxiliary heater must only make up the difference between incoming water temperatures and the desired supply temperature. If the solar heater is providing full-temperature water, this auxiliary remains off. If not, the auxiliary comes on just long enough to raise the solar heated water to the required temperature.

In addition to all these advantages, solar water heaters are a lot smaller than solar space heaters. The initial cost of a solar water heating system is lower and it can be built and operating within a very short time. All these aspects make solar water heaters ideal projects for learning how to build with the sun.

How Much Hot Water?

The smaller the amount of hot water you need, the smaller (and cheaper) your system can be. In the United States, the "average" daily consumption of water per person is about 28 gallons in public housing and 41 gallons in luxury apartments. But in Australia, the average is 10 gallons per person per day for residential purposes. In general, you can assume that hot water needs are about 30 percent of the water used at home each day. And there is high demand in the morning and evening—corresponding to bathing and dishwashing uses.

Small changes in household water usage will produce a significant reduction in hot water demand:

- using a basin of hot water rather than flow from the tap to rinse dishes
- taking showers rather than baths
- replacing steady-stream faucet nozzles with spray nozzles

- washing clothes in warm rather than hot water
- using a "suds saver" washing machine that recirculates hot wash water rather than disposing of it
- washing with cold water

These and other methods will reduce your hot water demand, simplifying the design of a solar water heater and reducing its size and cost.

If hot water were used steadily throughout the day, the demand could be met with a black rubber hose lying in the sun. By adjusting the rate of flow through the hose, you can get a steady stream of hot water. But hot water use is usually concentrated in the morning or evening. Hot water must be available after sunset and before sunrise. It must be *stored* during the daytime when sunlight is available but demand is low. So connect the hose to a storage tank where the heated water can collect. And insulate the tank to make sure the water stays hot overnight. To have hot water left on sunless days, your storage tank must be larger and better insulated. And an auxiliary heater is needed if hot water must always be available.

The greater the volume of water heated and the longer the storage time desired, the more complex and costly a solar water heating system becomes. If water conservation is practiced, however, long term storage would still be possible without an extremely large and expensive storage tank.

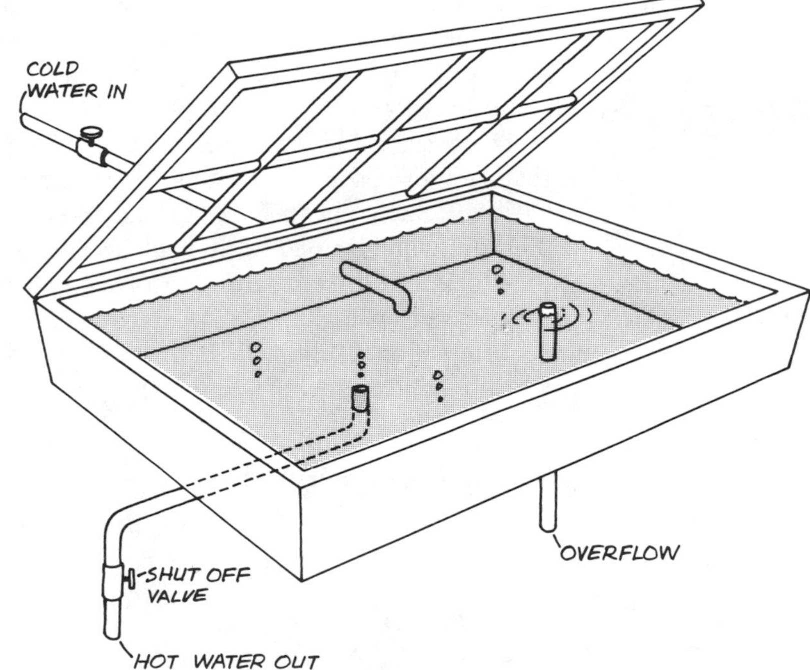

Shallow-trough solar water heater.

The degree of "user participation" also affects the design and cost of a solar water heating system. For example, an effective and inexpensive solar water heater is a shallow trough of water with a transparent cover, set in the sun. The heater is filled in the morning and the hot water drained for use in the afternoon or early evening. However, someone would have to fill the tank, put an insulating cover on it during cloudy weather, decide when the water was sufficiently hot, and drain the hot water as needed. But most people expect to get hot water with no more effort than turning

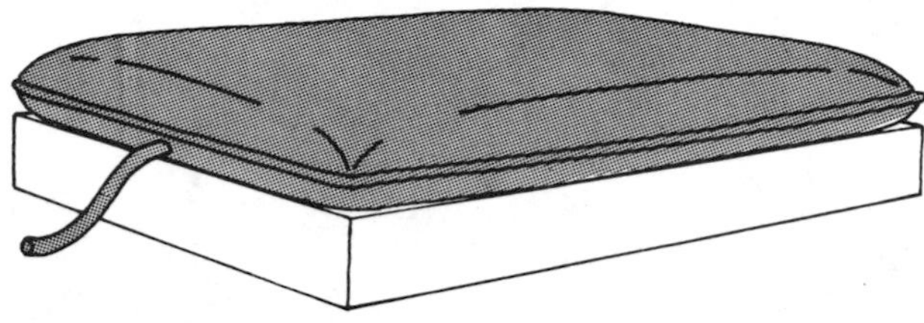

Black plastic bag water heater. This bag is filled in the morning and drained in the evening.

a faucet. This simple (and cheap!) solar water heater might be unacceptable to them as an alternative to their conventional water heaters.

There are a variety of solar water heaters that require somewhat less user participation. Such heaters are characteristic of those which gained fairly wide acceptance in Florida and Southern California before inexpensive natural gas became available. A brief discussion will help you decide which of these systems you prefer.

Storage-Type Water Heaters

There are several solar water heaters that are simple and inexpensive but far from automatic. All of them have a collection surface that is part of the storage tank.

In Japan there are water heaters that are little more than black plastic water-bed type bags set on a level platform. The bag is filled in the morning and drained in the evening. A variation is a wooden box lined with a plastic envelope that holds the water. Warmer water and higher collection efficiency are obtained if a transparent cover is placed over the bag. These horizontal, *flat-basin* water heaters are well-suited to the tropics—where the sun is high overhead all year.

A horizontal collector is less efficient at higher latitudes because the sun is lower in the sky (especially during winter). One solution is to use an aluminized surface to reflect more sunlight onto the water. If the supporting props are adjustable, the reflector can be set to the optimum angle for each month.

You can also tilt the heater itself. Flat metal tanks tilted for optimum collection have been built in Japan, Algeria, and India. Although construction details vary, these collectors are basically two sheets of metal sealed at the edges to form a container from 4 to 8 inches thick. The tank construction must be strong enough to prevent bulging under the extra water pressure caused by tilting the tank. The side toward the sun is painted black and covered with a sheet of glass.

What is most noteworthy about these two types of water heaters is that they function both as a collector of solar energy and as a

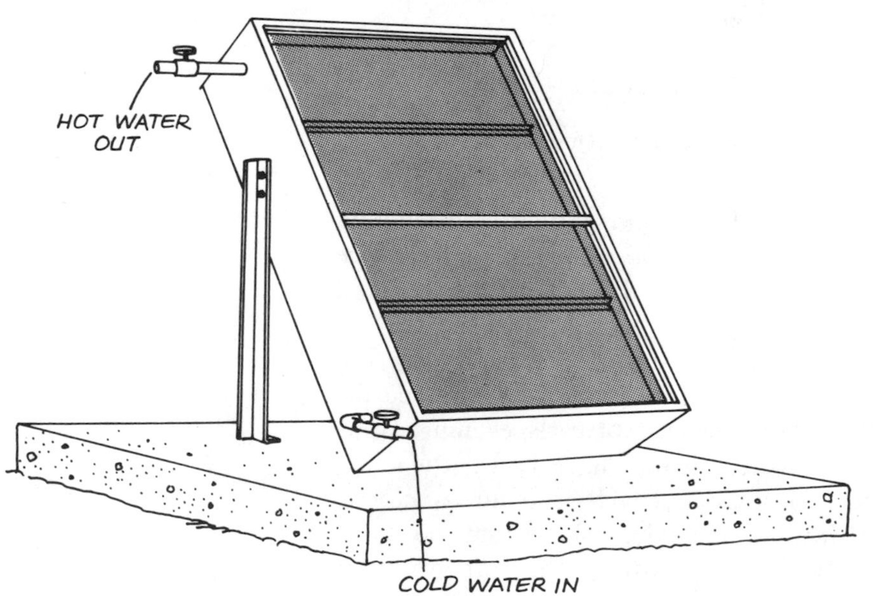

A tilted storage-type solar water heater.

storage unit for the heated water. In such *storage-type* water heaters, the size of the unit determines its storage capacity. Reducing the depth of the water also reduces storage capacity unless the surface area of the unit is increased. The shallower the tank, the quicker the entire volume of water heats up, but there will be fewer gallons of hot water per square foot of collection surface. The flat-basin water heater has an advantage on slightly overcast days—the basin can be filled only partially. Less water will be heated, but the water will be hotter.

The worst disadvantage of storage-type water heaters is their heat loss at night or during cloudy weather. The most effective response to this problem is to insulate the back and sides of the unit and use a removable insulating lid to cover the glazing or water when no sunlight is available. You can also put a reflecting surface on the underside of this lid and use it to gather additional sunlight when open. But such a solution does require more personal effort.

The Bread Box

The Bread Box is an efficient storage-type water heater that retains the virtues of low cost and simplicity of design. Variations of this water heater were used in the 1930's, but the improved version described here was developed by the folks at Zomeworks Corporation. A Bread Box has three basic parts: a water tank, the cover glass, and an insulated box.

The water tank can be made from a variety of tanks or drums. A common choice is the inside tank of a conventional domestic water heater minus the metal cover and insulation. By choosing an appropriate tank, you can fit the collection and storage capacities to your own hot water needs. The

The Bread Box—a simple and effective storage-type water heater.

tank is painted flat black and placed horizontally in the insulated box. Generally the long axis of the tank is oriented east to west, although a tilted tank is possible.

The basic Bread Box design calls for two sheets of glass, one on the south face of the box, and the other on the top. Zomeworks recommends double-glazing, even in mild climates, and triple-glazing in frigid ones. The more glass covers, the more careless you can be about promptly closing the box at night. At certain hours of the day, however, the glass can reflect most of the sun's rays away from the tank. You can avoid this problem by using instead a sheet of Filon or Sun-Lite arched above the tank. This solution also avoids the excessive weight of the glass.

The Bread Box minimizes heat loss from the stored hot water by means of an insulated box that encloses the tank at night and during cloudy weather. During the day, a top panel is raised and a side panel on the south face is lowered to expose the glass and tank to sunlight. The inside surfaces of the panels and the box itself are covered with a material such as aluminum foil that increases collection by reflecting sunlight around to the sides and back of the tank. When the panels are closed these surfaces reflect thermal radiation back to the tank.

The tank can be filled with water from either a pressurized or non-pressurized source. Once in the tank, the water is heated slowly but uniformly. Convection currents and conduction through the tank metal distribute the heat throughout the water, and little heat stratification occurs. In unpressurized systems, the hot water is nearly used up before the tank is refilled. In pressurized systems, cold replacement water is drawn into the tank as hot water is being drawn off and some mixing occurs. If dual tanks are used (see photo on the previous page), the Bread Box has less difficulty with mixing. Hot water is drawn from one tank while cold water flows into the other.

Freezing in cold weather is not much of a problem because of the large volume of warm water in the tank. However, the pipes must be protected. In severe climates, an electric heating element can be used to keep the tank from freezing, and heat tapes used on the pipes leading to and from the tanks.

Detailed design and construction plans for the Bread Box are available for $5 from Zomeworks Corporation. The plans include test data for a typical February day and also show how to hook up dual tanks. Because of its straightforward design, construction of a Bread Box does not require uncommon skills or tools. You can build one from new materials for about $200 or use recycled materials to reduce this cost. Either alternative puts the cost of a Bread Box well below that of other solar water heaters with similar performance.

Farallones Bread Box

In 1975, a Bread Box water heater was built and installed above the bathhouse at the Farallones Institute Rural Center in Occidental, California. This unit's 84-gallon tank sits under heavy panes of recycled storefront glass. The insulating cover can be adjusted to gain the maximum reflection of sunlight onto the tank. When empty, the entire unit weighs about 450 pounds. It required plenty of woman and manpower to hoist it onto the bathhouse roof.

During the winter of 1975-76 (a mild one), the Bread Box consistently produced water temperatures above 130°F. It's performance was sensitive to the amount of sunlight available but quite insensitive to the outdoor temperature. The 84 gallons of hot water have provided hot showers for an average of 15 students and staff through the use of a clever water-conserving showerhead: a hand-held, trigger-type nozzle that sprays water only when the trigger is depressed. Cold water is not drawn into the heater as hot water is drawn out. Instead, the tank is emptied during the day and refilled early the next morning.

Auxiliary heat is provided by sending the water through a copper coil bypass placed inside the stack of a wood-burning stove. This stove is frequently used to heat the bathhouse itself, and the extra hot water is an added dividend. With the stove lit, the bypass must be opened so that water can circulate through the coil and keep it from melting. But with any such auxiliary heating setup, the danger of overheating and consequent steam explosion cannot be overemphasized. To prevent such an occurrence, there should be a safety blow-out valve at the top of the collector.

Bread Box atop the bathhouse at Farallones Institute Rural Center.

Thermosiphoning Water Heaters

Less owner involvement is required with a thermosiphoning water heater, which exploits the natural buoyancy of the solar heated water to transport it to an insulated tank—away from the collection surface. Just as in a body of air, the warmest temperatures in an undisturbed body of water occur near the top. Because of small differences in density, warm water drifts to the top while cool water settles to the bottom. Consequently, if a storage tank is located *above* the collection surface, solar heated water will gather there. If the tank is *insulated,* this water will remain warm overnight. By contrast, the storage-type solar water heaters require someone to shut an insulating lid in the evening (and open it in the morning) to achieve the same result.

Perhaps the simplest thermosiphoning water heater is a model developed and used in the West Indies. The collector is a flat metal container painted black on the upper side and covered with glass. One end of this container bulges to form a large storage tank.

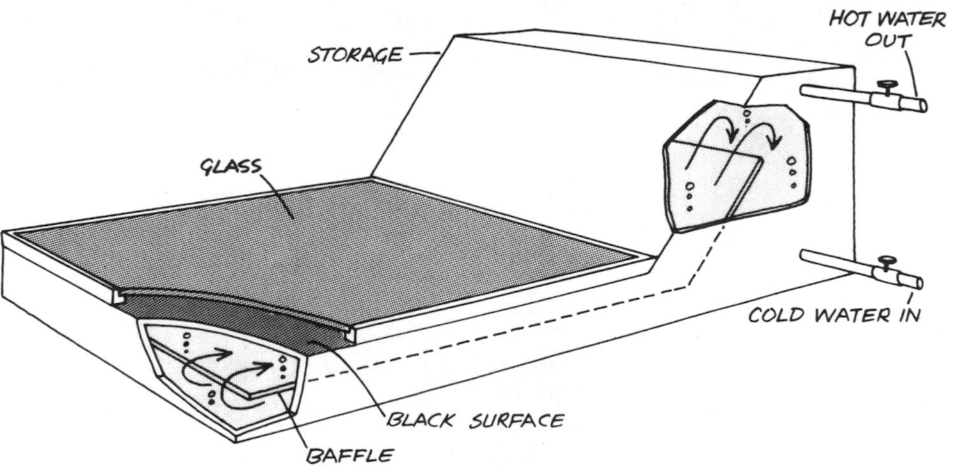

Thermosiphoning solar water heater developed in the West Indies.

A baffle plate across the mid-section of the container sets up a thermosiphoning circulation pattern (see diagram). Solar heated water from the collector flows up into the tank as shown, while cool water from the bottom of the tank returns to the collector. Because the storage tank has extra insulation, the solar heated water remains warm throughout the night.

Most thermosiphoning water heaters have a completely separate collector and storage tank. Although there are many different systems, virtually all are characterized by insulation-wrapped pipes connecting a tilted flat-plate collector with a well-insulated storage tank. When the water in the collector is heated by the sun, it rises, enters a pipe and flows into the top of the storage tank. Simultaneously, the cooler water at the bottom of the storage tank flows through another pipe leading back to the collector. As long as the sun shines, the water circulates and becomes warmer.

You can prevent reverse thermosiphoning by installing a check valve that closes when any reverse flow begins. But a height difference will accomplish the same thing. With the tank located at least a foot above the collector, the cool water settles to the base of the collector at night while the warm water remains up in the tank. However, reverse flow *can* occur if the collector and storage tank are on the same level and no check valve is used.

There is a wide variety of possible collector designs and piping patterns (see Chapter 7).

Because the pressure difference that drives thermosiphoning is small, a good collector design encourages an open flow of water. The parallel grid flow pattern (shown in the diagram) is an excellent choice for thermosiphoning water heaters.

The overall piping pattern should employ short lengths of pipe with large diameters and few joints or bends. Necessary joints and bends should be as gradual and smooth as possible to avoid interfering with circulation. For example, a 45 square foot collector with a height difference of 9 feet from the bottom of the collector to the hot water inlet at the top of the tank calls for 16 feet of 1-inch pipe. If these pipes are smaller in diameter, thermosiphoning will start later in the morning and end earlier in the afternoon because higher collector temperatures will be needed to drive the flow.

Insulation around the tank should be as thick as your pocketbook permits. Four inches of fiberglass isn't excessive. If possible, put the tank indoors. That way, the heat lost from it will flow to the rooms. But if the tank has to be outside, it should be shielded from the winds and lavishly insulated.

The piping between the collector and tank must also be well insulated in order to conserve the solar heat already collected and maintain the high temperature of the water coming from the collector. By locating the collector close to the tank and reducing the length of piping, you reduce the total surface area through which heat can be lost.

Because the thermal gradients between different levels of a thermosiphoning system are essential to circulation, the locations of the pipe connections to the tank are very important. The supply pipe to the bottom of the collector should feed from the coolest part of the tank—the bottom. Solar heated water from the collector should feed to the

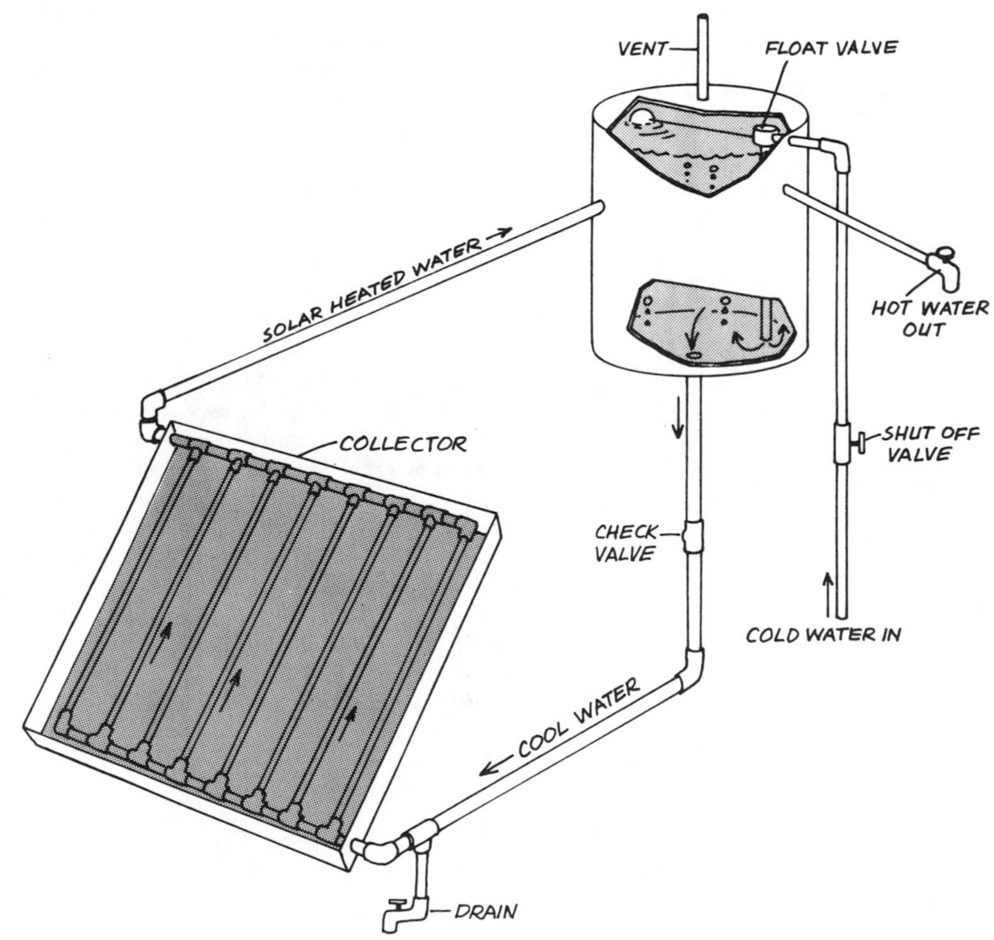

A thermosiphoning water heater with separate collector and storage tank.

Sizing the Collector

In good weather, an efficient collector will heat between one and two gallons of water per square foot per day. Based upon these figures, you can estimate the collector size from your daily hot water needs. For example, if a family of four uses 60 gallons of hot water per day, then 30 to 60 square feet of collector are needed. With an efficient collector in a good orientation producing 1.5 gallons per square foot per day, they would need 40 square feet.

A more exact estimate of collector size can be obtained with a little more math. First you need to know the daily heat demand H (in Btu/day), which you can calculate from the daily hot water demand D (in gallons/day) and the temperature difference ΔT *(in °F) between the hot tap water and the inlet cold water:*

$$H = 8.34 \times D \times \Delta T$$

For example, if the family needs 60 gallons of hot water at 120°F and the cold water comes from the city main at 55°F, the daily heat demand is

$$H = 8.34 \times 60 \times (120 - 55) = 32{,}526 \text{ Btu/day}.$$

The collector size C (in ft^2*) needed to satisfy this demand is:*

$$C = \frac{100 \times H}{I \times E}$$

where I is the average daily insolation (in $Btu/ft^2/day$*) and E is the average collector efficiency (in percent). To continue our example, suppose the family lives in a house with a south-facing roof pitched at an angle of 40° that receives 2000 Btu per square foot on a typical clear day. If the percentage of possible sunshine is 70 percent for their locale, that's 1400* $Btu/ft^2/day$*, on the average. With a collector efficiency of 45 percent, they need*

$$C = \frac{100 \times 32{,}526}{1{,}400 \times 45} = 52\ ft^2$$

of collector surface to supply their average daily needs.

The estimated collector size then gives an estimated collector cost if the costs per square foot are known. The size and cost of the tank follow directly from the daily hot water needs multiplied by the number of days of storage time desired. The total costs of the solar hot water heater can then be compared with the yearly savings in fuel to establish the payback period. For example, our model family saves the costs of 365 × 32,526 = 11.9 million Btu per year. At $6.00 per million Btu, that's more than $70 per year, and a $520 initial cost ($10 per square foot) can be paid back in about 7 years.

warmest part of the tank—the top. About half a day's storage volume should remain above the hot water inlet so that the hottest water can accumulate there. Water coming from the collector won't always be at peak temperature (on cloudy days or early in the morning) and will cool the hottest water if this inlet were located at the very top.

Hot water should always be drawn from the top of the tank, whether pressurized or not. If the inlet cold water comes from city mains or a well pump, the entire system will have about the same pressure and hot water can be supplied to parts of the house above the storage tank. However, more rugged collector construction is necessary because of the extra stresses on the piping. Non-pressurized systems, such as the one in the previous diagram, can supply hot water only to areas below the storage tank. But the stresses within these systems are much lower, and they are cheaper and easier to build.

Two day's storage capacity is generally the optimum. A larger tank will carry a household through longer cloudy periods (assuming enough collector area), but will require more time to reach the desired hot water temperature. The larger tank will also cost more. A smaller one, however, requires more frequent use of auxiliary heat. In general, the collector should be large enough to provide a single day's hot water needs under average conditions. Beyond this point, a larger collector provides an increasingly poor payback.

The collector should face south or as nearly south as possible. Because domestic hot water is required year round in about the same daily amounts, the collector should be tilted for about the same solar gain in all seasons. Relatively constant daily insolation strikes a south-facing collector tilted at an angle equal to the local latitude. Steeper tilts (up to latitude +10°) may be needed in areas with limited winter sunshine.

If the roof can support the collector, it is much less expensive to put it there than to build a separate structure. But true south orientation and latitude angle tilt may not be feasible on many houses. Fortunately, the loss in efficiency is fairly slight (10 to 15 percent from ideal) if the roof faces within 25° of true south and its tilt is within 15° of the latitude angle. This efficiency reduction can be easily recouped by making a proportionately larger collector.

For thermosiphoning solar water heaters, the choice of the collector location must allow placement of the storage tank at a higher level. For roof-mounted collectors, the storage tank may be placed under the roof ridge or even in a false chimney. Of course, the roof structure must be strong enough to support the weight of a full tank. One alternative is to build a collector support structure on the ground, detached from the house or leaning against it. The collector then feeds an elevated tank located beside it or inside the house. Another possible location is on the roof of a lean-to greenhouse built onto the south-facing side of a house. The greenhouse then provides flowers, food and solar heated air for the inhabitants while the collector supplies them with hot water!

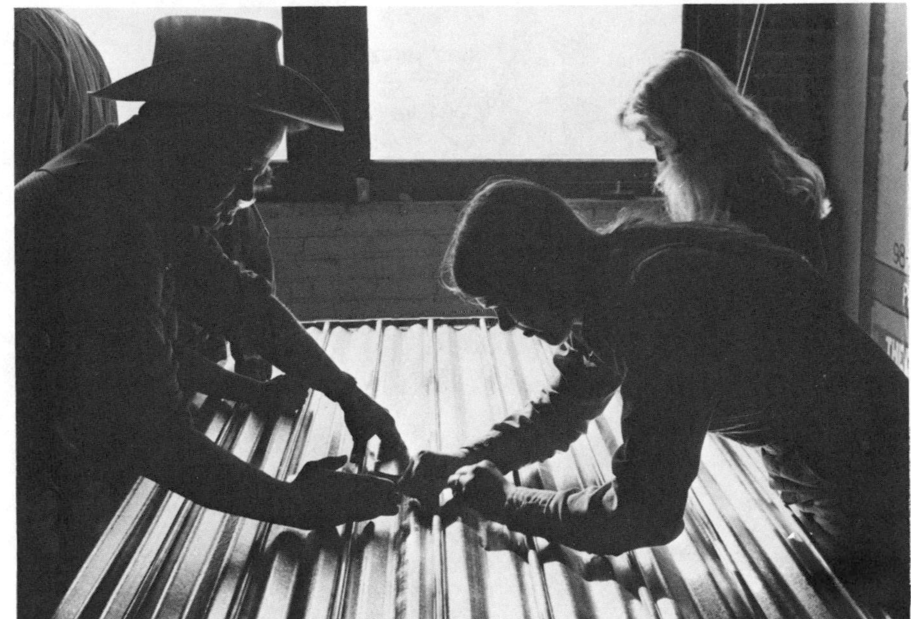

More than a thousand people have taken part in the ECOTOPE–RAIN water heater workshops.

ECOTOPE–RAIN Solar Water Heater

Over the past two years, Lee Johnson of RAIN Magazine and Ken Smith of the ECOTOPE Group have been showing people in the Pacific Northwest how to build their own solar water heaters. Those attending their popular 2-day workshops learn by building a thermosiphoning solar water heater from readily available materials and

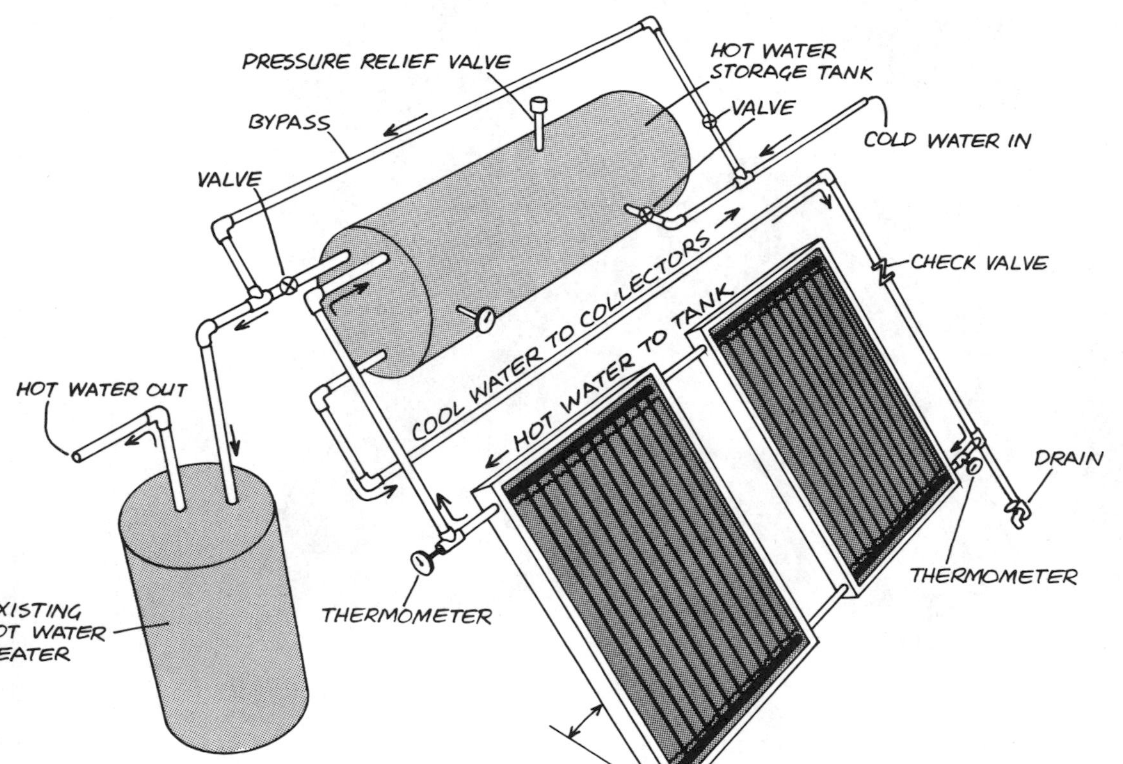

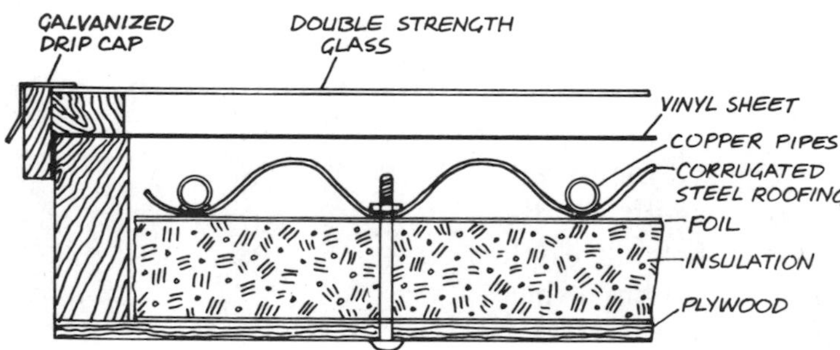

Construction details for ECOTOPE–RAIN solar water heater.

parts. More than a thousand people—housewives, handymen, and high school students—have returned from these workshops with the knowledge and enthusiasm to build their own water heater. Johnson and Smith have been able to perfect their design so that it is extremely simple, cheap, and easy to fabricate with basic tools and local materials.

Their standard solar water heater uses two collector panels—each with about 30 square feet of absorber surface. Corrugated galvanized steel forms the absorber plate, with ½ inch copper tubing wired or soldered to alternate corrugations as shown. The whole assembly is painted flat black and placed in a collector box built from 2 × 4's and a 4 × 8 sheet of plywood. Two inches of foil-faced styrofoam behind the absorber provide R-11 insulation. The transparent cover has two layers—a 4-mil sheet of vinyl plastic sitting beneath a layer of glass, Filon, or Kalwall Sun-Lite.

Johnson and Smith recommend using a 40 to 80 gallon recycled water heater tank (preferably glass-lined and insulated) as the storage tank. It rests on its side at least one foot above the top edge of the collectors. All the external piping connections are made with ¾-inch copper tubing insulated with at least 1 inch of fiberglass. Piping runs are kept as short as possible. A pressure relief valve on the tank guards against overpressures, and a check valve provides extra security against reverse thermosiphoning.

This solar water heater serves best as a preheater to an existing water heater having its own heat supply. The necessary connections are shown in the diagram. But tank temperatures easily hit 130°F on a sunny winter day, and this unit can serve as a stand-alone water heater. With a large storage tank, this solar water heater can provide the average hot water needs for 4 people in most areas of the United States. Freezing can be a problem in many areas, and Lee Johnson recommends using a hinged insulated lid that can be closed at night.

Step-by-step photo-illustrated instructions for building this solar water heater can be obtained for $3.00 from RAIN: Journal of Appropriate Technology (2270 NW Irving, Portland, OR 97210; attn: Solar Water Heater Workshop Manual). This manual has a detailed list of materials and tools needed and is a must for anyone contemplating such a project. The total cost of materials should fall between $200 and $250.

Integral Urban House Water Heater

The Farallones Institute also operates an urban center in Berkeley, California. An ongoing experiment in ecologically-sane living, this Integral Urban House is a renovated Victorian structure that houses 5 or 6 students and staff. In early 1975, Doug Daniels and Scott Matthews built a thermosiphoning solar water heater into the structure to supply most of the hot water needs.

Sitting on the roof of a greenhouse built onto the south side of this building is an 89 square foot collector. The absorber plate and the 3/8-inch tubing are copper—for efficient collection and long life. The only recycled material in the system is the collector glazing—sliding glass door seconds. A 120 gallon tank in the attic (15 feet above the collector) provides a storage capacity of 1-2 days. With proper maintenance, this glass-lined steel tank should last 15 to 20 years.

Flat-plate collector of the Integral Urban House water heater sits on the roof of a greenhouse addition.

Because Berkeley is situated in earthquake country, the tank (which weighs ¾ ton when full) had to be securely fastened to the strongest supports of the house. Its position is far from the collector, and 55 feet of 1½-inch copper pipe was required to link them. To reduce heat loss over this distance, the pipes are insulated with 3 to 4 inches of fiberglass wrap. The 10 inches of fiberglass insulation around the tank cost only $15—a good investment.

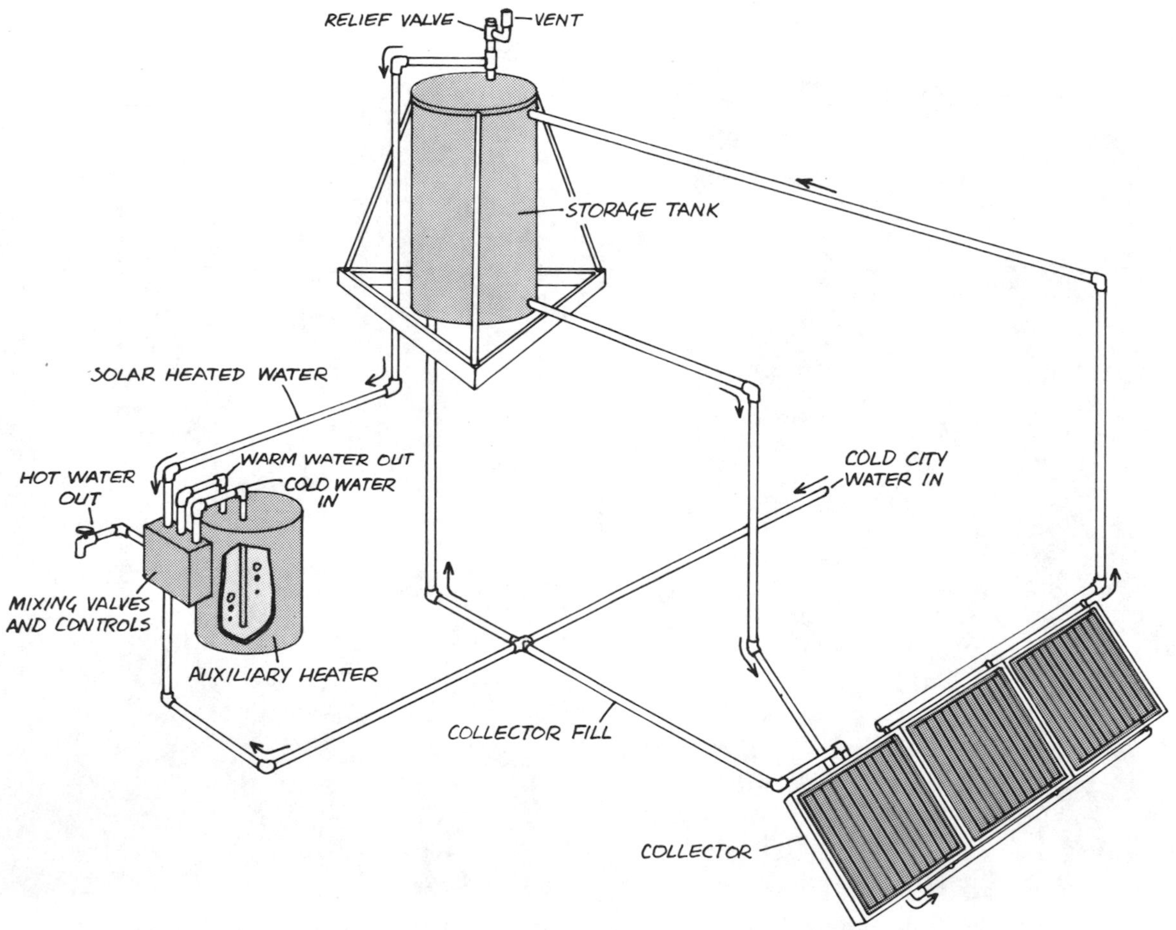

On a sunny day, the total volume of water cycles about 10 times between the collector and storage tank, and the incoming city water is heated from a temperature of 50-60°F to 140-160°F. Hot water for domestic use is drawn from the top of the tank and flows through a 30-gallon auxiliary heater (activated if the water is not sufficiently hot) to taps in the house. This solar water heater has provided about 95 percent of the hot water needs of the house. These needs have been kept to 70 gallons per day by a number of water conservation practices.

Daniels and Matthews figure the cost of materials for this system at about $1100, or $12.50 per square foot of collector. They estimate the time required to build and install it at 6 person-weeks. Detailed working drawings of this solar water heater are available for a modest $7.50 from the Farallones Institute (1516 Fifth Street, Berkeley, CA 94710). However, Daniels cautions that substantial changes in collector size or in the height difference between collector and storage can alter the system's performance unless pipe diameters are changed accordingly. If interested in building a similar water heater, you might contact the Berkeley Solar Group (3062 Shattuck Avenue, Berkeley, CA 94710), who developed the computer model for sizing this system.

The thermosiphoning solar water heating system in the Integral Urban House.

Protection Against Freezing

Virtually all thermosiphoning water heaters now operating are situated in warm climates where freezing temperatures are seldom encountered. But if even a single occurrence of freezing is possible, the collector must be protected from it. When air temperatures drop below 32°F, freezing water can burst the pipes or collector channels. Less obvious but no less worrisome is the freezing caused by nocturnal radiation. Copper pipes in collectors have frozen and burst on clear, windless nights when air temperatures never dipped below 40°F. The heat lost by thermal radiation was simply greater than that gained from the surrounding air. Safeguards must be included in a solar water heater designed for an area with any possibility of freezing.

There are four basic ways to protect a thermosiphoning collector against freezing:

1) heat the absorber plate;
2) cover the collector with movable insulation;
3) drain the collector; and
4) use an antifreeze solution in the collector piping.

The first method was used for the collector at the Integral Urban House, where freezing conditions occur about once every two years. A length of "heat tape" was fastened to the back of the absorber. This tape looks like an extension cord and has a small resistance to the flow of electricity. Under control of a thermostat, an electric current begins to heat the absorber when its temperature falls to 35°F. The heat tape is cheap and readily available at such outlets as Montgomery Ward and Sears. This solution works well in locations where freezing temperatures are a rarity. In more severe climates, the auxiliary heat would be required so often that electricity costs would be prohibitive.

Movable insulation is another cheap, effective alternative. But because of the severe consequences of even a single forgetful moment, an insulating cover should be automated somehow—even if only an alarm bell to warn of an impending freeze. And any moving mechanism must remain reliable through many seasons of harsh weather.

Draining the collector is difficult in a thermosiphoning system—but not impossible. First the collector is shut off from the tank; then a vent at the top and a drain at the bottom of the collector are opened, and the water is allowed to flow out. The collector is refilled by closing the drain, letting water in until all the air has escaped, and then reopening the shut-off valve to the tank. While not a difficult process to do by hand, it is difficult and expensive to automate. And manual operation requires a fair amount of time—especially on days of intermittent sun. In a draining system, the collector channels must be designed to drain completely—leaving no pockets of water behind. And vapor lock, caused by air pockets remaining after refilling, can be a problem.

The most effective means of protection against freezing is the use of an antifreeze-

and-water solution in the collector piping. But here we are confronted with yet another problem: the antifreeze cannot be allowed to mix with hot water going to the taps. A heat exchanger must be used to transfer solar heat from the antifreeze solution to the domestic hot water supply in the storage tank. One form of heat exchanger is a coil of copper tubing immersed in the storage tank. The coil is connected to the collector to form a single, closed flow loop. Some vent or pressure relief valve must be included in this circuit to guard against overpressures. Such a closed system can still be driven by thermosiphoning, or you may add a small pump to the circuit. A small percentage of the solar heat is lost because of the heat exchanger, but operation is relatively trouble-free and requires only occasional attention from the owner.

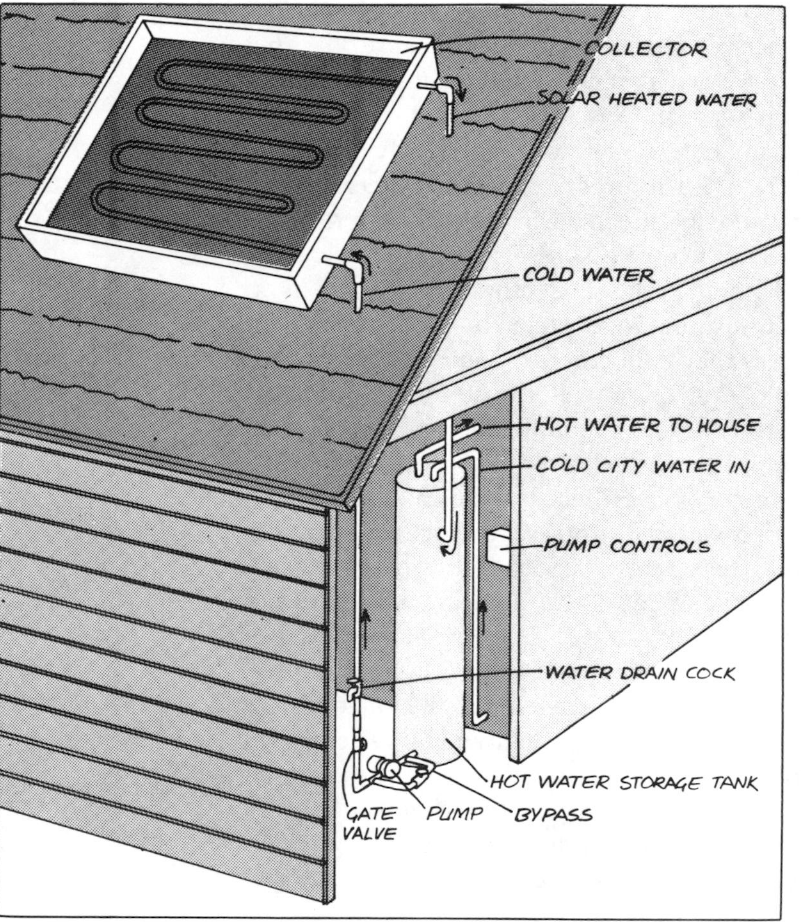

In a pumped system the collector can be located above the storage tank.

Instead of the coil of tubing inside the storage tank, you can use a smaller tank to hold the antifreeze solution. Or you might wrap the copper tubing around the *outside* of the storage tank and bond it there with a thermally conductive adhesive. Hardware and air-conditioning supply stores should be able to provide you with suitable adhesives. Because standard building codes often do not allow single walls between potable water and toxic substances like antifreeze, this latter approach may be your only alternative. At *all* costs, you must insure that antifreeze never enters the water supply.

Pumped Systems

The use of pumps can remove many of the architectural constraints of thermosiphoning water heaters. A pumped system is commonly used in cases where piping runs would be too long or there is no adequate position for an elevated tank. The penalty paid is the cost of the pump and its controls and the electricity needed to run them. But a collector has a higher efficiency with an assured steady flow of water, and you have much more freedom in the system layout. For instance, a pumped system can have a collector on the roof, a storage tank in the cellar, and the faucets on the first floor. The collector and its piping would form a single circulation loop tied into the hot water storage tank, either directly or through a heat exchanger.

The pump must be controlled so that circulation occurs only when the fluid in the collector is hotter than that in the tank. A differential thermostat, which has two sensors (one near the collector outlet and the other near the tank outlet) is the best control unit, but it may add $75-$125 to the overall system cost (1975 prices). A much cheaper alternative is a thermostat set to activate the pump when the collector reaches a preset temperature—say 130°F. Once again, the least expensive alternative is also the most time-consuming—switch the pump on and off yourself!

Modesto Solar Water Heater

To stimulate local interest in solar energy, Ecology Action of Modesto, California, designed and built a pumped solar water heater in 1975. Its flat-plate collector tilts at an angle of 48° (latitude +10°), leaning against an insulated wooden structure that houses the storage tank. The collector uses an aluminum ROLL-BOND absorber (discussed in Chapter 6) manufactured by Olin Brass. Covering the front face is a double layer of Dupont Tedlar, which was heat sealed to keep it from sagging or flapping in the wind.

Although built for balmy California, this unit would work well in much colder climes because an antifreeze solution is used as the heat transfer fluid. About 2½ gallons of propylene glycol and distilled water (to prevent corrosion—a must in areas with hard water) flows *up* through the collector and *down* through a heat exchanger inside the storage tank. This heat exchanger was fabricated from a 12 foot length of ¾-inch copper tubing by using a pipe bender. A small electric pump can circulate the liquid at two different flow rates.

Cold water flows from the well pump into the bottom of the tank and is heated by conduction from the heat exchanger. The hottest water collects at the top where it is drawn off for domestic uses. The 80 gallon tank was made from a recycled farm pressure tank that was cut open to install the heat exchanger and then welded shut. Locally manufactured cellulose fiber (from recycled newspaper) was used as the insulating material for both the tank and the collector.

On sunny January days, the system delivers hot water at 135°F when the outdoor air temperature is a nippy 50°F. During periods of partial cloudiness, the 2-speed pump is switched to its lower speed (2 gallons per minute instead of 20) so that

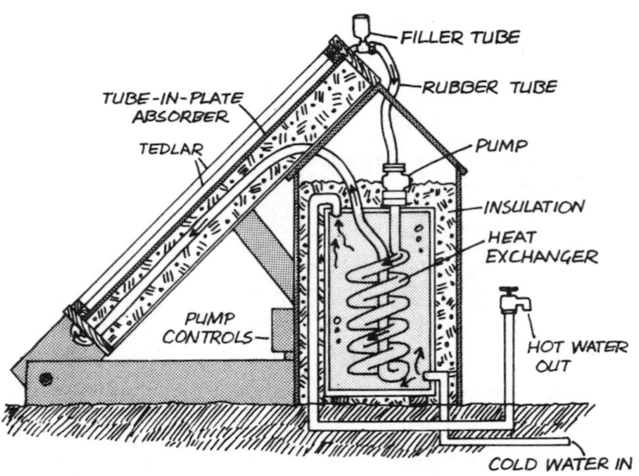

Pumped solar water heater built by Ecology Action of Modesto, California.

PERCENTAGE HEAT LOSSES FROM TYPICAL HOUSES			
	Type of House		
Heat Loss Through:	Uninsulated[1]	Moderately insulated[2]	Well insulated[3]
Roof	25	13	4
Windows and doors	25	27	15
Walls	25	15	24
Floor	5	5	2
Infiltration	20	40	55
Total	**100**	**100**	**100**

[1]A. Adams, *Your Energy-Efficient House.*
[2]E. Eccli, *Low-Cost Energy Efficient Shelter.*
[3]*Professional Builder* (November 1973).

water is heated substantially in a single pass through the collector. This feature of the system makes it especially appropriate for more frigid climates. The cost of this system was kept below $200 by using recycled materials whenever possible. Of the total cost, about $40 went for the absorber plate, $40 for the Tedlar, $25 for plywood, and $50 for the pump and controls.

Where to Go From Here

There are a few very readable works that can take you beyond the limits of this chapter. One is an excellent 25-page booklet, *Build Your Own Solar Water Heater,* published by the Florida Conservation Commission. This work gives detailed instructions for building a tube-type collector and using it with a solar water heater. It is well illustrated, and the information can be readily adapted to a number of other designs. Another booklet, available free from the Copper Development Association, is *How to Design and Build a Solar Swimming Pool Heater,* by Francis de Winter. Notwithstanding the obvious emphasis on copper and swimming pools, this book is a valuable aid to collector design. The economic analyses and cost evaluations are among the best ever done on the subject of solar heating.

ENERGY CONSERVATION MEASURES

The energy-efficient home requires much less energy—conventional or solar—to make it perform as well as, or better than, other homes. As discussed in Chapter 4, energy conservation is crucial for solar heated homes; the greater the heat loss, the larger and more expensive the collector and storage unit. People often think that energy conservation applies only to house heating. But summer ventilation and cooling are also important in many climates, and energy-conserving measures can provide savings on air conditioning equipment and power.

This section offers some tips on energy conservation in *existing homes.* It stresses cheap and effective methods of minimizing infiltration and conduction heat losses. These measures are an important prelude to the further step of refitting homes for solar space heating. Most of these measures will also reduce your summer cooling requirements.

Reducing Infiltration

Air infiltration can account for 20 to 55 percent of the total heat loss in existing homes. In those with some insulation, the heat loss from air infiltration exceeds conduction losses through the walls, ceiling and floor by up to 25 percent. Insulating older homes often requires a major overhaul, and tackling infiltration problems is the logical place to begin.

Measures for reducing infiltration include:

- general "tightening up" of the structure and foundations

- caulking and weatherstripping the doors and windows
- redesigning fireplace air flows
- creating foyer or vestibule entrances
- installing a vapor barrier in or on the walls
- creating windbreaks for the entrances and the entire house

To tighten up your house, start with the obvious defects. Close up cracks and holes in the foundation, and replace missing or broken shingles and siding. Hardened, cracked caulking on the outside of the house should be removed and replaced with fresh caulking. Be sure to refit obviously ill-fitting doors and windows. And plug up interior air leaks around mouldings, baseboards, and holes in the ceiling or floor.

Before you try to stem infiltration between windows or doors and their frames, make sure no air is leaking around the *outside* edges of the frames. Caulking will remedy any such problems and prevent water seepage during driving rainstorms. The caulking compounds with superior adhesive qualities are generally called sealants. An excellent reference, *Low-Cost, Energy-Efficient Shelter for the Owner & Builder,* recommends MONO sealant and some of the silicon sealants. It's worthwhile to obtain high quality sealants because caulking is a lot of work, and inferior compounds will decompose after a few years.

Weatherstripping is required to check infiltration where the edges of doors and windows meet (or don't quite meet!) their frames. Unlike high quality sealants, weatherstripping is readily available at the local hardware store. Different types are required for different applications. For example, gasket compression-type weatherstripping (black—with a fabric face, peel-off back, and adhesive coating) is best suited for hinged door and casement and awning type windows. For sliding windows, the spring bronze or felt-hair weatherstripping is most appropriate. Double-hung windows are more of a problem. Your best bet is to rely on storm windows or insulating shutters to reduce infiltration *and* conduction losses. Any unused windows can simply be caulked shut!

One obvious energy conservation step is to close a fireplace when not in use. If the fireplace is old and doesn't have a damper, you had best install one. You can get more heat from a fireplace by using a C-shaped tubular grate to cradle the burning wood. Cold air is drawn in at the bottom, warmed and delivered back into the room from the top by a thermosiphoning process. Another way to extract more heat is to install vents in the flue that reclaim heat from hot air rising up the chimney. You can also modify a fireplace to draw air directly from the basement or outdoors. Otherwise, the fire will continue to draw warm air from the rooms, effectively cooling the same living space you want to heat.

Plenty of cold outside air flows into a house every time you open a door, particularly if the door is on the windward side. However, a foyer or vestibule entrance will

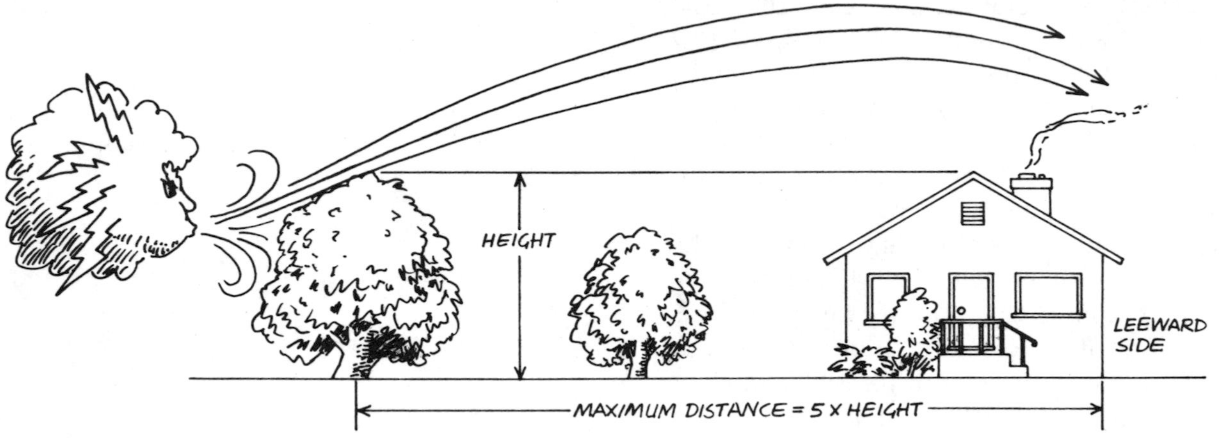

Tall, dense vegetation makes an effective windbreak.

eliminate this problem by creating an "air lock" effect. If a door opens into a hallway, another door can be hung about 4 feet into that hallway to make a foyer. If your entrance has no hallway, you can build two walls out from either side of the door, add a roof, and install a second door. This addition can be simple and inexpensive–you needn't insulate the vestibule walls, just weather-strip both doors. And be sure that the new doors open outwards for rapid exit in case of fire!

It may come as a surprise that air can infiltrate the walls themselves. Wind pressure forces air through the tiny cracks in the wall material (particularly wood siding). A good vapor barrier will eliminate this problem while fulfilling its primary purpose–that of maintaining comfortable indoor humidity. For older homes with no such vapor barrier, installing one is only practical if the inside surface of the walls is being removed for extensive rehabilitation or remodeling. Either foil-backed fiberglass insulation or polystyrene can be installed as an effective vapor barrier. If your home needs a vapor barrier but you have no intention of ripping your walls apart, certain high quality paints act as good vapor barriers. Ask for paint with low permeability.

If entrances and windows are exposed to prevailing winter winds, you may need wall or fence extensions, overhangs, baffles and plantings to deflect that wind. Such wind-breaks lower the wind's velocity and reduce infiltration. Vegetation should be dense and eventually reach as high as the house. The distance from the house to the wind-break, measured from the house's leeward side, should not exceed 5 times the building height. Local agricultural extension services can suggest the trees and shrubs best suited to your climate and the appropriate planting distance from the house.

Insulating the Ceiling, Walls and Floor

In most existing dwellings, the money spent on additional insulation is a very good investment. However, the *time* and *effort* involved in insulating an older house is substantially more than that required to reduce infiltration. So before you begin to rip out the siding, examine your house to determine how much additional insulation you need and how much work is required.

Publications recommended at the end of this section will help you to make this evaluation. This section serves as a general guide to insulation needs and the available types of insulation.

Thermal performance standards of buildings are determined by matching the insulation costs against fuel costs until the purchase price of the insulation and the projected cost of fuel bills reach a minimum. Previously, the amount of insulation yielding this minimum cost gave the walls a total thermal resistance, or R-value, of about 11. But with rising fuel costs it makes sense in many regions to increase the total resistance to as much as 19 in the walls and 25 to 30 in the ceilings. Moreover, installing extra insulation now will obviate the trouble of adding more in the future.

Attic insulation is the most crucial because substantial amounts of heat are lost in winter and gained in summer. A resistance of 20 to 30 can be obtained by applying thick blanket, batt, loose-fill, or poured insulation on the ceiling framing or directly on top of existing insulation. If the attic roof is too low, you can try to maneuver blanket insulation into place with a rake or have a contractor install blown insulation. If neither is possible, wait until re-roofing time.

The insulation of an existing stud wall is limited by the wall thickness. The three insulating materials which approach or exceed the desired resistance of 19 in a standard 2 X 4 stud wall are cellulose fiber, urethane, and urea formaldehyde. If you are fortunate enough to have 2 X 6 stud walls, you have many possible choices. Cellulose fiber, urea formaldehyde, and polystyrene beads can be blown into wood frame walls. However, holes will have to be bored in the interior wall between studs and above and below the firebreak. Later someone will have to patch the holes, providing you with an opportunity to use a vapor barrier paint. The other alternative—installing blanket or batt insulation—requires ripping out the interior walls. This makes sense only if the walls need renovation anyway. The table shows that mineral wool and fiberglass insulation will not provide the desired resistance of 19 in a 2 X 4 stud wall. Trying to compress these insulators to increase their resistance will have the opposite effect—compression reduces the air spaces needed to slow the flow of heat.

For masonry walls, one method is to blow loose-fill or foam insulation into the existing air spaces. This approach is possible if the plate for the ceiling rafters doesn't cover the concrete block cores or the cavity wall construction. Rigid board insulation can also be placed on the outside of a masonry wall and replastered or covered with siding.

Insulate floors and foundations last. Tacking foil-backed insulation supported by wire mesh to the underside of the floor (leaving a ½-inch air space) can provide a high resistance. If there isn't enough room to get under the house, seal the foundation but leave a few ventilation openings. For base-

R-VALUES OF COMMON INSULATORS

Insulation Material	R-Values		
	For one inch	Inside 2 x 4 stud wall*	Inside 2 x 6 stud wall*
Mineral Wool	3.0	13.7	19.7
Fiberglass	3.5	15.5	22.4
Polystyrene	4.0	17.2	25.2
Urethane	6.5	25.9	38.9
Vermiculite	2.5	11.9	16.9
Cellulose Fiber	4.5	18.9	27.9
Urea Formaldehyde	5.5	22.5	33.4

*includes insulating value of siding, sheathing, and air films.

SOURCE: E. Eccli, *Low-Cost Energy-Efficient Shelter*, 1976.

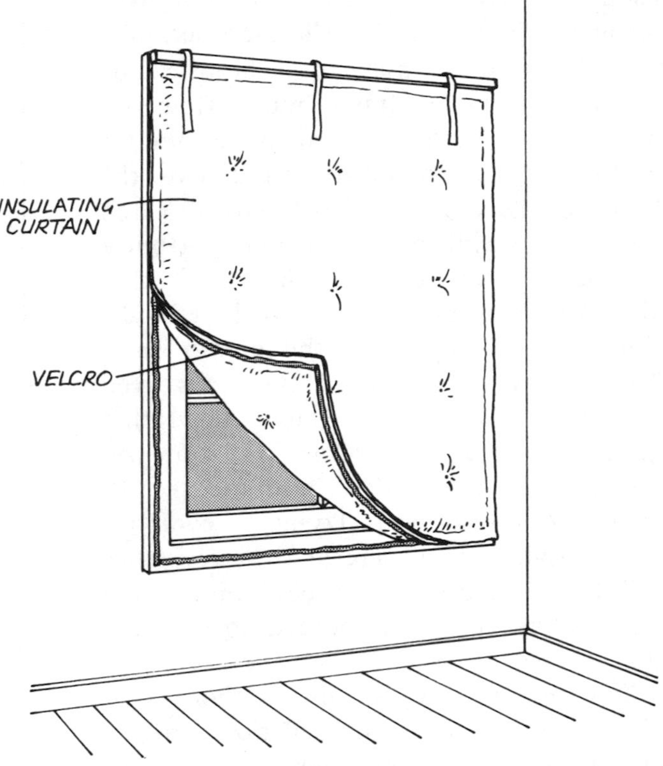

An insulating curtain that fastens tightly to the window frame.

ments being used as a living space, insulate the foundation walls all the way to the base of the foundation.Such "perimeter" insulating techniques were discussed briefly in Chapter 4.

Insulating the Windows

Even with all the foregoing improvements, you can still lose a lot of heat by conduction through the windows (see cover). A single-pane window loses heat about 20 times as quickly as a well-insulated wall of the same total area. There are a number of methods by which you can cut these losses—storm windows (discussed in Chapter 4), insulating curtains, and insulating shutters.

Insulating curtains can be made of tightly woven material lined with "fiberfill"—a loose stuffing or blanket-type insulating material—or other heavy material. They can be suspended from a curtain hanger or attached directly to the top window frame. In both cases you should fasten the curtain tightly to the sides and bottom of the window frame. Velcro fasteners (like those on down jackets and sleeping bags) allow a tight seal and permit you to open the curtain during times of winter sun. If you use a curtain hanger, put a cap over the top. These methods create a *dead space,* or pocket of cool air that can't be convected into the room. With a reflecting layer on the outer surface, such insulating curtains can also be used to reduce solar heat gains in summer.

Another idea is to insulate the windows with removable insulating shutters. With modern building materials and a little ingenuity, small miracles are possible. For example, you can make wooden shutters hinged on one side and filled with fiberglass or styrofoam insulation. Shutters should be weatherstripped where they meet the window frame and fastened tightly with a latch.

A recent innovation from Zomeworks (they never let up!) is the "Nightwall"—

styrofoam panels attached directly to the glass surface with magnetic clips. The clips are placed at intervals around the window or frame, with their mates affixed in corresponding locations on the styrofoam and recessed to allow a tight fit against the glass. The panels are light enough to be held firmly in place, but they can be readily removed and stored during the day. Get the styrofoam from your local building materials supply and Zomeworks will sell you the clips for a pittance.

On north windows, which provide indirect lighting and cross-ventilation in summer, the insulating panels or shutters can be left in place for most of the heating season. For east, west and especially south windows the panels should be put up only at night or during sunless winter periods. These same windows can be shuttered on sunny summer days, with the north windows left uncovered.

Where to go from here

You'll still need a way to apply the general information presented here to the specific needs of your house. One of the best guides is *In the Bank or up the Chimney? A Dollars and Cents Guide to Energy-Saving Home Improvements,* published by the Department of Housing and Urban Development. This booklet helps you determine the exact energy conservation measures your house requires and to estimate the costs. Do-it-yourself sections list materials, tools, installation procedures, and safety tips. Other sections tell how to deal with contractors and what preparatory work you can do yourself.

A second helpful publication is *Low Cost, Energy-Efficient Shelter for the Owner and Builder,* edited by Eugene Eccli, and published by Rodale Press. It contains very clear, in-depth discussions of energy conservation problems and possibilities for both new and existing homes. It also provides sources and cost figures for necessary (but often hard to locate) high-quality materials.

SOLAR HEATING AN EXISTING HOME

After you've made your house an energy-efficient "heat cocoon," you may want to solar heat it. Such a project is commonly called *retrofitting* a house for solar heating. A number of the houses discussed in Chapter 2—the Boulder House, the Desert Grassland Station and the Van Dresser House—were ordinary houses with solar heating added later.

To date, retrofitting has taken a back seat to the solar heating and cooling of *new* homes. Two of the obstacles to retrofitting are poor building orientation and the obstruction of winter sunlight by trees and other buildings. And since many old houses are drafty and poorly insulated, the owner must invest heavily in energy conservation measures before he or she can begin to think of solar heating. Furthermore, most systems suitable for existing houses cost several thou-

Nightwall—styrofoam shutter held in place by magnetic clips.

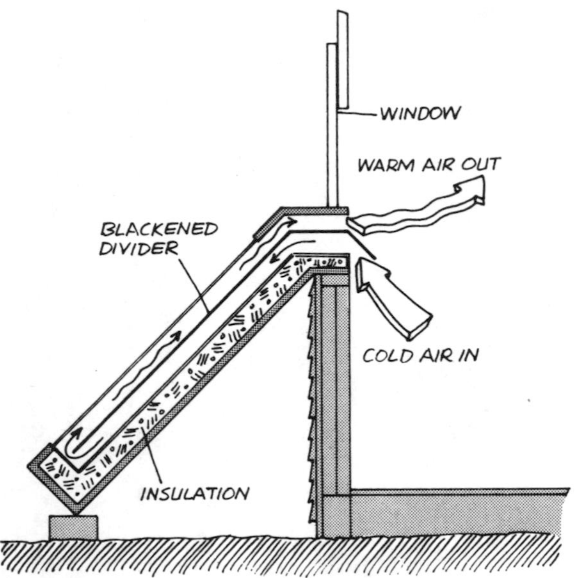

Window-box collector—daytime operation.

sand dollars—an investment rarely justified by fuel savings in the short run.

For houses without the physical impediments mentioned, however, there are a number of possibilities. Small scale "solar assist" projects like window-box or wall collectors may supply only 10 to 20 percent of a house's heat, but their extremely low cost justifies their use. Attached greenhouses that are well within the realm of do-it-yourselfers can supply up to 50 percent of the heating load. For those willing to build an additional room onto the south side of their house, an even greater percentage of the load can be supplied by other passive solar heating systems. Active or indirect systems have already received the lion's share of coverage in other publications, and scant mention will be made here of their potential for retrofitting. In general, they are too difficult to build and very expensive.

Window-Box and Wall Collectors

The construction of a "window-box" collector is a quick retrofitting project almost anyone can do. An insulated wooden box with a glazed top leans against the frame of a south-facing window (see diagram). The sun's rays are absorbed in a blackened sheet of plywood or other building board that divides the box along its midsection. The air above the black plywood is heated and flows into the house—drawing cooler replacement air from the house into the bottom part of the unit. As long as the sun shines, this window-box collector will help to warm the house. Actually, the window-box collector is a tilted version of the CNRS window collectors described in Chapter 4.

It's important to insulate the back, sides, and bottom of the collector box. It should be tightly sealed and fit tightly into the inside of the window frame. At night, the openings do not need to be closed off because the cold air settles to the bottom of the box and remains there. In summer the box can be removed and stored—or the openings closed off with dampers to keep the warm air out of the house.

Although the size of a window-box collector is arbitrary, its overall heat contribution is small unless it's significantly larger than the window area it replaces. A collector size of ¼ to ½ the house's floor area is usually required to supply about 50 percent of the heating load. It's fairly difficult to put that much collector area just outside your south windows. But the costs of a window-box collector are so minimal that it's always a sound investment.

If there are few windows on the south side of your house, you can put a collector on the outside walls. This thermosiphoning air heater is built by covering a portion of the wall with glass and cutting holes in the wall to let cool air into the collector at the bottom and warm air back into the house from the top. Old storm windows or fiberglass-reinforced polyester, such as Sun-Lite or Filon, make inexpensive glazing. These are held in place by an insulated box built of one-inch lumber and fitted around a 2 × 4 frame anchored to the wall. Concrete blocks and bricks under the box help support its weight. At intervals of 3 feet across the width of the collector, two 2" × 14" ducts should be cut through the wall at the top and bottom of the unit. A variety of dampers can be used to control daytime heat flow and prevent reverse thermosiphoning when there is no sun.

You can augment the solar energy collected by making a reflector from aluminum foil and plywood and placing it on the ground in front of the collector. Or when snow falls, shovel it around in front of the collector—it makes an excellent, inexpensive reflector. More detailed construction guidelines for window-box and wall collectors are available in the publication *Save Energy Save Money,* edited by Eugene Eccli. Other information in the manual makes it worthwhile reading.

Attached Greenhouses

Originally invented as a response to the demands of Tiberius Caesar for fresh cucumbers out of season, greenhouses were used in 19th century Europe as a source of winter heat and year-round food supply. A greenhouse was built onto the southern side of a dwelling and the excess solar heat collected by the greenhouse was vented into the rooms. Called "conservatories" in England, these greenhouses were built in sizes ranging from a small window unit to a large room-sized structure.

A century later, greenhouses have once again emerged as an ideal method of solar heating. And they provide many other benefits, too. The natural flow of heated air from a greenhouse has a higher relative humidity due to the vapors exhaled by the plants. Gone are the human discomfort and even respiratory ailments that result from the dry hot-air blasts of a forced-convection heating system. A greenhouse that produces food and house plants cheaply also makes a beautiful border for your living spaces. The heat losses from the wall, doors or windows it encompasses are reduced. And it can serve as a vestibule entrance—reducing the infiltration of outdoor air that comes with opening and closing a door. With an attached greenhouse, a dwelling can be extremely tight and still have a continuous flow of fresh air. Stuffy house air flows to the plants and returns to the house scrubbed of its carbon dioxide and other contaminants.

But first let's examine the heating potential of attached greenhouses. What, exactly, is their thermal performance? How much heat can they provide a home during the winter? How big must they be and what will they cost? The answers to these questions will vary with climate and needs, but a few examples will provide useful guidelines.

Solar Sustenance Project

Some of the most interesting solar greenhouses have been built as part of the Solar

A thermosiphoning wall collector added to the south side of a house.

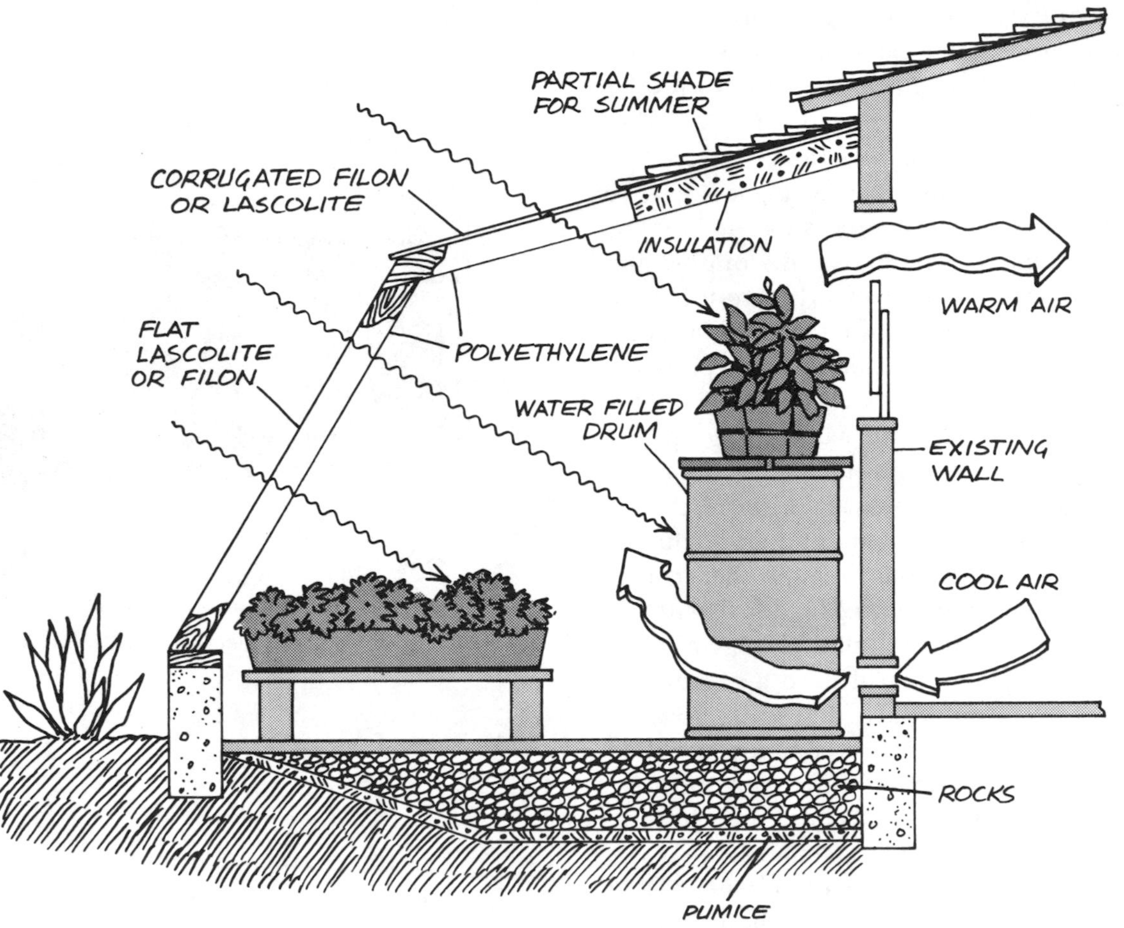

Cross section of Solar Sustenance Project greenhouse—daytime operation.

Sustenance Project, a program designed to supplement both the home heating and dietary needs of low-income families in mountainous northern New Mexico. Under the direction of Bill and Susan Yanda, lean-to greenhouses were built onto 11 private homes at altitudes ranging from 6000 to 8000 feet, where the growing season is as short as 90 days. The floor areas of these greenhouses varied from 100 to 360 square feet. In each case, an ordinary wood frame was built onto the southernmost side of the house. It was covered with Filon or Lascolite greenhouse-grade fiberglass sheeting. Both are unbreakable, have lifetimes of 20 years, and have the right spectral properties for growing plants. Directly under the fiberglass, a polyethylene sheet traps an insulating layer of air. The inner layer was made with Monsanto 602—an ultraviolet-resistant plastic sheeting with a lifetime of up to 5 years. Since the inner layer is fairly inexpensive and easy to replace, short lifetimes are no problem here.

In the nine greenhouses added to adobe structures, the adobe wall provides heat storage to modify the extremes of hot and cold. In the other two houses, including a mobile home, a storage wall was built with hollow masonry blocks filled with concrete and insulated on the outside. Extra thermal mass in many of the greenhouses comes from 55 gallon drums filled with water and stacked against the wall. In some of the greenhouses, the dirt floor was partially excavated and replaced with a layer of rocks to provide even more thermal mass. In others, the floors were covered with indigenous materials such as pumice, pea gravel, or flagstones obtained free.

The average cost of materials for each greenhouse was $1.86 per square foot of floor space. But Bill Yanda claims this figure would be \$2.50 at 1976 prices. This figure includes the cost of most of the flooring. All of the materials were bought at retail prices except for the Lascolite.

Although temperatures in northern New Mexico occasionally dip to −20°F, all of these uninsulated greenhouses remain warm enough to grow at least cool weather crops every month except January. They also contribute between 10 and 50 percent of household heating needs. Although a house loses heat to a greenhouse during cold nights (keeping the plants from freezing), there is an overall heat gain by the house during winter. Solar heated air at temperatures above 90°F collects at the apex of the greenhouse and flows into the house through the open doors and windows in the common wall.

Summer heat is allowed to escape through a high vent on the east wall. A low vent on the west wall aids this ventilation—particularly when the wind blows from the southwest. Most of the thermal mass is shaded from the high summer sun by a partial roof on the greenhouse. And a substantial portion of the most intense sunlight is reflected from the south facing side.

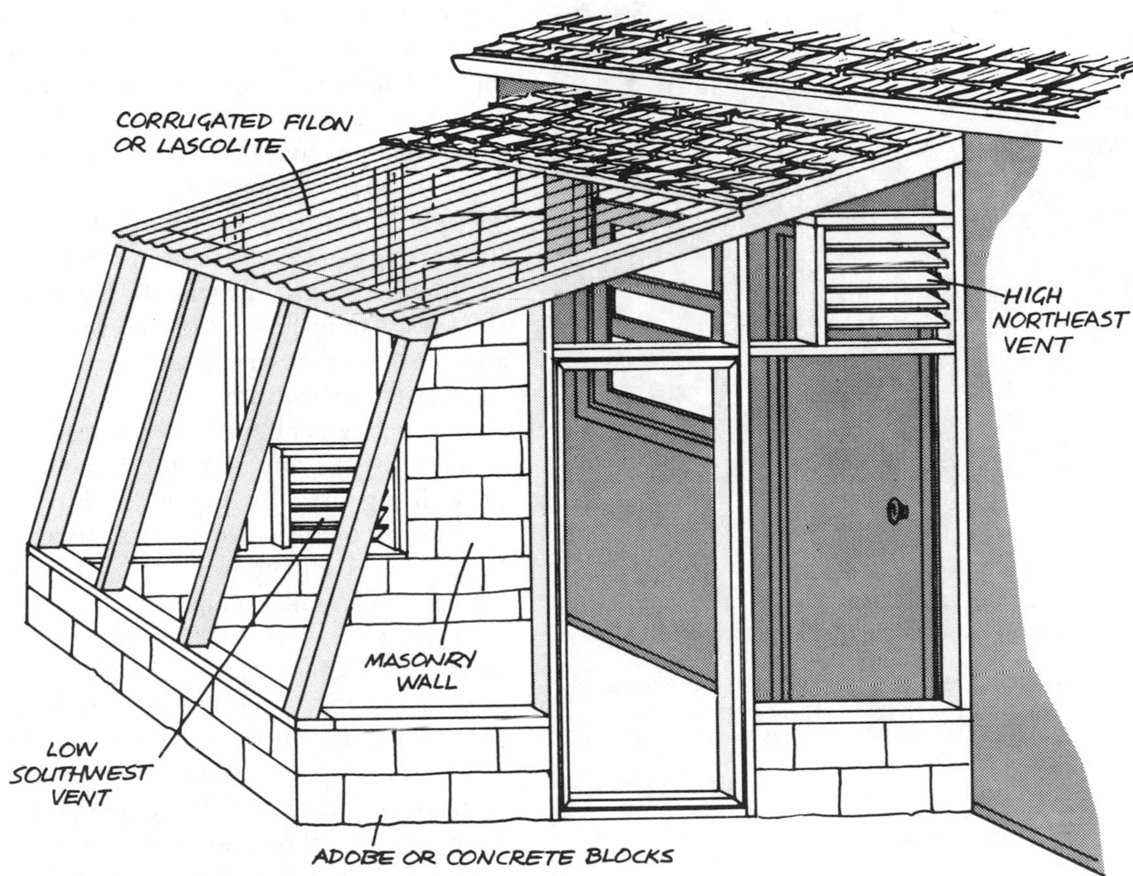

Construction details for a typical Solar Sustenance Project greenhouse.

Yanda Greenhouse

In 1974, Bill and Susan Yanda built a 160 square foot greenhouse onto their own adobe at Nambe. On its south face, the greenhouse has a 140 square foot section with two layers of plastic glazing tilted at 60° to the horizontal. Parts of the roof (which slopes at 23°) and the east and west walls are similarly glazed. The adobe wall of the Yandas' house provides the heat storage. It is supplemented by an adobe wall on the west face and an oil drum filled with water and placed on the greenhouse floor.

A door leading from the greenhouse to

Bill and Susan Yanda's greenhouse.

ECONOMICS OF AN ATTACHED SOLAR GREENHOUSE	
Southwestern Rocky Mountain Area (altitude 6000-7000 feet)	
Size of greenhouse: 160 square feet (floor area)	
Cost of greenhouse: 160 X $1.86/ft^2 = $297.60 (materials only)	
Value of plants:	
Food and seedlings	$320.61
Flowers	42.50
Value of supplemental heat:	
Greenhouse heats (2 X 140 ft^2) = 280 ft^2 of floor area or 12% of house;	
Average heating bill = $55/month for 7 month heating season;	
Therefore, value of solar heat is (0.12 X 7 X $55) =	46.20
Total income:	409.31
Expenses (seeds, fertilizer, water, etc.):	−48.00
Cost of greenhouse:	−297.60
Net profit after first year:	**$63.71**

SOURCE: Bill Yanda, "The Solar Greenhouse," in *Third Annual Life-Technics Conference*, 1974.

the kitchen is left open continuously during sunny winter days but remains closed at night. The average greenhouse temperature during the 1974-75 winter was 61°F, or 34° higher than the average outdoor temperature. When outdoor temperatures fell to −6°F, the greenhouse reached its low of 33°F. By putting rigid styrofoam insulation panels against the plastic at night, Bill reckons that the lowest temperatures inside could be kept above 50°F—well within the optimum range for winter horticulture.

Daytime greenhouse temperatures often averaged 80°F, providing the Yandas with a cozy living area. A supply of leafy green vegetables, onions, beets, peas, flowers and herbs was available well into the winter. Spring and fall temperatures remained between 53°F and 79°F—virtually identical to that of a conventional "cool" greenhouse in which nearly any plant can germinate and grow quickly. From February to May of 1975, the Yandas raised 450 seedlings for themselves and their neighbors – giving outdoor gardens a three-month head start.

During the winter of 1975-76 (admittedly a mild one), the house *never* returned heat to the greenhouse at night. The greenhouse acted only as an insulating buffer area and drew no energy from the home except through glass in the windows and door—which would have lost a lot of heat anyway. "With more storage mass, I believe the greenhouse will perform as well throughout a severe winter," Bill asserts. "With insulating blankets or panels on the plastic surfaces, it certainly will."

Economics of Greenhouses

During the winter of 1974 to 1975 (October 1 to May 1), the Yandas estimate that their greenhouse supplied 12 percent of the heat needed in their 2350 square foot adobe. Another greenhouse with 300 square feet of south-facing glazing provided about half of the heat requirements for a 1200 square foot home during that same period. The greenhouse attached to the mobile home reduced the owner's propane bill by 50 percent.

Working with these results and those from the other units in the project, Bill Yanda estimates that during the coldest months (December and January) 1 square foot of south-facing greenhouse glazing can heat about 1 square foot of floor area in the house. Performance over the entire heating season (October 1 to May 1) is better, with a ratio of 1 square foot of glazing needed to heat 2 square feet of floor area.

These ratios were used to calculate the fuel savings in Bill Yanda's chart "Economics of an Attached Solar Greenhouse," which was compiled from the performance of his own greenhouse. Quite remarkably, if a greenhouse is farmed intensively as Bill and Susan did, the value of the produce alone can pay back the cost of a similar owner-built greenhouse within *one* year. And only 2 hours of gardening time per week were required to achieve this output! Without gardening, it would take about 7 years for the greenhouse to pay for itself—even less if gas prices continue to spiral.

But what if you want your attached greenhouse to provide a larger portion of your heating needs? Using Yanda's 1:2 ratio (greenhouse glazing to house floor area) we've calculated the greenhouse floor and glazing areas needed to provide 40 and 60 percent of the heat for a 1000, 1500, or 2000 square foot house. Remember that this data is based on the winter climate of northern New Mexico, with its frequent sunny days, low temperatures, and only occasional lengthy storms. And the results will vary with the insulation and tightness of the house you are trying to heat. But they are useful as a rough estimate of the size and cost of a greenhouse suited to your needs.

To recover these costs through fuel savings alone can take many years, but certainly not the 10-20 years typical of many indirect solar heating systems. The cost can be recovered quickly from the extra food supplied by the greenhouse outside of the normal growing season. But the amount of food and other plants produced in a large greenhouse by such an intensive approach will far exceed the needs of a normal family. It would be necessary to market this food to neighbors, restaurants, plant shops or the local greengrocer. Or if you like the thermal benefits but just don't have a green thumb, you might rent out your gardening space.

There may not be enough property on the south side of your house to accomodate such a large greenhouse. In this case, your only alternatives are either to cut down on the number of Btu's needed inside your house or to squeeze more solar heat out of a small greenhouse. Most of the Solar Sustenance greenhouses were attached to drafty, poorly-insulated structures, so the solar heating percentages given here may be abnormally low. Beyond insulation, there are two basic ways to improve the performance of a greenhouse—increase its thermal mass, and get the solar heat inside the house quicker.

SAMPLE GREENHOUSE DIMENSIONS

Floor Area of House (ft^2)	Solar Heating Desired (%)	Greenhouse Glazing Area* (ft^2)	Greenhouse Floor Area (ft^2)	Greenhouse Cost† ($)
1000	40	200	229	571
1000	60	300	343	858
1500	40	300	343	858
1500	60	450	514	1285
2000	40	400	457	1142
2000	60	600	686	1715

*South-facing surface only, computed using Yanda's 1:2 rule of thumb.
†Based on Yanda's estimate of $2.50 per square foot of greenhouse floor area (1976 prices).

Greenhouse Variations

The attached greenhouses discussed so far

heat only the adjacent room or rooms. Interior partitions usually block the flow of solar heat to the rest of the house. Leaving doors open between rooms doesn't help much. Consequently, the room adjacent to the greenhouse can be toasty warm while the rest of the house remains cool on a sunny winter day. Warm air returns from this room to the greenhouse—driving its temperature and heat losses up.

There is an excellent solution to this problem in a house with a sealed cellar. Return air can be drawn into the greenhouse through a cellar window. The solar heated greenhouse air passes into the house through the upper half of an open window—just as before. But it now flows through a larger part of the house and down the back stairs to the cellar before returning to the greenhouse. Along the way, it releases more of its heat into the house. And the return air is *cooler*—keeping the greenhouse temperature and heat losses down.

Of course, you'll have to leave the door open to the room adjacent to the greenhouse if you want this added distribution. Or you might install registers in the floor and vents in the interior walls to aid the flow through the house. If gravity convection doesn't provide enough circulation for your tastes, haul out that window fan you've only been using in summer. Set it in the upper window and turn it on as needed to blow solar heated air into the room. Or you can rig up a thermostat to operate this fan only when the greenhouse gets above 70°F. This control might spare you a few chilly winter blasts on mornings when you just couldn't wait until the greenhouse got warm enough.

Yet another way to improve greenhouse performance is to feed the solar heated air into an existing forced-air distribution system. Lee Porter Butler used this approach in a greenhouse addition he designed for a 3300 square foot ranch style house in Selmer, Tennessee. Completed in August 1976, the 2230 square foot addition is designed to provide more than 90 percent of the household's heating needs. While the scope and cost of this project put it far beyond the realm of the do-it-yourself enthusiast, we include it here for its instructive value—the techniques and materials can be applied in smaller greenhouses.

The solar heat that penetrates the 1100 square feet of greenhouse glazing is absorbed and stored in the floor—not the back wall. The floor of this greenhouse is composed of a 4 inch concrete slab and 3 feet of earth above a 2¼-inch layer of polyurethane foam. A layer of brick covers the concrete and a 500 square foot swimming pool provides extra thermal mass!

About 25 percent of the solar heat gained by the greenhouse is stored directly in the floor and swimming pool. The rest is transferred to ducts beneath the concrete slab by a ¼ horsepower fan that draws solar heated air from the peak of the greenhouse. If the house needs heat, a series of dampers and ducts directs this warm air into the existing forced-air system, which distributes it through the house. When the house is sufficiently warm, a thermostat closes the dampers. If the greenhouse air is not warm enough, a second thermostat set a few degrees lower activates the existing furnace to supply the extra heat needed. In either case, cool room air sinks into return ducts which bring it back to the greenhouse.

With this kind of greenhouse, Butler suggests that 1 square foot of south-facing greenhouse glazing can heat about 4 square feet of house floor area. But he cautions that the house and greenhouse must be tightly built and well insulated, and that the actual performance may vary with climate. Nonetheless, his ratio is a significant improvement over Bill Yanda's 1:2 ratio. And Yanda himself thinks that a greenhouse with both high and low vents in the common wall can achieve a 1:3 ratio without using fans. He also recommends that people start with a small greenhouse and leave provisions to add on later.

To say that the *cost* of a greenhouse can vary tremendously is almost an understatement. The Solar Sustenance greenhouses may go for $2.50 per square foot, but a contractor-installed glass greenhouse with redwood frame and concrete slab floor can run as high as $20 per square foot. Fortunately, the owner-builder has a few alternatives. The first is to buy one of the many greenhouse kits now on the market. The pre-cut materials are shipped to you for on-site assembly. The April 1976 issue of *Popular Science* has an excellent article on greenhouse kits, including a list of suppliers and a comparison of the products they offer.

Another option is to design and build the greenhouse yourself from new or recycled materials. Obviously the choice of materials – wood or metal framing, glass or fiberglass – will be the determining factor in the final cost. Take the time to seek out old windows, wood framing, and other such materials that can be reused as part of your greenhouse. But be sure your construction and glazing details are tight. The cost of heat storage materials varies widely with the type and amount. Water-filled drums, adobe or concrete block walls, and crushed rock or insulated slab floors are the primary alternatives. Creative scrounging or do-it-yourself floor construction can keep the costs down. And when you're totalling up the costs of your prospective greenhouse, don't overlook its many potential benefits–both economic and aesthetic.

Hammond House

Jon Hammond, designer of the Winter's House described in Chapter 2, has also renovated his own wood-frame farmhouse in California's central valley. Using a combination of passive solar heating methods, he obtains most of the household heating needs. Much of the solar heating centers around a 576 square foot room added to the south side of the house. This room has a concrete slab floor and south wall that is almost

Drumwall in the Hammond House.

entirely glass. Just behind the glass sit 33 water-filled oil drums laid horizontally (as in Steve Baer's Zome) but stacked and held in place by means of small metal drum clips devised by Jon.

Instead of the reflecting panels that Baer uses for insulating the Drumwall at night, Jon installed insulating curtains. These were made from several layers of foil bonded to cloth and suspended from a pulley system between the drums and the glass. Later the pulley system failed and Jon discovered that the Drumwall performance was not seriously impaired. Temperatures inside the drums rarely exceed 85°F, and outdoor temperatures average about 50°F during winter, so heat losses are not severe.

Solar heat also enters the house through a 4 × 8 skylight installed in the south-facing roof of the original part of the house. The insulating lid of this skylight is operated with a rope and pulley system inside the house. To prevent heat loss during cold sunless weather and heat gain during the summer, the lid (with 3½ inches of fiberglass insulation) is pulled closed over the skylight. When solar heat is needed and available, the cover is raised so that its foil-coated underside reflects additional sunlight down into the rooms below.

It's difficult to estimate the percentages of solar heating provided by the Drumwall and skylight separately. But at a temperature of 85°F, a single barrel can store about 8000 Btu of useful solar heat. That's 264,000 Btu in all 33 barrels, and the concrete slab floor can store another 80,000 Btu with a 10°F temperature rise. Hammond claims that only half a cord was needed in his wood stove to provide backup heat during the (admittedly mild) winter of 1975-76. According to his rough estimates, about 80 percent of the home's heat comes from the sun.

It's also difficult to separate the cost of the solar heating device from the overall costs of renovation—adding the south room, residing, re-roofing and insulating the house, and putting insulating shutters on the windows. All told, the entire renovation cost the Hammonds $10,000. The barrels for the Drumwall came free of charge, but lately barrels have been hard to find in California. In some places they cost $25 apiece.

Some of these passive solar heating methods could be used in an existing home without building a special addition onto the house. A skylight can be installed in just about any south-facing roof section. A reflective, insulating lid is always a good idea, but be sure to brace it properly against possible strong gusts of wind. For houses with a ceiling and an attic, some shaft or clerestory would be needed to get the sunlight down into the living space. And remember that skylights must be caulked and sealed very carefully to prevent rain leakage.

If you can't cope with all that direct sun coming in your south windows, there are a number of ways to put some storage mass between you and the glass. The Kalwall storage tubes mentioned in Chapter 5 require much less space than the water-filled oil drums that Baer and Hammond have been using. These 12 or 18 inch diameter tubes of Sun-Lite stand vertically just behind the south glass. For a 25°F temperature rise, a single 12 inch tube (more suitable for home applications than the 18 inch tube) can store almost 10,000 Btu. If you're queasy about the possibilities of leaks you can wall the tubes in (and provide a drain!). Heat is then extracted by drawing cool room air through this enclosure with a fan and sending it to the rooms or through an existing forced-air heating system. More information on this "Solar Battery" and the water storage tubes can be obtained from the Solar Components Division of Kalwall Corporation (P.O. Box 237, Manchester, N.H. 03105).

Other Passive Retrofits

If you're already thinking of building an addition onto the south side of a house, you should seriously consider solar heating it. A veritable plethora of direct methods and integrated systems can be used to provide solar heat for such an addition. If you're clever, you can also supply heat for the rest of the house. Begin by putting in a concrete slab floor—or at least some massive interior walls.

Then go to work collecting sunlight. With a concrete slab and a lot of glass on the south wall you can easily heat the room itself through most of the winter. A good rule of thumb is to allow a cubic foot of exposed concrete for every square foot of south facing glass. That's 2 square feet of concrete in a 6 inch slab and 3 square feet in a 4 inch slab. Again, make sure all the concrete gets a lot of direct sun.

You're fortunate if your house already has a concrete floor slab to absorb the solar heat coming through the windows. The slab should *not* be covered with thick carpets. To get the best use of the concrete, you should leave it bare and paint it black. But if this is aesthetically offensive, asphalt tile or thin carpets aren't too bad.

If the house doesn't have much exposed thermal mass, your best bet is water-filled drums or other suitable containers placed so that sunlight falls directly on them. They should be located to catch most of the sunlight coming in between 10:00 a.m. and 2:00 p.m. when the sun is most intense. But they won't do much good in a dark corner of the room. Be sure to use watertight containers and to seal them well.

Some designers worry that Sun-Lite tubes are generally prone to leaks and might not be able to withstand earthquake activity. Where there is a moderate seismic hazard, Jon Hammond advocates the use of locally manufactured 16-gauge corrugated steel culverts as storage containers. Welded to a base plate and stood on end, they make an excellent support and containment structure for a water wall.

Whatever method of containment you use, the water wall can serve the dual purpose of heating and cooling if movable insulation is installed on the south glass. Just as in Steve

SUN-LITE FIBERGLASS TUBES

	Specifications	
Tube Diameter	12 in	18 in
Base Diameter	13 in	19 in
Height	8 ft	10 ft
Volume	6.3 ft^3	17.7 ft^3
Weight of Contained Water	392 lbs	1102 lbs
Heat Storage Capacity	392 Btu/°F	1102 Btu/°F
Price (unpackaged)*	$22	$37

*Prices (F.O.B. Manchester, NH) effective July, 1975. Packaging costs vary according to quantity shipped.

SOURCE: Kalwall Corporation

Baer's house, the water will absorb excess heat during the summer day, only to release it to the cool sky when the panels are removed at night.

Summary

Thus far, we have examined only a few of the promising alternatives for solar heating an existing home. There are many more excellent projects than we could ever hope to cover here. Already, several of your neighbors may have harnessed the sun for their own domestic use. But retrofitting has had surprisingly little press coverage. Media people seem to prefer expensive new demonstration homes as examples of solar heating. They ignore more practical solutions that are occurring right now on roofs and in backyards across the land.

The available resource material on retrofitting is also very sparse. Though several books could be written on the subject, they would probably be out of date by the time they reached the bookstores. Bill Yanda and his co-workers have made some of the best efforts to fill this void. With his wife Susan, he has written an 18-page booklet *An Attached Solar Greenhouse* that explains the principles, site requirements, construction and operation of a greenhouse built onto the side of an existing dwelling. This bi-lingual (Spanish and English) publication includes photographs and illustrations; it may be ordered from the publisher. The last word on this subject is a full-length book titled *The Food and Heat-Producing Solar Greenhouse* that Bill has written with Richard C. Fisher. There are also a number of retrofitting ideas in *Low-Cost Energy-Efficient Shelter for the Owner and Builder,* by Eugene Eccli. Both of these books should be available at your local bookstore.

FURTHER READING

Adams, Anthony. *Your Energy-Efficient House; Building and Remodeling Ideas.* Charlotte, Vermont: Garden Way Publishing, 1975.

Andrassy, Stella. "Solar Water Heaters." in *Proceedings of the United Nations Conference on New Sources of Energy, Rome, 1961.* New York: United Nations, 1964.

Brace Research Institute. *How to Build a Solar Water Heater.* Montreal: McGill University, 1965.

Close, D.J. "The Performance of Experimental Solar Hot Water Heaters with Natural Circulation." in *Solar Energy,* Vol. 2, Nos. 3-4, July and October 1958.

Department of Housing and Urban Development. *In the Bank or up the Chimney? A Dollars and Cents Guide to Energy-Saving Home Improvements.* Washington: U.S. Department of Housing and Urban Development, 1975. (available for $1.95 from Chilton Book Company, Radnor PA).

Eccli, Eugene, ed. *Low-Cost Energy-Efficient Shelter for the Owner and Builder.* Emmaus, Pennsylvania: Rodale Press, 1976.

Eccli, Eugene and Sandy. *Save Energy, Save Money.* Washington: Institute on Energy Conservation and the Poor; Office of Economic Opportunity, 1974 (available from the National Center for Community Action, 1711 Connecticut Avenue, NW, Washington DC 20009).

Florida Conservation Foundation, Inc. *Build Your Own Solar Water Heater.* Winter Park: Florida Conservation Foundation, Inc., 1976 (available for a minimum donation of $2.50 from the Environmental Information Center, 935 Orange Avenue, Winter Park, Florida 32789).

Garg, H.P. "Year Round Performance Studies on a Built-in Storage Type Solar Water Heater at Jodhpur, India." in *Solar Energy,* Vol. 17, pp. 167-172, 1975.

Morse, R.N. "Solar Water Heaters." in *World Symposium on Applied Solar Energy, Phoenix, Arizona, 1955.* Menlo Park, California: Stanford Research Institute, 1956.

Stone, Greg. "Greenhouse Kits for All-Season Gardening." in *Popular Science,* April 1976.

Whillier, Austin. "Effect of Materials and Construction Details on the Thermal Performance of Solar Water Heaters." in *Solar Energy,* Vol. 9, No. 1, January 1965.

de Winter, Francis. *How to Design and Build a Solar Swimming Pool Heater.* New York: Copper Development Association, Inc., 1975 (available free from the publisher, 405 Lexington Ave., New York, NY 10017).

Yanda, William F. "The Solar Greenhouse." *Third Annual Life-Technics Conference,* October 12-13, 1974 (available from the New Mexico Solar Energy Association, P.O. Box 2004, Santa Fe, NM).

Yanda, William F. and Richard C. Fisher. *The Food and Heat-Producing Solar Greenhouse.* Santa Fe, New Mexico: John Muir Publications, 1977.

Yanda, William F. and Susan. *An Attached Solar Greenhouse.* Santa Fe, New Mexico: The Lightning Tree, 1976 (available for $1.75 from publisher, P.O. Box 1837, Santa Fe, NM 87501).

8
Perspectives

The Social and Cultural Imperatives of an Emerging Solar Age

An Australian aboriginal energy spirit at the hub of night and day.

Solar energy, in the last analysis, has always been the basis not only of civilization, but of life; from the primeval sun-basking plankton to modern man harvesting his fields and burning coal and oil beneath his boilers, solar energy has provided the ultimate moving force. But its direct utilization at a higher level of technology is a new phenomenon, and rich with new potentialities at this stage of human affairs.

Peter van Dresser,
Landscape, Spring 1956

There are still many obstacles blocking the widespread use of the sun's energy for heating and cooling. They include financial constraints, the lack of good equipment, construction problems, building code restrictions, and legal difficulties. Most of these problems are *non-technical* in nature, having to do with solar energy's impact upon and acceptance by society as a whole. Whenever a new building method bursts upon the scene, financial institutions and the building trades are understandably conservative until that method has proved itself. But with the costs of fossil fuels rising sharply, and with increasing local and Federal incentives for solar heating and cooling, the rapid development of this vastly neglected power source is imminent.

Financial Constraints

The greatest barrier to the immediate use of solar energy for heating and cooling is the high initial cost of the apparatus. Depending upon size and complexity, a solar heating system can add $2000 to $10,000 to the price of a new house. A complete system fitted to an existing house costs slightly more. Financing such an initial expenditure is particularly difficult when building costs are already burdensome, interest rates high, and mortgage money difficult to obtain. One of the principal reasons that people decide against using solar energy is that they just cannot obtain the money to finance these costs.

Compounding these financing difficulties is the fact that an auxiliary heating system must be provided in most solar-heated homes. People prefer complete heating systems rather than systems which provide only 50 to 90 percent of their heating needs. But 100 percent solar heating systems are usually far too large to be practical, so the additional expense of a conventional heating system must also be borne.

Over the long run, the use of solar energy for heating and cooling of buildings is often competitive with the use of fossil fuels or electricity. Several analyses have compared the costs of solar energy with those of conventional methods of heating. George Löf of Colorado State University and Richard Tybout of Ohio State University have completed some of the most extensive research on this subject. In their detailed studies, the initial costs of an indirect solar heating system were amortized over a twenty year period at six percent interest, and fuel costs

COSTS OF SPACE HEATING IN DOLLARS PER MILLION BTU*					
City	Solar Heating§		Conventional Heating Costs		
	Collector ($2/ft²)	Collector ($4/ft²)	Electric Heat†	Gas Heat‡	Oil Heat‡
Santa Maria	1.10	1.59	4.28	1.52	1.91
Albuquerque	1.60	2.32	4.63	0.95	2.44
Phoenix	2.05	3.09	5.07	0.85	1.89
Omaha	2.45	2.98	3.25	1.12	1.56
Boston	2.50	3.02	5.25	1.85	2.08
Charleston	2.55	3.56	4.22	1.03	1.83
Seattle	2.60	3.82	2.29	1.96	2.36
Miami	4.05	4.64	4.87	3.01	2.04

*Values in table are the costs (1970 prices) per million Btu of useful heat delivered to the house.

§Costs based on optimized solar heating system in a house that loses 25,000 Btu per degree day. Capital cost of system amortized over 20 years at 6 percent interest.

†Assuming 30,000 kwh used per year.

‡Values reflect cost of fuel only.

SOURCE: Tybout and Löf, "Solar House Heating," *Natural Resources Journal,* April 1970.

were assumed to remain constant. Their results are promising, as shown in the table, but may be too optimistic. In the seven cities studied, solar heating costs were always lower than electric heating, and often lower than gas or oil heating costs. However, the solar heating system costs used in the study seem to be unrealistically low. The lowest realistic estimate of *installation* costs, which include the cost of the collectors and the controls, pumps, piping, heat storage and labor, is about 10 dollars per square foot of collector, based on a system with 500 square feet of collector surface. The simplest owner-built systems cost half this amount, while more technologically complex systems cost twice this amount. But the fact that fossil fuel prices will continue to rise makes solar energy a sound investment.

All too frequently, however, financial institutions disregard the ever increasing operating costs of a conventionally-heated home and focus upon the large initial costs of a solar heating system. Home-financing plans generally discourage such an investment even though lowered heating bills over the lifetime of the system make it a sound buy. Some lending institutions are easing the situation by the use of *life-cycle* costing methods, which compare the higher initial costs to lower operating and maintenance costs. These methods emphasize the lowest monthly home-owning costs—mortgage payments *plus* utilities. Lenders should allow higher monthly mortgage payments if monthly energy costs are lower. Most progressive lending institutions are doing just that.

Part of the problem, of course, is that solar energy is not yet an established alternative. Banks are reluctant to fund an expensive addition that they consider unproved and prone to failure. Frequently they require that an extensive conventional heating system also be installed before granting any mortgage at all. As more solar heating systems come into general use, however, and bear out the claims of lowered heating costs, loans for these systems should become more readily available.

With the growing shortages of fuel, however, arguments against a big cash outlay are losing their clout while the *availability* of fuel becomes the real issue. Some people who find themselves without fuel are deciding that this shortage is reason enough for using solar energy—that the actual costs of the system are less important. The attitude toward alternative sources of energy is changing dramatically as this realization strikes home. More and more the old American battle cry of *self-reliance* is being heard throughout the land. With its Project Independence, the nation is striving toward freedom from foreign energy sources. In the same way, some people are trying to free themselves from a similar energy bind through more direct control of their energy supplies.

Lack of Equipment

A major difficulty in the design and manufacture of indirect solar heating and cooling systems is the necessary combination of low cost, good performance, and durability. The designer should have a thorough understanding of the principles of solar energy and of the pitfalls discovered in the past. Even then, many things can go wrong with such complex systems, and many architectural and engineering firms hesitate to invest extra time and money in their design. Firms that do tackle such projects often delay their selection of components, since they hope that new technologies and increasing mass production

Life-Cycle Costing

Life-cycle costing is an estimating method that includes the future costs of energy consumption, maintenance, and repair in the economic comparison of several alternatives. These future costs can make an initially cheaper system costlier over the life cycle of the system. Life-cycle costing methods make such costs visible at the outset—and they include the economic impact of interest rates and inflation. They are ideally suited for comparing the costs of solar heating and cooling with those of conventional methods.

In order to obtain consistent cost comparisons among several alternatives, all the costs of each system (over a selected "life-cycle") are reduced to total costs over a unit of time—usually the first year. The purchase price, installation costs, and operating and maintenance expenses are all included.

Allowances are also made for prevailing interest rates and inflation. Future savings such as lower fuel costs are discounted to "present-value" dollars—the amount of money that, if invested today, would grow to the value of the savings in the intervening years. And if the annual operating and maintenance expenses can be predicted to grow at some steady inflation rate, the present-value total of those expenditures over the life cycle of the system, P_e, can be calculated using the following equation:

$$P_e = A\frac{R(R^n - 1)}{R - 1}$$

where

$$R = \frac{1 + g}{1 + i}$$

and i and g are the fractional rates of interest and inflation. In this equation, the current annual expense, A, is multiplied by a factor which accounts for the number of years in the life cycle, n, and the rate at which the annual expense A is expected to increase. An example will help to illustrate its use.

EXAMPLE: The cost of oil is 40 cents a gallon at present and it will increase at about 8% per year. What is the present value of the expenditures for 1 gallon of oil each year for the next 30 years?

SOLUTION: Assuming an interest rate of 6%, the ratio R equals 1.08 ÷ 1.06, or 1.0189. Applying the equation, we find the present value of 30 gallons of oil expended over the next 30 years is \$16.25:

$$P_e = \$0.40 \times \frac{1.0189 \times (1.0189^{30} - 1)}{1.0189 - 1}$$

$$= \$0.40 \times 40.63 = \$16.25$$

To get the life-cycle costs of a system, the purchase and installation prices are added to the present value of the total operating and maintenance costs.

For example, an owner-builder might want to compare the life-cycle costs of 2 × 4 insulated stud walls to those of 2 × 6 insulated stud walls. He estimates that the 2 × 4 walls cost \$7140 to build but lose 48.4 million Btu per year; the 2 × 6 walls cost \$7860 and lose 34.8 million Btu. Assuming a gallon of oil produces 100,000 Btu of useful heat, the house will require 484 or 348 gallons of oil per year, depending upon the wall construction. Over a 30-year life cycle, the operating costs will be (\$16.25 × 484) = \$7865 for the 2 × 4 stud walls and (\$16.25 × 348) = \$5655 for the 2 × 6 walls, in present value dollars. Maintenance expenses are equal for the two alternatives and hence are ignored. Adding operating and installation costs, the owner-builder finds that the 30-year life-cycle costs of the two alternatives are:

2 × 4 walls: \$7140 + \$7865 = \$15,005

2 × 6 walls: \$7860 + \$5655 = \$13,515

Life cycle costing indicates that the "more expensive" wall is almost \$1500 cheaper when future costs are figured in the initial price. Similar costing practices can be used to compare the costs of solar heating and cooling with conventional approaches.

will bring the costs of solar apparatus down.

Several hundred manufacturers are either seriously involved in solar energy or hesitantly putting money into development programs, in the belief that they can solve these cost problems. They are understandably reluctant, however, to tool up assembly lines before an adequate demand develops. On the other hand, a bull market for solar components will not develop until high quality components become available at reasonable prices.

One of the most appealing aspects of solar heating and cooling is that custom design and on-site construction is usually a cheaper alternative than buying manufactured collectors. Many of the present solar heated homes, such as the Mathew House and the Thomason houses, were built in just this way. Although labor costs for on-site construction are generally quite high, much of the work can be done by the average home handyman. If the designs are kept simple to reduce possible failures, such personal projects can be both less expensive and more rewarding. For owner-built systems, flat-plate collectors are not the only alternative. Direct methods or integrated solar heating and cooling systems are often cheaper, more efficient, and more reliable. Such systems are even more appropriate for custom design and on-site construction.

Construction Problems

There are other potential problems centered in the housing construction industry and in the laws which regulate it. The industry has a record of slow adaptation to change, particularly when the change means higher construction costs. There are thousands of builders and the building industry is consequently very fragmented, with 90 percent of all construction work done by companies that produce fewer than 100 units per year. The profit margin in this industry is small, and innovation is a risk which few builders will take.

But the fragmented nature of the construction industry is essential to the localized industries that are springing up around solar energy. Even if a few manufacturers achieve low-cost solar collectors, interstate transportation costs will remain relatively high, adding one or two dollars per square foot to collector costs. Glass, in particular, is best fabricated locally and installed at the construction site. On-site construction and local fabrication of components will always be a good alternative.

Fortunately, some contractors and developers are installing solar equipment in order to evoke interest in their recent housing developments. Developers are often having great difficulty obtaining oil and gas for new houses, and are turning to solar energy for heating and cooling purposes. In actual fact, most builders and contractors are surprisingly enthusiastic about solar homes.

Building codes often dampen the enthusiasm of a builder or homeowner wanting to include solar energy in the design of a house. Fire codes, for example, are likely to hinder the spread of solar heating and cooling, particularly in indirect systems where high temperatures are the rule. Some fire codes will not allow the use of large quantities of paraffin (an excellent heat storage material) inside of buildings. Unbreakable plastics and fiberglass are alternatives to glass for use in cover plates, but such products are frowned upon because they are less fire-resistant than glass. Insulation on the back sides of solar collectors are usually close to the absorber plate, which can become very hot (up to 350° F) in some applications.

Still other building codes may limit the use of solar energy. Zoning limitations may restrict the height of your house so that it cannot receive enough of the sun's rays through the surrounding trees. Health codes may also apply when using mixtures of ethylene glycol and water as a heat transfer medium. The glycol cannot be allowed to pass near the domestic water supply. Plumbing and electrical codes may add hundreds of dollars to the cost of a system just by restricting your choice of materials. Generally, building codes restrict the use of indirect systems more than direct methods or integrated systems.

Legal Difficulties

A user of solar energy must be certain that the sun's energy will not be taken away from him by the actions of other people. Architectural agreements allowing for unobstructed exposure to the sun's rays will be necessary as more homes rely on solar energy for their heating and cooling. Laws are needed to guarantee that neighboring construction and vegetation will not reduce the solar energy striking a building. Until such laws are commonplace, however, solar houses may be subject to rude shadows from new buildings or trees.

Another potential problem is the reflection of sunlight off the large expanses of glass-covered collectors into the eyes of pedestrians, motorists, and neighbors. Reflections from tilted collectors may occur when the sun is relatively high in the sky, and the discomfort experienced would be similar to that of facing directly into the sun.

Vertical wall collectors, on the other hand, pose no more reflection problems than large all-glass buildings. Glare from such buildings has rarely been dangerous or discomforting. The most severe reflections off vertical glass surfaces occur when the sun is low in the sky, and the effect is not as bad as driving into the sunset.

Non reflective cover plates, which even increase the collector efficiency, are now being developed. In the near future, the best collectors will reflect very little light as compared to today's flat-plate varieties. After all, if less sunlight is reflected, more is available to be absorbed by the collector!

The possibility of vandalism of glass surfaces used for solar collection has been of some concern to owners and designers of solar buildings. Vertical wall collectors and windows can be easy targets for the projectiles

Reflections from a rooftop collector can be a hazard to passers-by.

of neighborhood pranksters. Tilted collectors on rooftops are less vulnerable, and they can be arranged in a sawtooth fashion to minimize the amount of surface area exposed to public abuse.

Alternatives to ordinary glass, such as plastics and tempered glass, can be used in demanding situations. Lexan, an expensive tough plastic also used for the canopies of fighter jets, was used for cover plates in a project at the Grover Cleveland Junior High School in South Boston, where schools have recently been targets of scorn. Over the 35-year history of solar building in the U.S., however, vandalism has not been much of a problem.

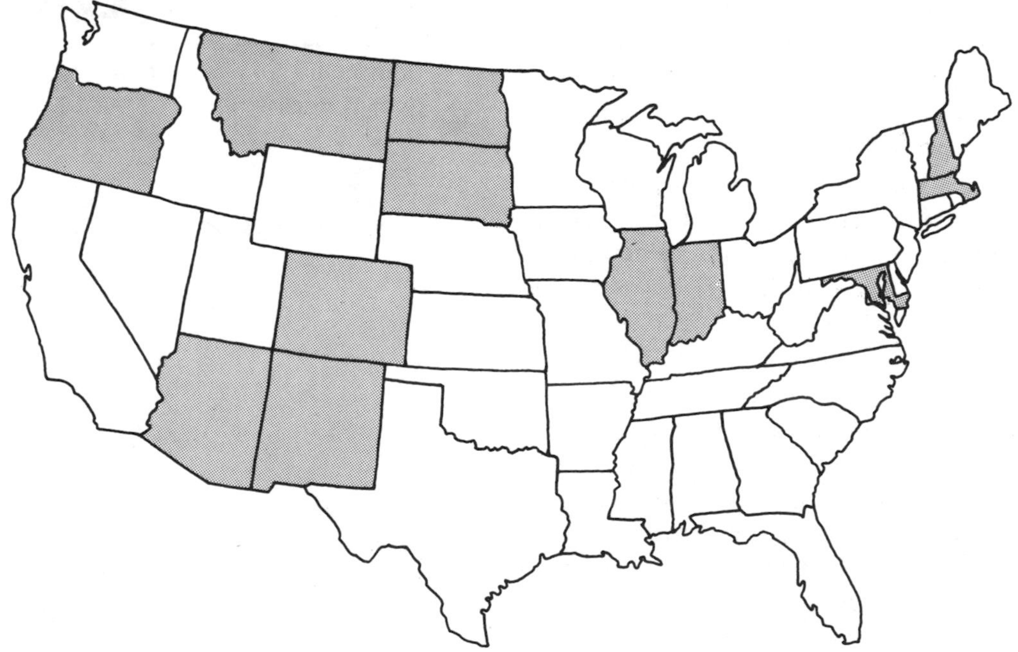

States offering property or income tax incentives for solar heating and cooling apparatus (1975).

Government Incentives

Local governments are slowly beginning to acknowledge the benefits of solar energy in their tax structures. The extra work stimulated by solar energy, which requires local labor to build and install the components, can be a boon to local economies. Annual cash outflows for gas and oil from energy-poor areas like New England amount to billions of dollars that can be saved through the use of solar energy. And reductions in the level of pollution result from lowered consumption of fossil fuels for heating and cooling. However, local tax structures tend to discourage solar heating and cooling, because the installation costs mean higher property taxes for the homeowner. Lowering these taxes to encourage the use of solar energy is a desirable goal, but assessment of property is often a local matter and change comes slowly.

Indiana took the lead in modifying this situation by offering a property tax incentive for using solar components in both new and existing buildings. Assessed property values are determined by subtracting the value of the solar heating system (up to a limit of $2000) from the total valuation. By 1975, eleven other states had followed Indiana's lead, while Arizona and New Mexico were offering income tax incentives.

At the Federal level, legislation authorizing

income tax incentives for insulation or solar heating and cooling equipment is always being introduced. The Solar Tax Incentive Act, which passed the House and is before the Senate Finance Committee at this writing, would provide a tax credit of up to $400 on the first $1000 spent for solar heating or cooling apparatus. Smaller credits would be allowed for more expensive systems. Such credits or deductions would be a direct subsidy similar to oil depletion allowances granted the oil industry.

Low interest loans for solar energy systems are often proposed for both homeowners and small contractors. The Department of Housing and Urban Development (HUD) is preparing to insure home improvement loans of up to $5000 toward the installation of solar heating and cooling systems in new or existing buildings. They have already developed solar energy design criteria for FHA financed homes. In 1976, HUD awarded grants totalling more than a million dollars for the installation of solar heating and cooling equipment in 143 new and existing dwellings. Although these grants are not available to the average homeowner, they are a strong indication of the Federal interest in promoting solar energy.

The importance of government incentive and development programs cannot be underestimated. Other energy suppliers are already receiving massive Federal support. Solar energy must overcome many other obstacles, apart from its competition with established, government-supported energy suppliers such as the nuclear industry. The importance of solar energy to global and national welfare is more than adequate justification for the American people to demand better-than-equal treatment for its promotion.

But more than technological innovation and government incentive will be needed to make solar heating and cooling a universal reality. In a larger sense, the nation's energy future rests in the personal choices of its people. The high-energy practices learned in an age of plentiful fossil fuels cannot be supported by a solar economy. Only nuclear power, with all its attendant ills, can possibly satisfy these demands. We can enjoy clean, inexhaustible solar energy much sooner if we accept the small changes in our lives needed to use the simple methods advocated in this book. In choosing these approaches, we take a cue from centuries of a more intimate relationship between humanity and nature—prior to the advent of cheap fossil energy. Instead of exaggerated wants, people had simple needs that could be supplied by the energy and materials around them. They interacted with their climates to take full advantage of the natural heating and cooling available. These attitudes of conservation must once again become our standards. A new solar age will occur when we can forego our high-energy ways of life and return to our place in the sun.

Appendix

The sun's position in the sky is described by two angular measurements, the solar altitude θ and the solar azimuth ϕ. As explained in Chapter 3, the solar altitude is the angle of the sun above the horizon. The azimuth is its angular deviation from true south.

The exact calculation of θ and ϕ depends upon three variables: the latitude L, the declination δ, and the hour angle H. Latitude is the angular distance of the observer north or south of the equator—it can be read from any good map. Declination is a measure of how far north or south of the equator the sun has moved. At the summer solstice $\delta = +23\frac{1}{2}°$, while at the winter solstice $\delta = -23\frac{1}{2}°$ in the Northern Hemisphere; at both equinoxes, $\delta = 0°$. This quantity varies from month to month and can be read directly from the graph below.

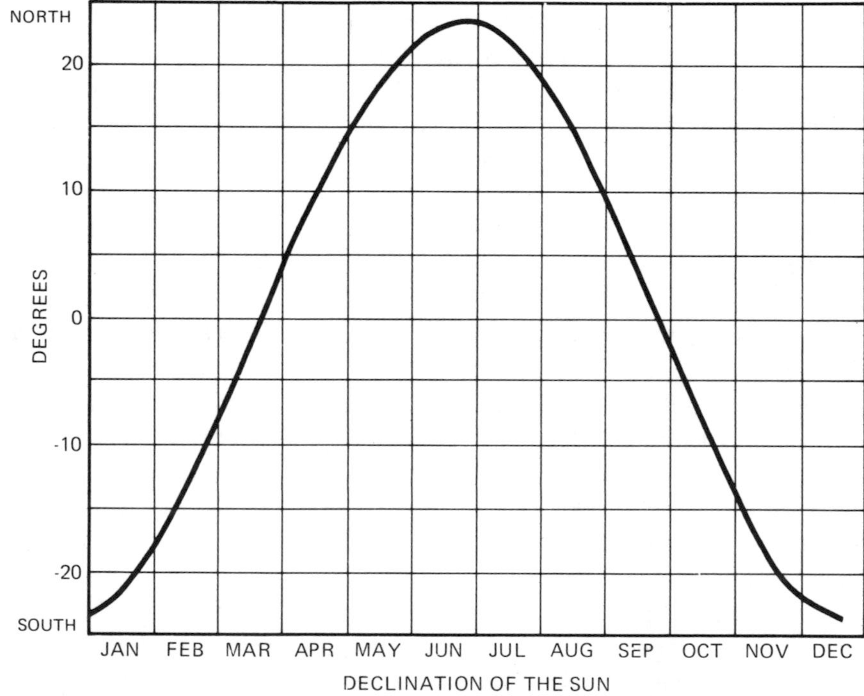

DECLINATION OF THE SUN

The hour angle H depends on Local Solar Time, which is the time that would be read from a sundial oriented south. Solar Time is measured from solar noon—the moment when the sun is highest in the sky. At different times of the year, the lengths of solar days (measured from solar noon to solar noon) are slightly different from days measured by a clock running at a uniform rate. Local Solar Time is calculated taking this difference into account. There is also a correction if the observer is not on the standard time meridian for his time zone.

To correct local standard time (read from an accurate clock) to Local Solar Time, three steps are necessary:

1) If daylight savings time is in effect, subtract one hour.
2) Determine the longitude of the locality and the longitude of the standard time meridian (75° for Eastern ST, 90° for Central ST, 105° for Mountain ST, 120° for Pacific ST, 135° for Yukon ST, 150° for Alaska-Hawaii ST). Multiply the difference in longitudes by 4 minutes/degree. If the locality is east of the standard meridian, add the correction minutes; if it's west, subtract them.
3) Add the equation of time (from the next graph) for the date in question. The result is the Local Solar Time.

Once you know the Local Solar Time, you can obtain the hour angle H from:

$$H = 0.25 \times (\text{number of minutes from solar noon}).$$

From the latitude L, declination δ and hour angle H, the solar altitude θ and azimuth ϕ follow after a little trigonometry:

$$\sin\theta = \cos L \cos\delta \cos H + \sin L \sin\delta \,;$$
$$\sin\phi = \cos\delta \sin H/\cos\theta \,.$$

Example: Determine the altitude and azimuth of the sun in Abilene, Texas on December 1, when it is 1:30 p.m. (CST). First we need to calculate the Local Solar Time. It is not daylight savings time, so no correction for that is needed. Looking at a map we see that Abilene is on the 100°W meridian, or 10 degrees west of the standard meridian—90°W. We subtract the 4 X 10 = 40 minutes from local time; 1:30 – 0:40 = 12:50 p.m. From the equation of time for December 1, we must *add* about 11 minutes. 12:50 + 0:11 = 1:01 Local Solar Time, or 61 minutes past solar noon. Consequently, the hour angle is H = 0.25 x 61 or about 15°. The latitude of Abilene is read from the same map: L = 32°, and the declination for December 1 is $\delta = -22°$.

We have come thus far with maps, graphs, and the back of an envelope, but now we need a pocket calculator or a table of trigonometric functions:

$$\begin{aligned}\sin\theta &= \cos(32°)\cos(-22°)\cos(15°) + \sin(32°)\sin(-22°)\\ &= 0.85 \times 0.93 \times 0.97 + 0.53 \times (-0.37)\\ &= 0.76 - 0.20 = 0.56\end{aligned}$$

Then $\theta = \arcsin(0.56) = 34.12°$ above the horizon. Similarly:

$$\begin{aligned}\sin\phi &= \cos(-22°)\sin(15°)/\cos(34.12°)\\ &= (0.93 \times 0.26)/0.83 = 0.29 \,.\end{aligned}$$

Then $\phi = \arcsin(0.29) = 16.85°$ west of true south. At 1:30 p.m. on December 1 in Abilene Texas, the solar altitude is 34.12° and the azimuth is 16.85° west.

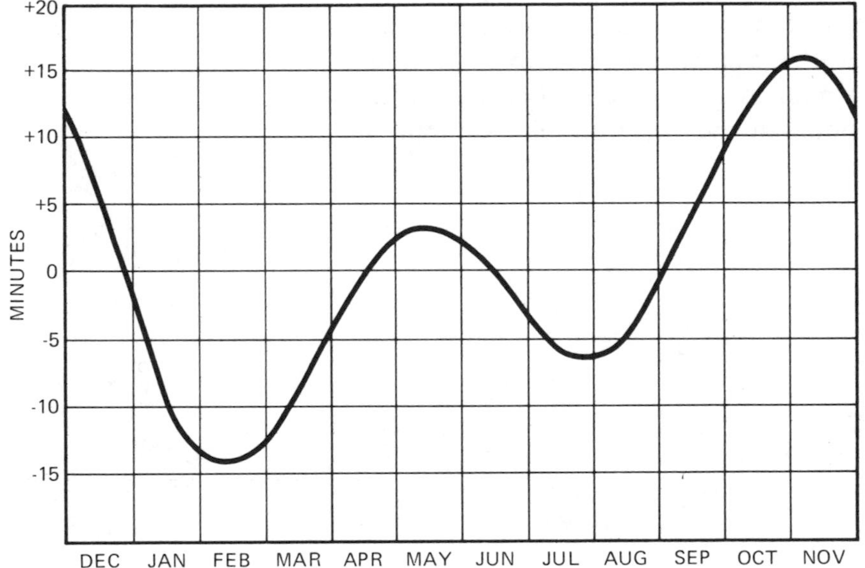

Sun Path Diagrams

In applications where strict accuracy is superfluous, solar angles can be quickly determined with sun path diagrams. In these diagrams, the sun's path across the sky vault is represented by a curve projected onto a horizontal plane (see diagram below). The horizon appears as a circle with the observation point at its center. Equally-spaced concentric circles represent the altitude angles, θ, at 10° intervals, and equally-spaced radial lines represent the azimuth angles, ϕ, at the same intervals.

The elliptical curves running horizontally are the projection of the sun's path on the 21st day of each month; they are designated by two Roman numerals for the two months when the sun follows approximately this same path. A grid of vertical curves indicate the hours of the day in Arabic numerals.

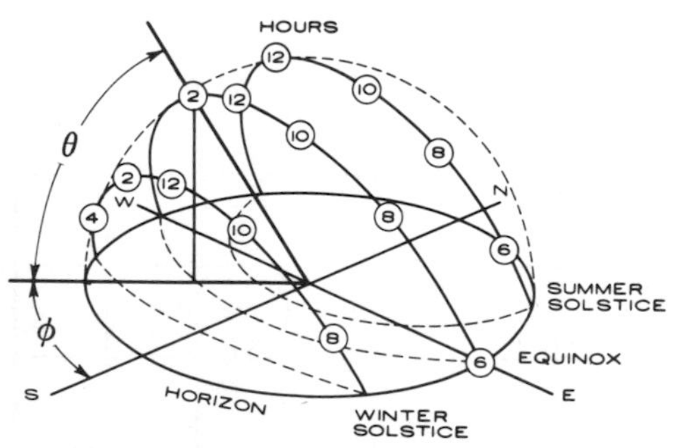

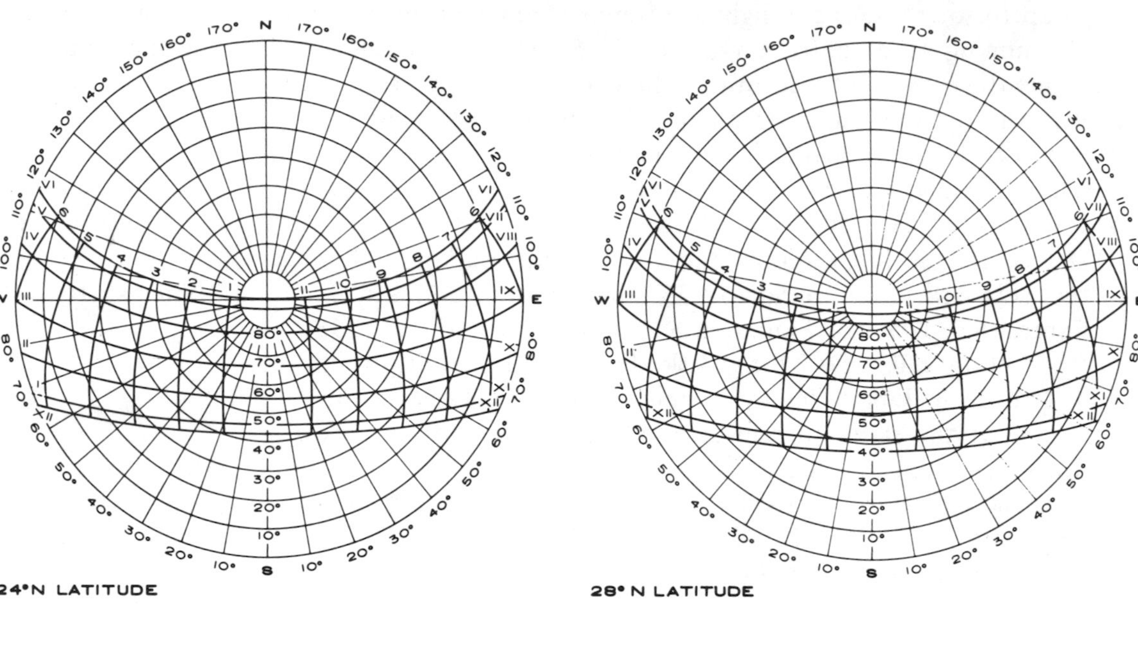

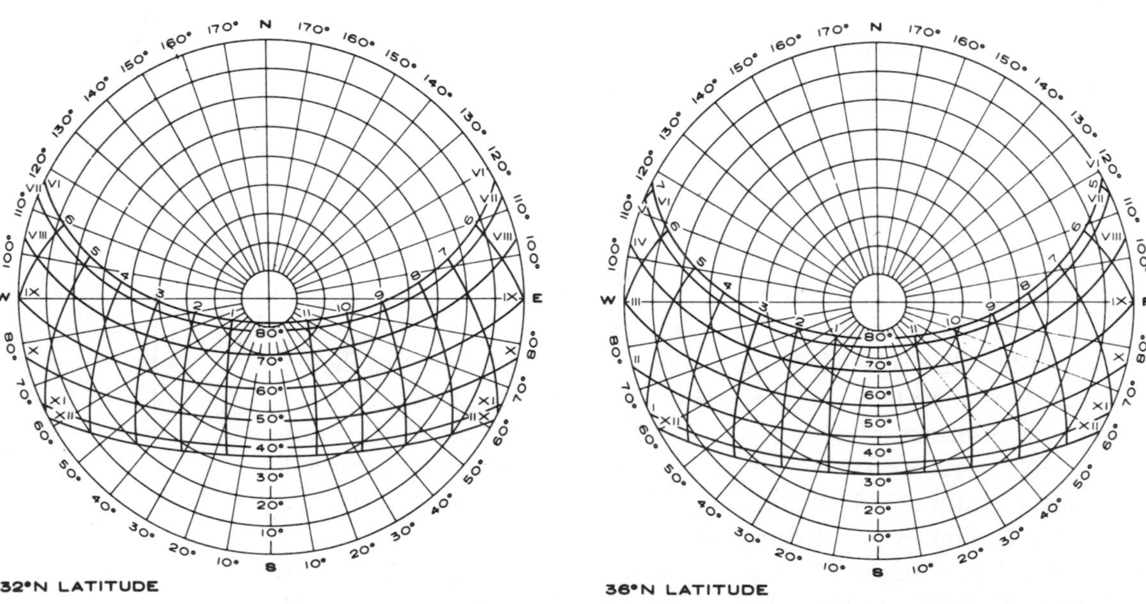

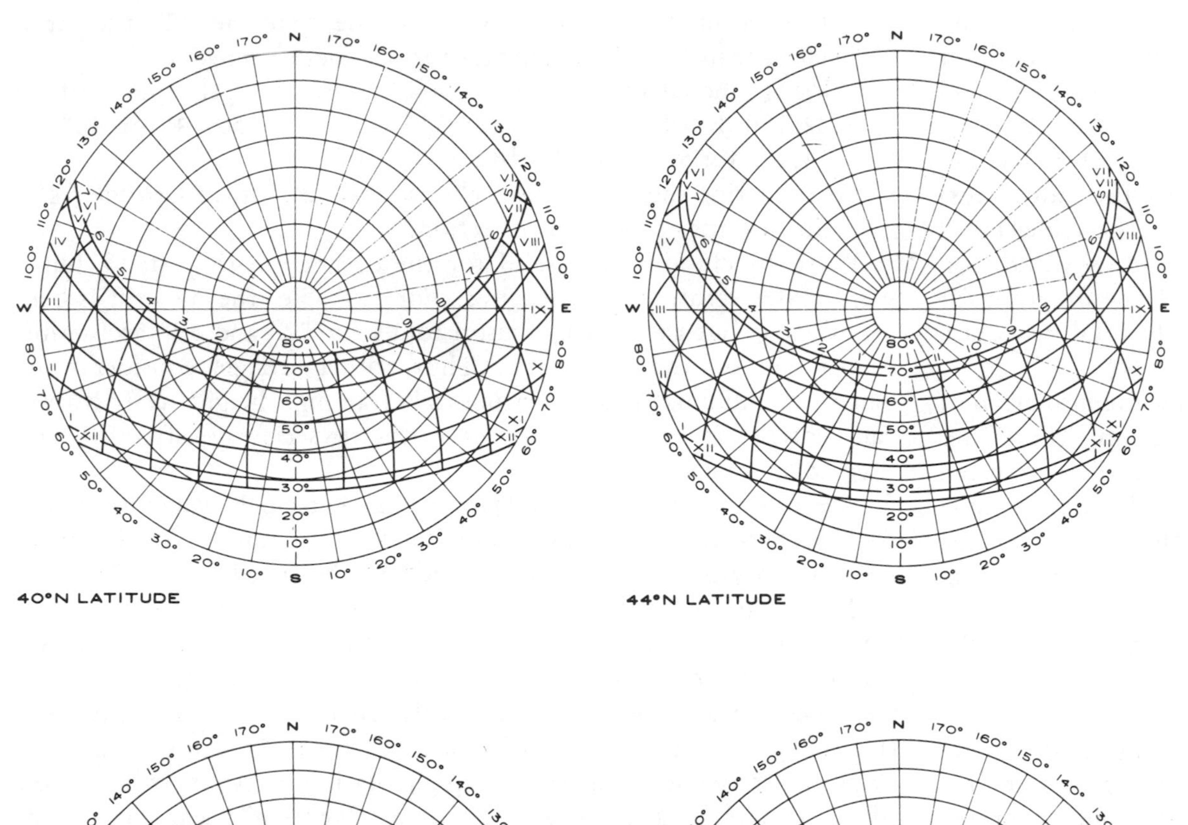

40°N LATITUDE

44°N LATITUDE

The shading mask protractor provided below can be used to construct shading masks characteristic of various shading devices. The bottom half of the protractor is used for constructing the segmental shading masks characteristic of horizontal devices (such as overhangs), as explained on pages 90-91. The upper half, turned around so that the 0° arrow points down (south) is used to construct the radial shading masks characteristic of vertical devices. These masks can be superimposed on the appropriate sun path diagram to determine the times when a surface will be shaded by these shading devices.

SOURCE: Ramsey, C. G. and H. R. Sleeper, *Architectural Graphic Standards.* New York: John Wiley & Sons, 1972.

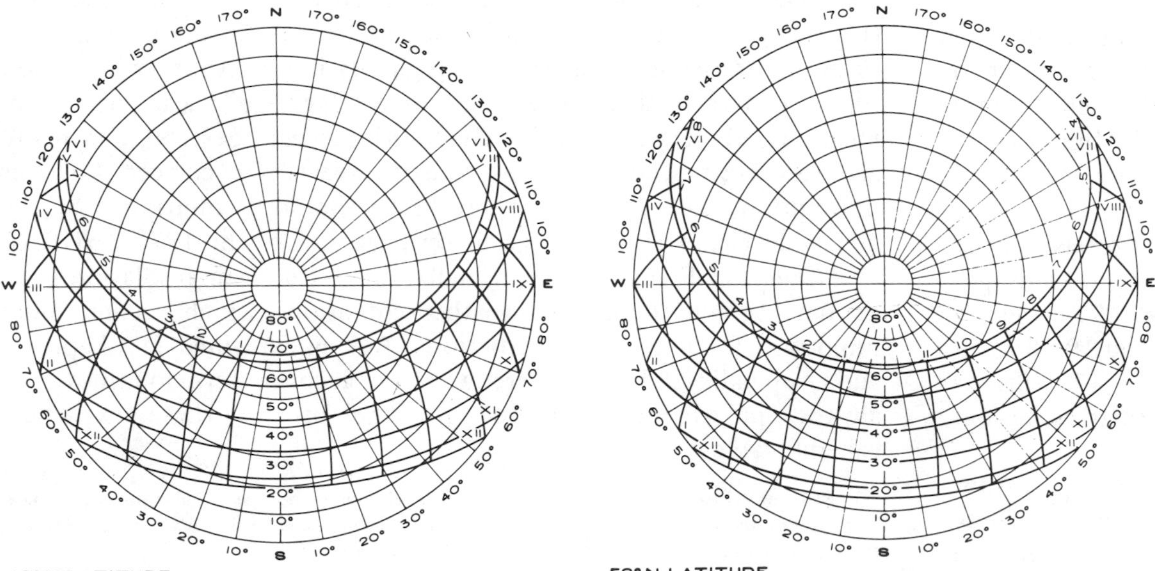

48°N LATITUDE

52°N LATITUDE

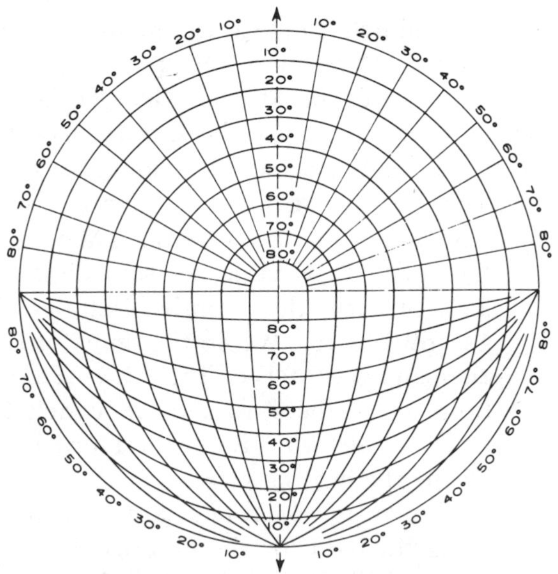

SHADING MASK PROTRACTOR

ASHRAE has developed tables that give the clear day insolation on tilted and south facing surfaces–such as those commonly used for solar collectors. For North latitudes L equal to 24°, 32°, 40°, 48°, and 56°, insolation values are given for south facing surfaces with tilt angles equal to L-10°, L, L+10°, L+20°, and 90° (vertical). Values are also given for the direct normal (perpendicular to the sun's rays) radiation and the insolation on a horizontal surface. The values listed in these tables are the sum of the direct solar and diffuse sky radiation hitting each surface on an average cloudless day. Data are given for the 21st day of each month; both hourly and daily total insolation are provided.

A brief examination of the 24°N table reveals that the insolation of south facing surfaces is symmetrical about solar noon. The values given for 8 a.m. are the same as those for 4 p.m., and they are listed concurrently. Moving from left to right on any fixed time line, you encounter values of: the solar altitude and azimuth (in degrees); the direct normal radiation and the insolation on a horizontal surface (in Btu/hr/ft²); and the insolation of the five south facing surfaces discussed above (in Btu/hr/ft²). Below these hourly data are values of the daily total insolation for each of these surfaces (in Btu/ft²). An example will help to illustrate the use of these tables.

Example: Determine the optimum tilt angle for a flat plate collector located in Atlanta, Georgia (32°N latitude). Select the tilt angle so as to maximize the surface insolation for the following three periods: a) heating season, b) cooling season, and c) the full year.

1) The heating season in Atlanta lasts from October through April; the cooling season from May to September.
2) Using the 32°N table, we sum the surface daily totals for the 22° tilt for the months October through April, and get 14,469 Btu/ft². We do the same for the 32°, 42°, 52°, and 90° tilts, and get totals of 15,142; 15,382; 15,172; and 10,588.
3) Comparing these totals, we conclude that the 42° tilt, or latitude plus 10°, is the best orientation for solar collection during the heating season.
4) A similar set of totals is generated for the cooling season, using the data for the months May through September. These are 11,987 Btu/ft² for 22°; 11,372 for 32°; 10,492 for 42°; 9,320 for 52°; and 3,260 for 90° tilt.
5) Comparing these totals, we conclude that the 22° tilt, or latitude minus 10°, is the best for summer cooling.
6) Using the data for the whole year, we get totals of: 24,456 Btu/ft² for 22°; 26,514 for 32°; 25,874 for 42°; 24,492 for 52°; and 13,848 for 90° tilt.
7) Comparing these totals, we choose the 32° tilt, or latitude, as the best for year-round collection.

These conclusions are useful for the designer as they stand, but a little closer scrutiny is instructive. For example, the 42° tilt is best for heating, but the heating season totals for 32° and 52° are within 2 percent of the 42° total. Thus, other design considerations (such as building layout, structural framing, height restrictions) can enter the decision process without seriously affecting the final collector efficiency.

The Clear Day Insolation Data are an extremely valuable design tool, but their limitations should be kept in mind. For instance, there is no ground reflection included in the listed values. This can lead one to underestimate the clear day insolation on a vertical surface. In the example above, the heating season total for a 90° surface is about 30 percent below the 42° maximum. In reality, the insolation on a vertical surface is only 10 to 20 percent lower than this maximum during the heating season because of the contribution of ground reflection–especially at higher latitudes. Another limitation of these data is their assumption of an "average" clear day. Many locations are clearer than this (high altitudes and deserts), and many are less clear (industrial and dusty areas). To correct for this assumption, the numbers in these tables should be multiplied by the area-wide clearness factors given on page 394 of the 1972 ASHRAE *Handbook of Fundamentals.* Finally, the Clear Day Insolation Data do not account for cloudy weather conditions, which become quite important for long term predictions. But as explained in Chapter 3 and Appendix 2.4, there are ways to account for cloudiness. As always, judgement is needed for the use of these extremely valuable tables.

SOURCE: Morrison, C. A. and E. A. Farber. "Development and Use of Solar Insolation Data in Northern Latitudes for South Facing Surfaces" in Symposium on *Solar Energy Applications.* New York: ASHRAE, 1974. Reprinted by permission.

SOLAR POSITION AND INSOLATION, 24°N LATITUDE

DATE	SOLAR TIME		SOLAR POSITION		BTUH/SQ. FT. TOTAL INSOLATION ON SURFACES						
	AM	PM	ALT	AZM	NORMAL	HORIZ.	SOUTH FACING SURFACE ANGLE WITH HORIZ. 14	24	34	54	90
JAN 21	7	5	4.8	65.6	71	10	17	21	25	28	31
	8	4	16.9	58.3	239	83	110	126	137	145	127
	9	3	27.9	48.8	288	151	188	207	221	228	176
	10	2	37.2	36.1	308	204	246	268	282	287	207
	11	1	43.6	19.6	317	237	283	306	319	324	226
	12		46.0	0.0	320	249	296	319	332	336	232
	SURFACE DAILY TOTALS				2766	1622	1984	2174	2300	2360	1766
FEB 21	7	5	9.3	74.6	158	35	44	49	53	56	46
	8	4	22.3	67.2	263	116	135	145	150	151	102
	9	3	34.4	57.6	298	187	213	225	230	228	141
	10	2	45.1	44.2	314	241	273	286	291	287	168
	11	1	53.0	25.0	321	276	310	324	328	323	185
	12		56.0	0.0	324	288	323	337	341	335	191
	SURFACE DAILY TOTALS				3036	1998	2276	2396	2446	2424	1476
MAR 21	7	5	13.7	83.8	194	60	63	64	62	59	27
	8	4	27.2	76.8	267	141	150	152	149	142	64
	9	3	40.2	67.9	295	212	226	229	225	214	95
	10	2	52.3	54.8	309	266	285	288	283	270	120
	11	1	61.9	33.4	315	300	322	326	320	305	135
	12		66.0	0.0	317	312	334	339	333	317	140
	SURFACE DAILY TOTALS				3078	2270	2428	2456	2412	2298	1022
APR 21	6	6	4.7	100.6	40	7	5	4	4	3	2
	7	5	18.3	94.9	203	83	77	70	62	51	10
	8	4	32.0	89.0	256	160	157	149	137	122	16
	9	3	45.6	81.9	280	227	227	220	206	186	41
	10	2	59.0	71.8	292	278	282	275	259	237	61
	11	1	71.1	51.6	298	310	316	309	293	269	74
	12		77.6	0.0	299	321	328	321	305	280	79
	SURFACE DAILY TOTALS				3036	2454	2458	2374	2228	2016	488
MAY 21	6	6	8.0	108.4	86	22	15	10	9	9	5
	7	5	21.2	103.2	203	98	85	73	59	44	12
	8	4	34.6	98.5	248	171	159	145	127	106	15
	9	3	48.3	93.6	269	233	224	210	190	165	16
	10	2	62.0	87.7	280	281	275	261	239	211	22
	11	1	75.5	76.9	286	311	307	293	270	240	34
	12		86.0	0.0	288	322	317	304	281	250	37
	SURFACE DAILY TOTALS				3032	2556	2447	2286	2072	1800	246
JUN 21	6	6	9.3	111.6	97	29	20	12	12	11	7
	7	5	22.3	106.8	201	103	87	73	58	41	13
	8	4	35.5	102.6	242	173	158	142	122	99	16
	9	3	49.0	98.7	263	234	221	204	182	155	18
	10	2	62.6	95.0	274	280	269	253	229	199	18
	11	1	76.3	90.8	279	309	300	283	259	227	19
	12		89.4	0.0	281	319	310	294	269	236	22
	SURFACE DAILY TOTALS				2994	2574	2422	2230	1992	1700	204
JUL 21	6	6	8.2	109.0	81	23	16	11	10	9	6
	7	5	21.4	103.8	195	98	85	73	59	44	13
	8	4	34.8	99.2	239	169	157	143	125	104	16
	9	3	48.4	94.5	261	231	221	207	187	161	18
	10	2	62.1	89.0	272	278	270	256	235	206	21
	11	1	75.7	79.2	278	307	302	287	265	235	32
	12		86.6	0.0	280	317	312	298	275	245	36
	SURFACE DAILY TOTALS				2932	2526	2412	2250	2036	1766	246
AUG 21	6	6	5.0	101.3	35	7	5	4	4	4	2
	7	5	18.5	95.6	186	82	76	69	60	50	11
	8	4	32.2	89.7	241	158	154	146	134	118	16
	9	3	45.9	82.9	265	223	222	214	200	181	39
	10	2	59.3	73.0	278	273	275	268	252	230	58
	11	1	71.6	53.2	284	304	309	301	285	261	71
	12		78.3	0.0	286	315	320	313	296	272	75
	SURFACE DAILY TOTALS				2864	2408	2402	2316	2168	1958	470
SEP 21	7	5	13.7	83.8	173	57	60	60	59	56	26
	8	4	27.2	76.8	248	136	144	146	143	136	62
	9	3	40.2	67.9	278	205	218	221	217	206	93
	10	2	52.3	54.8	292	258	275	278	273	261	116
	11	1	61.9	33.4	299	291	311	315	309	295	131
	12		66.0	0.0	301	302	323	327	321	306	136
	SURFACE DAILY TOTALS				2878	2194	2342	2366	2322	2212	992
OCT 21	7	5	9.1	74.1	138	32	40	45	48	50	42
	8	4	22.0	66.7	247	111	129	139	144	145	99
	9	3	34.1	57.1	284	180	206	217	223	221	138
	10	2	44.7	43.8	301	234	265	277	282	279	165
	11	1	52.5	24.7	309	268	301	315	319	314	182
	12		55.5	0.0	311	279	314	328	332	327	188
	SURFACE DAILY TOTALS				2868	1928	2198	2314	2364	2346	1442
NOV 21	7	5	4.9	65.8	67	10	16	20	24	27	29
	8	4	17.0	58.4	232	82	108	123	135	142	124
	9	3	28.0	48.9	282	150	186	205	217	224	172
	10	2	37.3	36.3	303	203	244	265	278	283	204
	11	1	43.8	19.7	312	236	280	302	316	320	222
	12		46.2	0.0	315	247	293	315	328	332	228
	SURFACE DAILY TOTALS				2706	1610	1962	2146	2268	2324	1730
DEC 21	7	5	3.2	62.6	30	3	7	9	11	12	14
	8	4	14.9	55.3	225	71	99	116	129	139	130
	9	3	25.5	46.0	281	137	176	198	214	223	184
	10	2	34.3	33.7	304	189	234	258	275	283	217
	11	1	40.4	18.2	314	221	270	295	312	320	236
	12		42.6	0.0	317	232	282	308	325	332	243
	SURFACE DAILY TOTALS				2624	1474	1852	2058	2204	2286	1808

SOLAR POSITION AND INSOLATION, 32°N LATITUDE

DATE	SOLAR TIME		SOLAR POSITION		BTUH/SQ. FT. TOTAL INSOLATION ON SURFACES						
							SOUTH FACING SURFACE ANGLE WITH HORIZ.				
	AM	PM	ALT	AZM	NORMAL	HORIZ.	22	32	42	52	90
JAN 21	7	5	1.4	65.2	1	0	0	0	0	1	1
	8	4	12.5	56.5	203	56	93	106	116	123	115
	9	3	22.5	46.0	269	118	175	193	206	212	181
	10	2	30.6	33.1	295	167	235	256	269	274	221
	11	1	36.1	17.5	306	198	273	295	308	312	245
	12		38.0	0.0	310	209	285	308	321	324	253
	SURFACE DAILY TOTALS				2458	1288	1839	2008	2118	2166	1779
FEB 21	7	5	7.1	73.5	121	22	34	37	40	42	38
	8	4	19.0	64.4	247	95	127	136	140	141	108
	9	3	29.9	53.4	288	161	206	217	222	220	158
	10	2	39.1	39.4	306	212	266	278	283	279	193
	11	1	45.6	21.4	315	244	304	317	321	315	214
	12		48.0	0.0	317	255	316	330	334	328	222
	SURFACE DAILY TOTALS				2872	1724	2188	2300	2345	2322	1644
MAR 21	7	5	12.7	81.9	185	54	60	60	59	56	32
	8	4	25.1	73.0	260	129	146	147	144	137	78
	9	3	36.8	62.1	290	194	222	224	220	209	119
	10	2	47.3	47.5	304	245	280	283	278	265	150
	11	1	55.0	26.8	311	277	317	321	315	300	170
	12		58.0	0.0	313	287	329	333	327	312	177
	SURFACE DAILY TOTALS				3012	2084	2378	2403	2358	2246	1276
APR 21	6	6	6.1	99.9	66	14	9	6	6	5	3
	7	5	18.8	92.2	206	86	78	71	62	51	10
	8	4	31.5	84.0	255	158	156	148	136	120	35
	9	3	43.9	74.2	278	220	225	217	203	183	68
	10	2	55.7	60.3	290	267	279	272	256	234	95
	11	1	65.4	37.5	295	297	313	306	290	265	112
	12		69.6	0.0	297	307	325	318	301	276	118
	SURFACE DAILY TOTALS				3076	2390	2444	2356	2206	1994	764
MAY 21	6	6	10.4	107.2	119	36	21	13	13	12	7
	7	5	22.8	100.1	211	107	88	75	60	44	13
	8	4	35.4	92.9	250	175	159	145	127	105	15
	9	3	48.1	84.7	269	233	223	209	188	163	33
	10	2	60.6	73.3	280	277	273	259	237	208	56
	11	1	72.0	51.9	285	305	305	290	268	237	72
	12		78.0	0.0	286	315	315	301	278	247	77
	SURFACE DAILY TOTALS				3112	2582	2454	2284	2064	1788	469
JUN 21	6	6	12.2	110.2	131	45	26	16	15	14	9
	7	5	24.3	103.4	210	115	91	76	59	41	14
	8	4	36.9	96.8	245	180	159	143	122	99	16
	9	3	49.6	89.4	264	236	221	204	181	153	19
	10	2	62.2	79.7	274	279	268	251	227	197	41
	11	1	74.2	60.9	279	306	299	282	257	224	56
	12		81.5	0.0	280	315	309	292	267	234	60
	SURFACE DAILY TOTALS				3084	2634	2436	2234	1990	1690	370
JUL 21	6	6	10.7	107.7	113	37	22	14	13	12	8
	7	5	23.1	100.6	203	107	87	75	60	44	14
	8	4	35.7	93.6	241	174	158	143	125	104	16
	9	3	48.4	85.5	261	231	220	205	185	159	31
	10	2	60.9	74.3	271	274	269	254	232	204	54
	11	1	72.4	53.3	277	302	300	285	262	232	69
	12		78.6	0.0	279	311	310	296	273	242	74
	SURFACE DAILY TOTALS				3012	2558	2422	2250	2030	1754	458
AUG 21	6	6	6.5	100.5	59	14	9	7	6	6	4
	7	5	19.1	92.8	190	85	77	69	60	50	12
	8	4	31.8	84.7	240	156	152	144	132	116	33
	9	3	44.3	75.0	263	216	220	212	197	178	65
	10	2	56.1	61.3	276	262	272	264	249	226	91
	11	1	66.0	38.4	282	292	305	298	281	257	107
	12		70.3	0.0	284	302	317	309	292	268	113
	SURFACE DAILY TOTALS				2902	2352	2388	2296	2144	1934	736
SEP 21	7	5	12.7	81.9	163	51	56	56	55	52	30
	8	4	25.1	73.0	240	124	140	141	138	131	75
	9	3	36.8	62.1	272	188	213	215	211	201	114
	10	2	47.3	47.5	287	237	270	273	268	255	145
	11	1	55.0	26.8	294	268	306	309	303	289	164
	12		58.0	0.0	296	278	318	321	315	300	171
	SURFACE DAILY TOTALS				2808	2014	2288	2308	2264	2154	1226
OCT 21	7	5	6.8	73.1	99	19	29	32	34	36	32
	8	4	18.7	64.0	229	90	120	128	133	134	104
	9	3	29.5	53.0	273	155	198	208	213	212	153
	10	2	38.7	39.1	293	204	257	269	273	270	188
	11	1	45.1	21.1	302	236	294	307	311	306	209
	12		47.5	0.0	304	247	306	320	324	318	217
	SURFACE DAILY TOTALS				2696	1654	2100	2208	2252	2232	1588
NOV 21	7	5	1.5	65.4	2	0	0	0	1	1	1
	8	4	12.7	56.6	196	55	91	104	113	119	111
	9	3	22.6	46.1	263	118	173	190	202	208	176
	10	2	30.8	33.2	289	166	233	252	265	270	217
	11	1	36.2	17.6	301	197	270	291	303	307	241
	12		38.2	0.0	304	207	282	304	316	320	249
	SURFACE DAILY TOTALS				2406	1280	1816	1980	2084	2130	1742
DEC 21	8	4	10.3	53.8	176	41	77	90	101	108	107
	9	3	19.8	43.6	257	102	161	180	195	204	183
	10	2	27.6	31.2	288	150	221	244	259	267	226
	11	1	32.7	16.4	301	180	258	282	298	305	251
	12		34.6	0.0	304	190	271	295	311	318	259
	SURFACE DAILY TOTALS				2348	1136	1704	1888	2016	2086	1794

SOLAR POSITION AND INSOLATION, 40°N LATITUDE

DATE	SOLAR TIME		SOLAR POSITION		BTUH/SQ. FT. TOTAL INSOLATION ON SURFACES						
	AM	PM	ALT	AZM	NORMAL	HORIZ.	SOUTH FACING SURFACE ANGLE WITH HORIZ. 30	40	50	60	90
JAN 21	8	4	8.1	55.3	142	28	65	74	81	85	84
	9	3	16.8	44.0	239	83	155	171	182	187	171
	10	2	23.8	30.9	274	127	218	237	249	254	223
	11	1	28.4	16.0	289	154	257	277	290	293	253
	12		30.0	0.0	294	164	270	291	303	306	263
	SURFACE DAILY TOTALS				2182	948	1660	1810	1906	1944	1726
FEB 21	7	5	4.8	72.7	69	10	19	21	23	24	22
	8	4	15.4	62.2	224	73	114	122	126	127	107
	9	3	25.0	50.2	274	132	195	205	209	208	167
	10	2	32.8	35.9	295	178	256	267	271	267	210
	11	1	38.1	18.9	305	206	293	306	310	304	236
	12		40.0	0.0	308	216	306	319	323	317	245
	SURFACE DAILY TOTALS				2640	1414	2060	2162	2202	2176	1730
MAR 21	7	5	11.4	80.2	171	46	55	55	54	51	35
	8	4	22.5	69.6	250	114	140	141	138	131	89
	9	3	32.8	57.3	282	173	215	217	213	202	138
	10	2	41.6	41.9	297	218	273	276	271	258	176
	11	1	47.7	22.6	305	247	310	313	307	293	200
	12		50.0	0.0	307	257	322	326	320	305	208
	SURFACE DAILY TOTALS				2916	1852	2308	2330	2284	2174	1484
APR 21	6	6	7.4	98.9	89	20	11	8	7	7	4
	7	5	18.9	89.5	206	87	77	70	61	50	12
	8	4	30.3	79.3	252	152	153	145	133	117	53
	9	3	41.3	67.2	274	207	221	213	199	179	93
	10	2	51.2	51.4	286	250	275	267	252	229	126
	11	1	58.7	29.2	292	277	308	301	285	260	147
	12		61.6	0.0	293	287	320	313	296	271	154
	SURFACE DAILY TOTALS				3092	2274	2412	2320	2168	1956	1022
MAY 21	5	7	1.9	114.7	1	0	0	0	0	0	0
	6	6	12.7	105.6	144	49	25	15	14	13	9
	7	5	24.0	96.6	216	214	89	76	60	44	13
	8	4	35.4	87.2	250	175	158	144	125	104	25
	9	3	46.8	76.0	267	227	221	206	186	160	60
	10	2	57.5	60.9	277	267	270	255	233	205	89
	11	1	66.2	37.1	283	293	301	287	264	234	108
	12		70.0	0.0	284	301	312	297	274	243	114
	SURFACE DAILY TOTQLS				3160	2552	2442	2264	2040	1760	724
JUN 21	5	7	4.2	117.3	22	4	3	3	2	2	1
	6	6	14.8	108.4	155	60	30	18	17	16	10
	7	5	26.0	99.7	216	123	92	77	59	41	14
	8	4	37.4	90.7	246	182	159	142	121	97	16
	9	3	48.8	80.2	263	233	219	202	179	151	47
	10	2	59.8	65.8	272	272	266	248	224	194	74
	11	1	69.2	41.9	277	296	296	278	253	221	92
	12		73.5	0.0	279	304	306	289	263	230	98
	SURFACE DAILY TOTALS				3180	2648	2434	2224	1974	1670	610

DATE	SOLAR TIME		SOLAR POSITION		BTUH/SQ. FT. TOTAL INSOLATION ON SURFACES						
	AM	PM	ALT	AZM	NORMAL	HORIZ.	SOUTH FACING SURFACE ANGLE WITH HORIZ. 30	40	50	60	90
JUL 21	5	7	2.3	115.2	2	0	0	0	0	0	0
	6	6	13.1	106.1	138	50	26	17	15	14	9
	7	5	24.3	97.2	208	114	89	75	60	44	14
	8	4	35.8	87.8	241	174	157	142	124	102	24
	9	3	47.2	76.7	259	225	218	203	182	157	58
	10	2	57.9	61.7	269	265	266	251	229	200	86
	11	1	66.7	37.9	275	290	296	281	258	228	104
	12		70.6	0.0	276	298	307	292	269	238	111
	SURFACE DAILY TOTALS				3062	2534	2409	2230	2006	1728	702
AUG 21	6	6	7.9	99.5	81	21	12	9	8	7	5
	7	5	19.3	90.0	191	87	76	69	60	49	12
	8	4	30.7	79.9	237	150	150	141	129	113	50
	9	3	41.8	67.9	260	205	216	207	193	173	89
	10	2	51.7	52.1	272	246	267	259	244	221	120
	11	1	59.3	29.7	278	273	300	292	276	252	140
	12		62.3	0.0	280	282	311	303	287	262	147
	SURFACE DAILY TOTALS				2916	2244	2354	2258	2104	1894	978
SEP 21	7	5	11.4	80.2	149	43	51	51	49	47	32
	8	4	22.5	69.6	230	109	133	134	131	124	84
	9	3	32.8	57.3	263	167	206	208	203	193	132
	10	2	41.6	41.9	280	211	262	265	260	247	168
	11	1	47.7	22.6	287	239	298	301	295	281	192
	12		50.0	0.0	290	249	310	313	307	292	200
	SURFACE DAILY TOTALS				2708	1788	2210	2228	2182	2074	1416
OCT 21	7	5	4.5	72.3	48	7	14	15	17	17	16
	8	4	15.0	61.9	204	68	106	113	117	118	100
	9	3	24.5	49.8	257	126	185	195	200	198	160
	10	2	32.4	35.6	280	170	245	257	261	257	203
	11	1	37.6	18.7	291	199	283	295	299	294	229
	12		39.5	0.0	294	208	295	308	312	306	238
	SURFACE DAILY TOTALS				2454	1348	1962	2060	2098	2074	1654
NOV 21	8	4	8.2	55.4	136	28	63	72	78	82	81
	9	3	17.0	44.1	232	82	152	167	178	183	167
	10	2	24.0	31.0	268	126	215	233	245	249	219
	11	1	28.6	16.1	283	153	254	273	285	288	248
	12		30.2	0.0	288	163	267	287	298	301	258
	SURFACE DAILY TOTALS				2128	942	1636	1778	1870	1908	1686
DEC	8	4	5.5	53.0	89	14	39	45	50	54	56
	9	3	14.0	41.9	217	65	135	152	164	171	163
	10	2	20.7	29.4	261	107	200	221	235	242	221
	11	1	25.0	15.2	280	134	239	262	276	283	252
	12		26.6	0.0	285	143	253	275	290	296	263
	SURFACE DAILY TOTALS				1978	782	1480	1634	1740	1796	1646

SOLAR POSITION AND INSOLATION, 48°N LATITUDE

DATE	SOLAR TIME AM	SOLAR TIME PM	SOLAR POSITION ALT	SOLAR POSITION AZM	BTUH/SQ. FT. TOTAL INSOLATION ON SURFACES NORMAL	HORIZ.	SOUTH FACING SURFACE ANGLE WITH HORIZ. 38	48	58	68	90
JAN 21	8	4	3.5	54.6	37	4	17	19	21	22	22
	9	3	11.0	42.6	185	46	120	132	140	145	139
	10	2	16.9	29.4	239	83	190	206	216	220	206
	11	1	20.7	15.1	261	107	231	249	260	263	243
	12		22.0	0.0	267	115	245	264	275	278	255
	SURFACE DAILY TOTALS				1710	596	1360	1478	1550	1578	1478
FEB 21	7	5	2.4	72.2	12	1	3	4	4	4	4
	8	4	11.6	60.5	188	49	95	102	105	106	96
	9	3	19.7	47.7	251	100	178	187	191	190	167
	10	2	26.2	33.3	278	139	240	251	255	251	217
	11	1	30.5	17.2	290	165	278	290	294	288	247
	12		32.0	0.0	293	173	291	304	307	301	258
	SURFACE DAILY TOTALS				2330	1080	1880	1972	2024	1978	1720
MAR 21	7	5	10.0	78.7	153	37	49	49	47	45	35
	8	4	19.5	66.8	236	96	131	132	129	122	96
	9	3	28.2	53.4	270	147	205	207	203	193	152
	10	2	35.4	37.8	287	187	263	266	261	248	195
	11	1	40.3	19.8	295	212	300	303	297	283	223
	12		42.0	0.0	298	220	312	315	309	294	232
	SURFACE DAILY TOTALS				2780	1578	2208	2228	2182	2074	1632
APR 21	6	6	8.6	97.8	108	27	13	9	8	7	5
	7	5	18.6	86.7	205	85	76	69	59	48	21
	8	4	28.5	74.9	247	142	149	141	129	113	69
	9	3	37.8	61.2	268	191	216	208	194	174	115
	10	2	45.8	44.6	280	228	268	260	245	223	152
	11	1	51.5	24.0	286	252	301	294	278	254	177
	12		53.6	0.0	288	260	313	305	289	264	185
	SURFACE DAILY TOTALS				3076	2106	2358	2266	2114	1902	1262
MAY 21	5	7	5.2	114.3	41	9	4	4	4	3	2
	6	6	14.7	103.7	162	61	27	16	15	13	10
	7	5	24.6	93.0	219	118	89	75	60	43	13
	8	4	34.7	81.6	248	171	156	142	123	101	45
	9	3	44.3	68.3	264	217	217	202	182	156	86
	10	2	53.0	51.3	274	252	265	251	229	200	120
	11	1	59.5	28.6	279	274	296	281	258	228	141
	12		62.0	0.0	280	281	306	292	269	238	149
	SURFACE DAILY TOTALS				3254	2482	2418	2234	2010	1728	982
JUN 21	5	7	7.9	116.5	77	21	9	9	8	7	5
	6	6	17.2	106.2	172	74	33	19	18	16	12
	7	5	27.0	95.8	220	129	93	77	59	39	15
	8	4	37.1	84.6	246	181	157	140	119	95	35
	9	3	46.9	71.6	261	225	216	198	175	147	74
	10	2	55.8	54.8	269	259	262	244	220	189	105
	11	1	62.7	31.2	274	280	291	273	248	216	126
	12		65.5	0.0	275	287	301	283	258	225	133
	SURFACE DAILY TOTALS				3312	2626	2420	2204	1950	1644	874
JUL 21	5	7	5.7	114.7	43	10	5	5	4	4	3
	6	6	15.2	104.1	156	62	28	18	16	15	11
	7	5	25.1	93.5	211	118	89	75	59	42	14
	8	4	35.1	82.1	240	171	154	140	121	99	43
	9	3	44.8	68.8	256	215	214	199	178	153	83
	10	2	53.5	51.9	266	250	261	246	224	195	116
	11	1	60.1	29.0	271	272	291	276	253	223	137
	12		62.6	0.0	272	279	301	286	263	232	144
	SURFACE DAILY TOTALS				3158	2474	2386	2200	1974	1694	956
AUG 21	6	6	9.1	98.3	99	28	14	10	9	8	6
	7	5	19.1	87.2	190	85	75	67	58	47	20
	8	4	29.0	75.4	232	141	145	137	125	109	65
	9	3	38.4	61.8	254	189	210	201	187	168	110
	10	2	46.4	45.1	266	225	260	252	237	214	146
	11	1	52.2	24.3	272	248	293	285	268	244	169
	12		54.3	0.0	274	256	304	296	279	255	177
	SURFACE DAILY TOTALS				2898	2086	2300	2200	2046	1836	1208
SEP 21	7	5	10.0	78.7	131	35	44	44	43	40	31
	8	4	19.5	66.8	215	92	124	124	121	115	90
	9	3	28.2	53.4	251	142	196	197	193	183	143
	10	2	35.4	37.8	269	181	251	254	248	236	185
	11	1	40.3	19.8	278	205	287	289	284	269	212
	12		42.0	0.0	280	213	299	302	296	281	221
	SURFACE DAILY TOTALS				2568	1522	2102	2118	2070	1966	1546
OCT 21	7	5	2.0	71.9	4	0	1	1	1	1	1
	8	4	11.2	60.2	165	44	86	91	95	95	87
	9	3	19.3	47.4	233	94	167	176	180	178	157
	10	2	25.7	33.1	262	133	228	239	242	239	207
	11	1	30.0	17.1	274	157	266	277	281	276	237
	12		31.5	0.0	278	166	279	291	294	288	247
	SURFACE DAILY TOTALS				2154	1022	1774	1860	1890	1866	1626
NOV 21	8	4	3.6	54.7	36	5	17	19	21	22	22
	9	3	11.2	42.7	179	46	117	129	137	141	135
	10	2	17.1	29.5	233	83	186	202	212	215	201
	11	1	20.9	15.1	255	107	227	245	255	258	238
	12		22.2	0.0	261	115	241	259	270	272	250
	SURFACE DAILY TOTALS				1668	596	1336	1448	1518	1544	1442
DEC 21	9	3	8.0	40.9	140	27	87	98	105	110	109
	10	2	13.6	28.2	214	63	164	180	192	197	190
	11	1	17.3	14.4	242	86	207	226	239	244	231
	12		18.6	0.0	250	94	222	241	254	260	244
	SURFACE DAILY TOTALS				1444	446	1136	1250	1326	1364	1304

SOLAR POSITION AND INSOLATION, 56°N LATITUDE

DATE	SOLAR TIME AM	SOLAR TIME PM	SOLAR POSITION ALT	SOLAR POSITION AZM	BTUH/SQ. FT. TOTAL INSOLATION ON SURFACES NORMAL	HORIZ.	SOUTH FACING SURFACE ANGLE WITH HORIZ. 46	56	66	76	90
JAN 21	9	3	5.0	41.8	78	11	50	55	59	60	60
	10	2	9.9	28.5	170	39	135	146	154	156	153
	11	1	12.9	14.5	207	58	183	197	206	208	201
	12		14.0	0.0	217	65	198	214	222	225	217
	SURFACE DAILY TOTALS				1126	282	934	1010	1058	1074	1044
FEB 21	8	4	7.6	59.4	129	25	65	69	72	72	69
	9	3	14.2	45.9	214	65	151	159	162	161	151
	10	2	19.4	31.5	250	98	215	225	228	224	208
	11	1	22.8	16.1	266	119	254	265	268	263	243
	12		24.0	0.0	270	126	268	279	282	276	255
	SURFACE DAILY TOTALS				1986	740	1640	1716	1742	1716	1598
MAR 21	7	5	8.3	77.5	128	28	40	40	39	37	32
	8	4	16.2	64.4	215	75	119	120	117	111	97
	9	3	23.3	50.3	253	118	192	193	189	180	154
	10	2	29.0	34.9	272	151	249	251	246	234	205
	11	1	32.7	17.9	282	172	285	288	282	268	236
	12		34.0	0.0	284	179	297	300	294	280	246
	SURFACE DAILY TOTALS				2586	1268	2066	2084	2040	1938	1700
APR 21	5	7	1.4	108.8	0	0	0	0	0	0	0
	6	6	9.6	96.5	122	32	14	9	8	7	6
	7	5	18.0	84.1	201	81	74	66	57	46	29
	8	4	26.1	70.9	239	129	143	135	123	108	82
	9	3	33.6	56.3	260	169	208	200	186	167	133
	10	2	39.9	39.7	272	201	259	251	236	214	174
	11	1	44.1	20.7	278	220	292	284	268	245	200
	12		45.6	0.0	280	227	303	295	279	255	209
	SURFACE DAILY TOTALS				3024	1892	2282	2186	2038	1830	1458
MAY 21	4	8	1.2	125.5	0	0	0	0	0	0	0
	5	7	8.5	113.4	93	25	10	9	8	7	6
	6	6	16.5	101.5	175	71	28	17	15	13	11
	7	5	24.8	89.3	219	119	88	74	58	41	16
	8	4	33.1	76.3	244	163	153	138	119	98	63
	9	3	40.9	61.6	259	201	212	197	176	151	109
	10	2	47.6	44.2	268	231	259	244	222	194	146
	11	1	52.3	23.4	273	249	288	274	251	222	170
	12		54.0	0.0	275	255	299	284	261	231	178
	SURFACE DAILY TOTALS				3340	2374	2374	2188	1962	1682	1218
JUN 21	4	8	4.2	127.2	21	4	2	2	2	2	1
	5	7	11.4	115.3	122	40	14	13	11	10	8
	6	6	19.3	103.6	185	86	34	19	17	15	12
	7	5	27.6	91.7	222	132	92	76	57	38	15
	8	4	35.9	78.8	243	175	154	137	116	92	55
	9	3	43.8	64.1	257	212	211	193	170	143	98
	10	2	50.7	46.4	265	240	255	238	214	184	133
	11	1	55.6	24.9	269	258	284	267	242	210	156
	12		57.5	0.0	271	264	294	276	251	219	164
	SURFACE DAILY TOTALS				3438	2562	2388	2166	1910	1606	1120
JUL 21	4	8	1.7	125.8	0	0	0	0	0	0	0
	5	7	9.0	113.7	91	27	11	10	9	8	6
	6	6	17.0	101.9	169	72	30	18	16	14	12
	7	5	25.3	89.7	212	119	88	74	58	41	15
	8	4	33.6	76.7	237	163	151	136	117	96	61
	9	3	41.4	62.0	252	201	208	193	173	147	106
	10	2	48.2	44.6	261	230	254	239	217	189	142
	11	1	52.9	23.7	265	248	283	268	245	216	165
	12		54.6	0.0	267	254	293	278	255	225	173
	SURFACE DAILY TOTALS				3240	2372	2342	2152	1926	1646	1186
AUG 21	5	7	2.0	109.2	1	0	0	0	0	0	0
	6	6	10.2	97.0	112	34	16	11	10	9	7
	7	5	18.5	84.5	187	82	73	65	56	45	28
	8	4	26.7	71.3	225	128	140	131	119	104	78
	9	3	34.3	56.7	246	168	202	193	179	160	126
	10	2	40.5	40.0	258	199	251	242	227	206	166
	11	1	44.8	20.9	264	218	282	274	258	235	191
	12		46.3	0.0	266	225	293	285	269	245	200
	SURFACE DAILY TOTALS				2850	1884	2218	2118	1966	1760	1392
SEP 21	7	5	8.3	77.5	107	25	36	36	34	32	28
	8	4	16.2	64.4	194	72	111	111	108	102	89
	9	3	23.3	50.3	233	114	181	182	178	168	147
	10	2	29.0	34.9	253	146	236	237	232	221	193
	11	1	32.7	17.9	263	166	271	273	267	254	223
	12		34.0	0.0	266	173	283	285	279	265	233
	SURFACE DAILY TOTALS				2368	1220	1950	1962	1918	1820	1594
OCT 21	8	4	7.1	59.1	104	20	53	57	59	59	57
	9	3	13.8	45.7	193	60	138	145	148	147	138
	10	2	19.0	31.3	231	92	201	210	213	210	195
	11	1	22.3	16.0	248	112	240	250	253	248	230
	12		23.5	0.0	253	119	253	263	266	261	241
	SURFACE DAILY TOTALS				1804	688	1516	1586	1612	1588	1480
NOV 21	9	3	5.2	41.9	76	12	49	54	57	59	58
	10	2	10.0	28.5	165	39	132	143	149	152	148
	11	1	13.1	14.5	201	58	179	193	201	203	196
	12		14.2	0.0	211	65	194	209	217	219	211
	SURFACE DAILY TOTALS				1094	284	914	986	1032	1046	1016
DEC 21	9	3	1.9	40.5	5	0	3	4	4	4	4
	10	2	6.6	27.5	113	19	86	95	101	104	103
	11	1	9.5	13.9	166	37	141	154	163	167	164
	12		10.6	0.0	180	43	159	173	182	186	182
	SURFACE DAILY TOTALS				748	156	620	678	716	734	722

The quantity of solar radiation actually available for use in heating is difficult to calculate exactly. Most of this difficulty is due to the many highly variable factors that influence the radiation available at a collector location. But most of these factors can be treated by statistical methods using long-term averages of recorded weather data.

The least modified and therefore most usable solar radiation data is available from the U.S. Weather Bureau. Some of these data, averaged over a period of many years, have been published in the *Climatic Atlas of the United States* in the form of tables or maps. A selection of these average data is reprinted here for your convenience. They are taken to be a good indicator of future weather trends. More recent and complete information may be obtained from the National Weather Records Center in Asheville, North Carolina, as advised in Chapter 3.

As one example, daily insolation has been recorded at more than 80 weather stations across the United States. The available data have been averaged over a period of more than 30 years; these averages are summarized in the first 12 (one for each month) contour maps—"Mean Daily Solar Radiation." Values are given in langleys, or calories per square centimeter; multiply by 3.69 to convert to Btu per square foot. These figures represent the monthly average of the daily total of direct, diffuse, and reflected radiation on a horizontal surface. Trigonometric conversions (as explained later in this Appendix) must be applied to these data to convert them to the insolation on vertical or tilted surfaces.

Other useful information include the Weather Bureau records of the amount of sunshine, which is listed as the "hours of sunshine" or the "percentage of possible sunshine." A device records the cumulative total hours each day when there is enough direct solar radiation to "cast a shadow." This number of hours is then divided by the total hours from sunrise to sunset to get the percentage of possible sunshine. Monthly averages of this percentage are provided in the next 12 contour maps—"Mean Percentage of Possible Sunshine." These values can be taken as the average portion of the daytime hours each month when the sun is not obscured by clouds.

Also included in each of these 12 maps is a table of the average number of hours between sunrise and sunset for that month. You can multiply this number by the mean percentage of possible sunshine to obtain the mean number of hours of sunshine for a particular month and location. A table at the end of this section lists the mean number of hours of sunshine for selected locations across the United States.

These national maps are useful for getting an overview or approximation of the available solar radiation at a particular spot. For many locations, they may be the only way of finding a particular value. As a rule, however, they should be used only when other more local data are unavailable. Many local factors can have significant effect, so care and judgement are important when using interpolated data from these national weather maps.

SOURCE: Environmental Science Services Administration, *Climatic Atlas of the United States*. Washington, U.S. Department of Commerce, 1968.

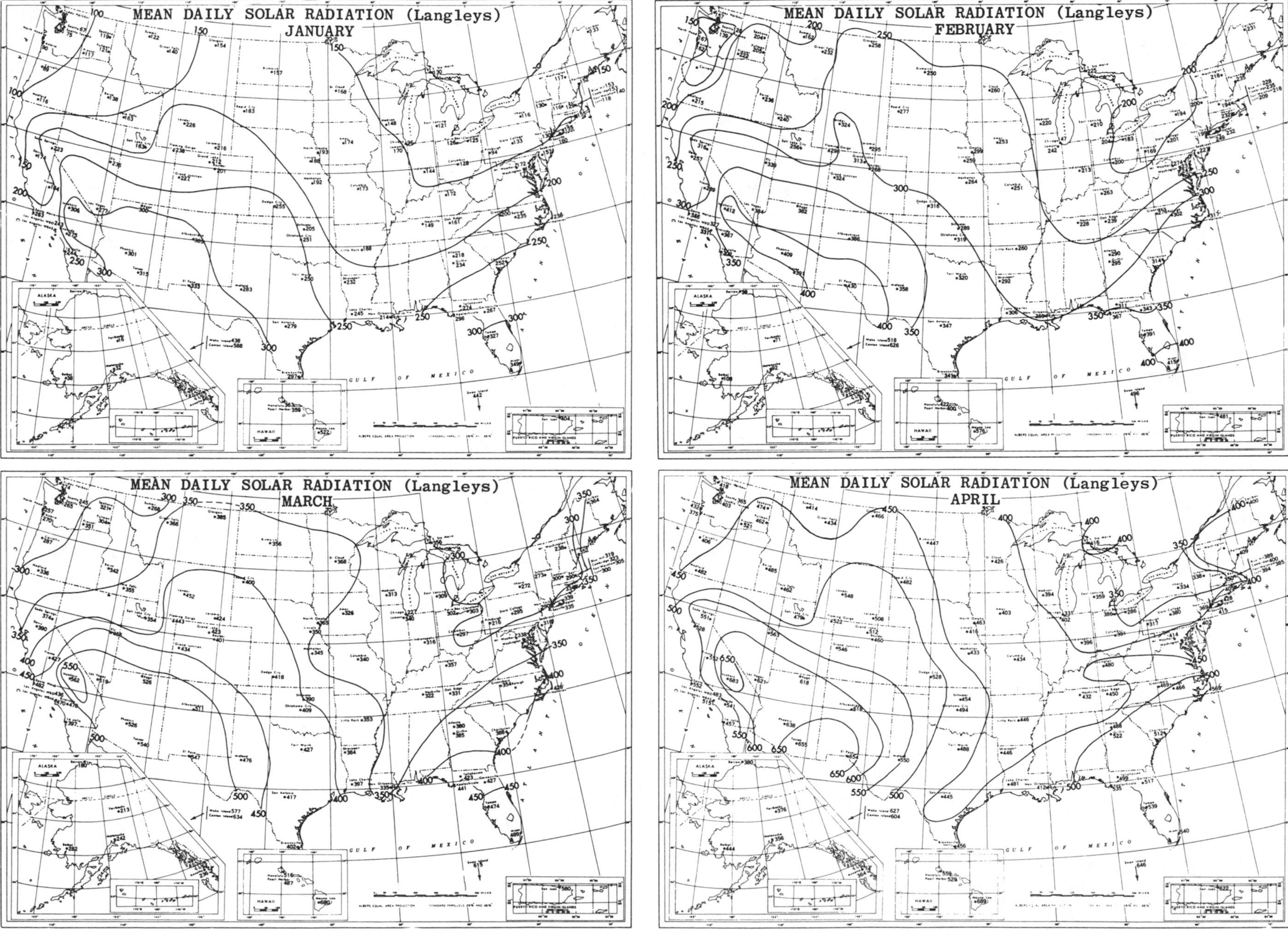
MEAN DAILY SOLAR RADIATION (Langleys)
JANUARY
MEAN DAILY SOLAR RADIATION (Langleys)
FEBRUARY
MEAN DAILY SOLAR RADIATION (Langleys)
MARCH
MEAN DAILY SOLAR RADIATION (Langleys)
APRIL

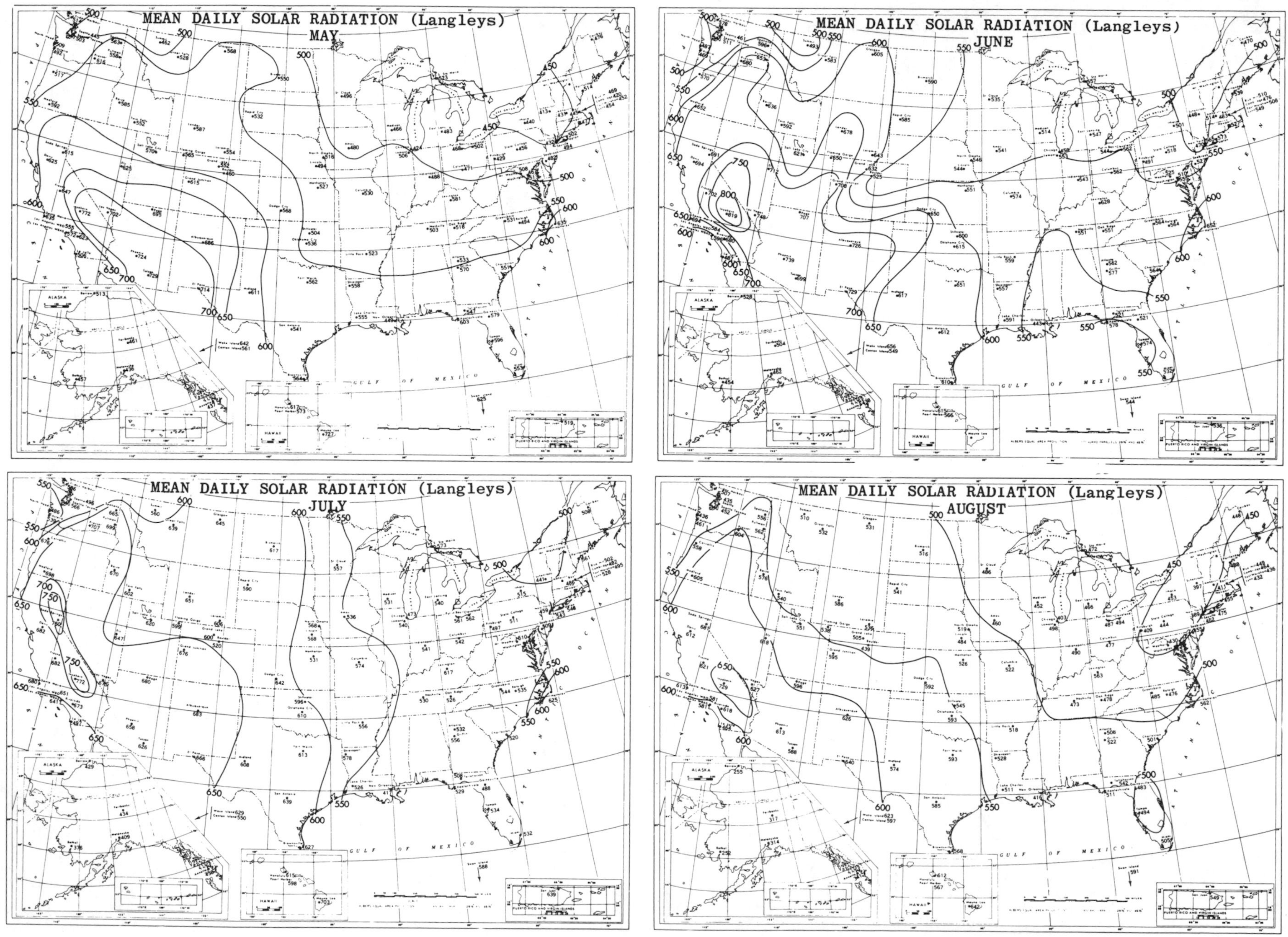
MEAN DAILY SOLAR RADIATION (Langleys)
MAY
MEAN DAILY SOLAR RADIATION (Langleys)
JUNE
MEAN DAILY SOLAR RADIATION (Langleys)
JULY
MEAN DAILY SOLAR RADIATION (Langleys)
AUGUST
GULF OF MEXICO

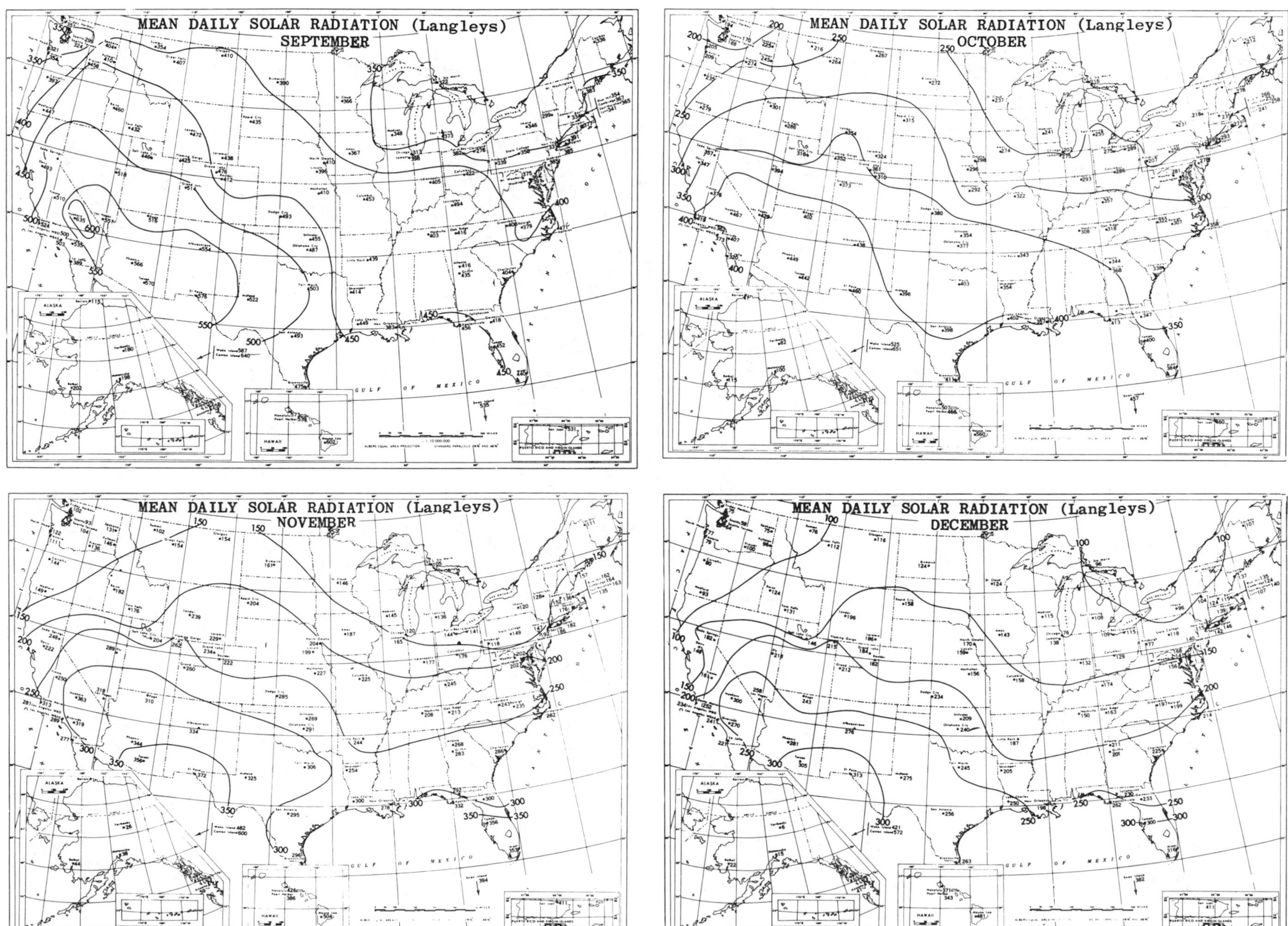
MEAN DAILY SOLAR RADIATION (Langleys)
SEPTEMBER
MEAN DAILY SOLAR RADIATION (Langleys)
OCTOBER
MEAN DAILY SOLAR RADIATION (Langleys)
NOVEMBER
MEAN DAILY SOLAR RADIATION (Langleys)
DECEMBER
GULF OF MEXICO

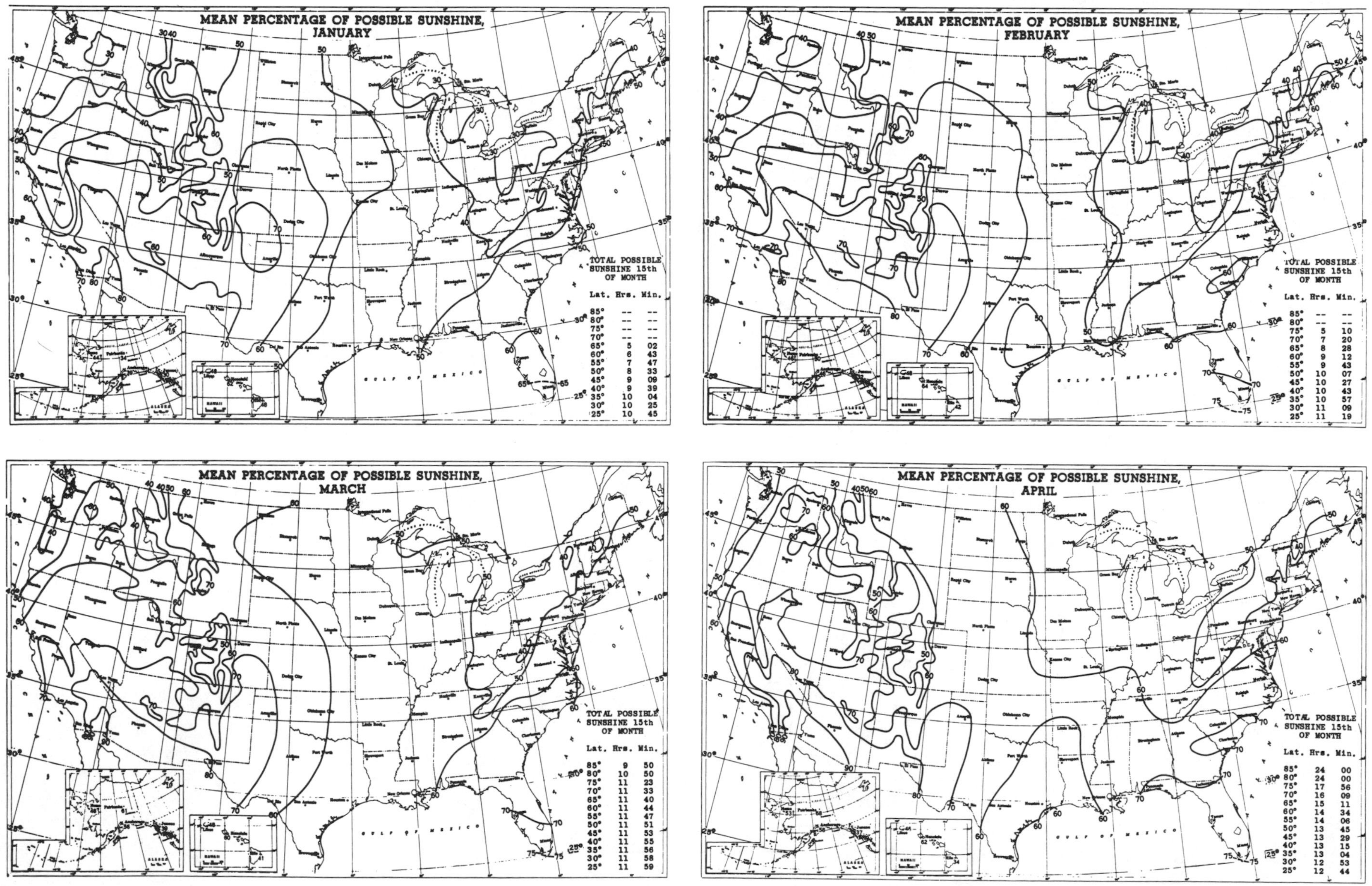
MEAN PERCENTAGE OF POSSIBLE SUNSHINE, JANUARY
TOTAL POSSIBLE SUNSHINE 15th OF MONTH

Lat.	Hrs.	Min.
85°	--	--
80°	--	--
75°	--	--
70°	--	--
65°	5	02
60°	6	43
55°	7	47
50°	8	33
45°	9	09
40°	9	39
35°	10	04
30°	10	25
25°	10	45

GULF OF MEXICO
ALASKA
HAWAII
MEAN PERCENTAGE OF POSSIBLE SUNSHINE, FEBRUARY
TOTAL POSSIBLE SUNSHINE 15th OF MONTH

Lat.	Hrs.	Min.
85°	--	--
80°	--	--
75°	5	10
70°	7	20
65°	8	28
60°	9	12
55°	9	43
50°	10	07
45°	10	27
40°	10	43
35°	10	57
30°	11	09
25°	11	19

MEAN PERCENTAGE OF POSSIBLE SUNSHINE, MARCH
TOTAL POSSIBLE SUNSHINE 15th OF MONTH

Lat.	Hrs.	Min.
85°	9	50
80°	10	50
75°	11	23
70°	11	33
65°	11	40
60°	11	44
55°	11	47
50°	11	51
45°	11	53
40°	11	55
35°	11	56
30°	11	58
25°	11	59

MEAN PERCENTAGE OF POSSIBLE SUNSHINE, APRIL
TOTAL POSSIBLE SUNSHINE 15th OF MONTH

Lat.	Hrs.	Min.
85°	24	00
80°	24	00
75°	17	56
70°	16	09
65°	15	11
60°	14	34
55°	14	06
50°	13	45
45°	13	29
40°	13	15
35°	13	04
30°	12	53
25°	12	44

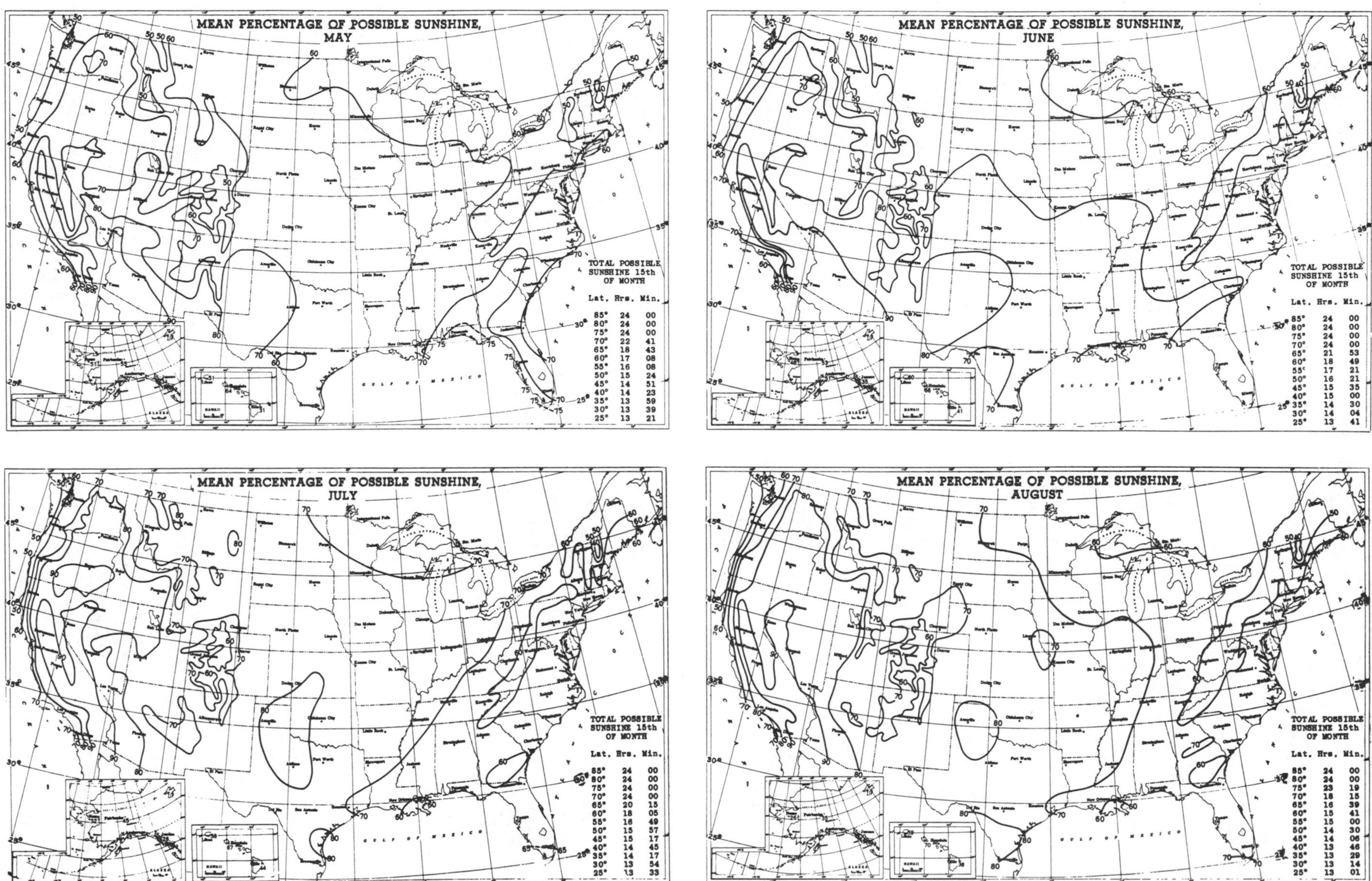
MEAN PERCENTAGE OF POSSIBLE SUNSHINE, MAY
TOTAL POSSIBLE SUNSHINE 15th OF MONTH
Lat. Hrs. Min.
85° 24 00
80° 24 00
75° 24 00
70° 22 41
65° 18 43
60° 17 08
55° 16 08
50° 15 24
45° 14 51
40° 14 23
35° 13 59
30° 13 39
25° 13 21
GULF OF MEXICO
MEAN PERCENTAGE OF POSSIBLE SUNSHINE, JUNE
TOTAL POSSIBLE SUNSHINE 15th OF MONTH
Lat. Hrs. Min.
85° 24 00
80° 24 00
75° 24 00
70° 24 00
65° 21 53
60° 18 49
55° 17 21
50° 16 21
45° 15 35
40° 15 00
35° 14 30
30° 14 04
25° 13 41
MEAN PERCENTAGE OF POSSIBLE SUNSHINE, JULY
TOTAL POSSIBLE SUNSHINE 15th OF MONTH
Lat. Hrs. Min.
85° 24 00
80° 24 00
75° 24 00
70° 24 00
65° 20 15
60° 18 05
55° 16 49
50° 15 57
45° 15 17
40° 14 45
35° 14 17
30° 13 54
25° 13 33
MEAN PERCENTAGE OF POSSIBLE SUNSHINE, AUGUST
TOTAL POSSIBLE SUNSHINE 15th OF MONTH
Lat. Hrs. Min.
85° 24 00
80° 24 00
75° 23 19
70° 18 15
65° 16 39
60° 15 41
55° 15 00
50° 14 30
45° 14 06
40° 13 46
35° 13 29
30° 13 14
25° 13 01

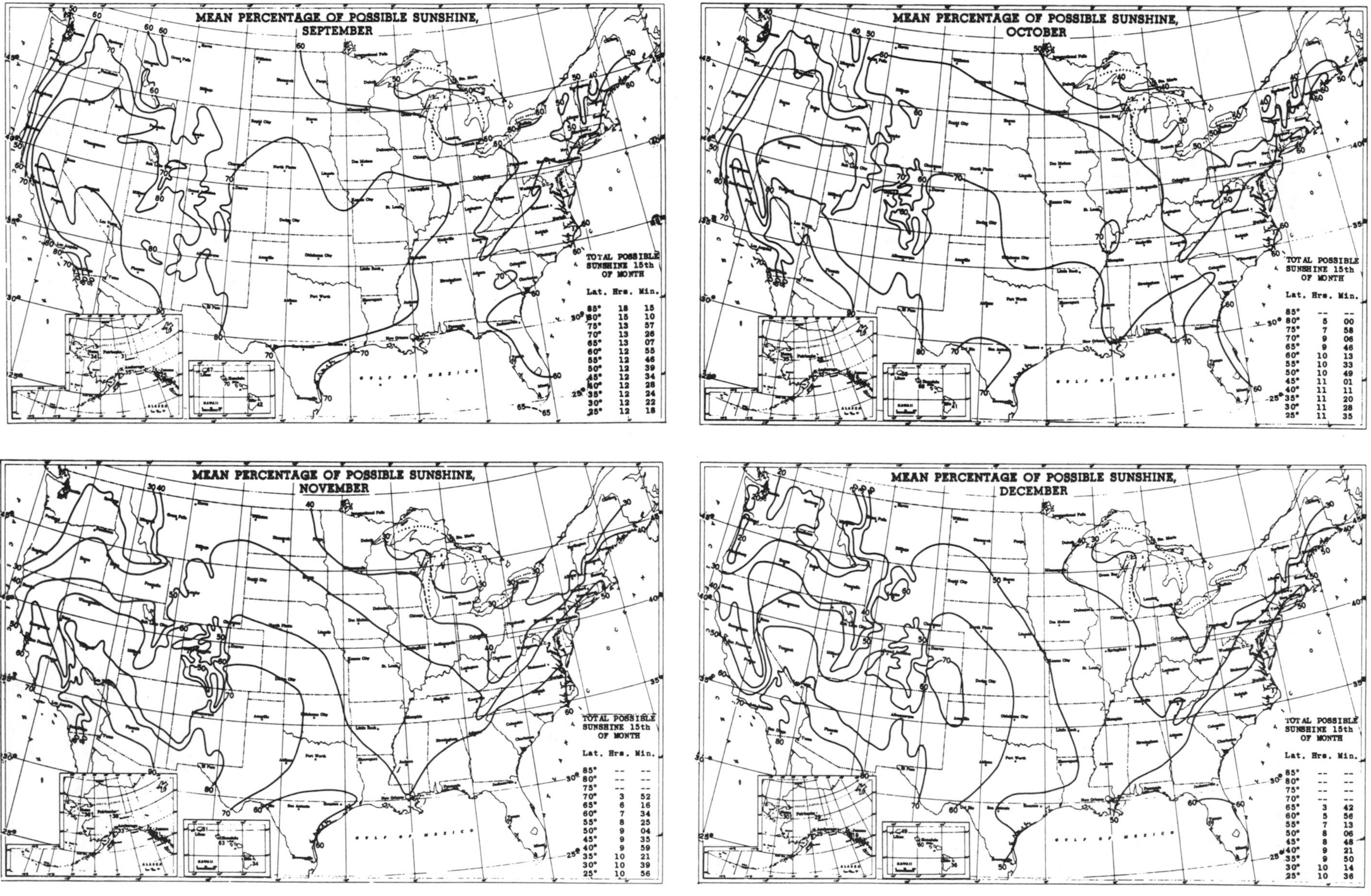
MEAN PERCENTAGE OF POSSIBLE SUNSHINE, SEPTEMBER
TOTAL POSSIBLE SUNSHINE 15th OF MONTH
Lat. Hrs. Min.
85° 18 15
80° 15 10
75° 13 57
70° 13 26
65° 13 07
60° 12 55
55° 12 46
50° 12 39
45° 12 34
40° 12 28
35° 12 24
30° 12 22
25° 12 18
MEAN PERCENTAGE OF POSSIBLE SUNSHINE, OCTOBER
TOTAL POSSIBLE SUNSHINE 15th OF MONTH
Lat. Hrs. Min.
85° -- --
80° 5 00
75° 7 58
70° 9 06
65° 9 46
60° 10 13
55° 10 33
50° 10 49
45° 11 01
40° 11 11
35° 11 20
30° 11 28
25° 11 35
MEAN PERCENTAGE OF POSSIBLE SUNSHINE, NOVEMBER
TOTAL POSSIBLE SUNSHINE 15th OF MONTH
Lat. Hrs. Min.
85° -- --
80° -- --
75° -- --
70° 3 52
65° 6 16
60° 7 34
55° 8 25
50° 9 04
45° 9 35
40° 9 59
35° 10 21
30° 10 39
25° 10 56
MEAN PERCENTAGE OF POSSIBLE SUNSHINE, DECEMBER
TOTAL POSSIBLE SUNSHINE 15th OF MONTH
Lat. Hrs. Min.
85° -- --
80° -- --
75° -- --
70° -- --
65° 3 42
60° 5 56
55° 7 13
50° 8 06
45° 8 48
40° 9 21
35° 9 50
30° 10 14
25° 10 36
GULF OF MEXICO

MEAN NUMBER OF HOURS OF SUNSHINE

STATE AND STATION	YEARS	JAN.	FEB.	MAR.	APR.	MAY	JUNE	JULY	AUG.	SEPT.	OCT.	NOV.	DEC.	ANNUAL
ALA. BIRMINGHAM	30	138	152	207	248	293	294	269	265	244	234	182	136	2662
MOBILE	22	157	158	212	253	301	289	249	259	235	254	195	146	2708
MONTGOMERY	30	160	168	227	267	317	311	288	290	260	250	200	156	2894
ALASKA ANCHORAGE	19	78	114	210	254	268	288	255	184	128	96	68	49	1992
FAIRBANKS	20	54	120	224	302	319	334	274	164	122	85	71	36	2105
JUNEAU	29	71	102	171	200	230	251	193	161	123	67	60	51	1680
NOME	27	72	109	193	226	285	297	204	146	142	101	67	42	1884
ARIZ. PHOENIX	30	248	244	314	346	404	404	377	351	334	307	267	236	3832
PRESCOTT	14	222	230	293	323	378	392	323	305	315	286	254	228	3549
TUCSON	13	255	266	317	350	399	394	329	329	335	317	280	258	3829
YUMA	30	258	266	337	365	419	420	404	380	351	330	285	262	4077
ARK. FT. SMITH	30	146	156	202	234	268	303	321	305	261	230	174	147	2747
LITTLE ROCK	30	143	158	213	243	291	316	321	316	265	251	181	142	2840
CALIF. EUREKA	30	120	138	180	209	247	261	244	205	195	164	127	108	2198
FRESNO	29	153	192	283	330	389	418	435	406	355	306	221	144	3632
LOS ANGELES	30	224	217	273	264	292	299	352	336	295	263	249	220	3284
RED BLUFF	15	156	186	246	302	366	396	438	407	341	277	199	154	3468
SACRAMENTO	30	134	169	255	300	367	405	437	406	347	283	197	122	3422
SAN DIEGO	30	216	212	262	242	261	253	293	277	255	234	236	217	2958
SAN FRANCISCO	30	165	182	251	281	314	330	300	272	267	243	198	156	2959
COLO. DENVER	30	207	205	247	252	281	311	321	297	274	246	200	192	3033
GRAND JUNCTION	30	169	182	243	265	314	350	349	311	291	255	198	168	3095
PUEBLO	30	224	217	261	271	299	340	349	318	290	265	225	211	3270
CONN. HARTFORD	30	141	166	206	223	267	285	299	268	220	193	137	136	2541
NEW HAVEN	30	155	178	215	234	274	291	309	284	238	215	157	154	2704
D. C. WASHINGTON	30	138	160	205	226	267	288	291	264	233	207	162	135	2576
FLA. APALACHICOLA	26	193	195	233	274	328	296	273	259	236	263	216	175	2941
JACKSONVILLE	30	192	189	241	267	296	260	255	248	199	205	191	170	2713
KEY WEST	30	229	238	285	296	307	273	277	269	236	237	226	225	3098
LAKELAND	7	204	186	222	251	285	268	252	242	203	209	212	198	2732
MIAMI	30	222	227	266	275	280	251	267	263	216	215	212	209	2903
PENSACOLA	30	175	180	232	270	311	302	278	284	249	265	206	166	2918
TAMPA	30	223	220	260	283	320	275	257	252	232	243	227	209	3001
GA. ATLANTA	25	154	165	218	266	309	304	284	285	247	241	188	160	2821
MACON	30	177	178	235	279	321	314	292	295	253	236	202	168	2950
SAVANNAH	30	175	173	229	274	307	279	267	256	212	216	197	167	2752
HAWAII HILO	7	153	135	161	112	106	158	184	134	137	153	106	131	1670
HONOLULU	30	227	202	250	255	276	280	293	290	279	257	221	211	3041
LIHUE	10	171	162	176	176	211	246	246	236	246	210	170	161	2411
IDAHO BOISE	30	116	144	218	274	322	352	412	378	311	232	143	104	3006
POCATELLO	30	111	143	211	255	300	338	380	347	296	230	145	108	2864
ILL. CAIRO	15	124	160	218	254	298	324	345	336	279	254	181	145	2918
CHICAGO	30	126	142	199	221	274	300	333	299	247	216	136	118	2611
MOLINE	18	132	139	189	214	255	279	337	300	251	214	130	123	2563
PEORIA	30	134	149	198	229	273	303	336	299	259	222	149	122	2673
SPRINGFIELD	30	127	149	193	224	282	304	346	312	266	225	152	122	2702
IND. EVANSVILLE	30	123	145	199	237	294	322	342	318	274	236	156	120	2766
FT. WAYNE	30	113	136	191	217	281	310	342	306	242	210	120	102	2570
INDIANAPOLIS	30	118	140	193	227	278	313	342	313	265	222	139	118	2668
TERRE HAUTE	24	125	148	189	231	274	302	341	305	253	235	150	122	2675
IOWA BURLINGTON	19	148	165	217	241	284	315	353	327	270	243	175	147	2885
CHARLES CITY	22	137	157	190	226	258	285	336	290	241	207	130	115	2572
DES MOINES	30	155	170	203	236	276	303	346	299	263	227	156	136	2770
SIOUX CITY	30	164	177	216	254	300	320	363	320	270	236	160	146	2926
KAN. CONCORDIA	30	180	172	214	243	281	315	348	308	249	245	189	172	2916
DODGE CITY	30	205	191	249	268	305	335	359	335	290	266	218	198	3219
TOPEKA	18	159	160	193	215	260	287	310	304	263	229	173	149	2702
WICHITA	30	187	186	233	254	291	321	350	325	277	245	206	182	3057
KY. LOUISVILLE	30	115	135	188	221	283	303	324	295	256	219	148	114	2601
LA. NEW ORLEANS	30	160	158	213	247	292	287	260	269	241	260	200	157	2744
SHREVEPORT	19	151	172	214	240	298	332	339	322	289	273	208	177	3015
MAINE EASTPORT	22	133	151	196	201	245	248	275	260	205	175	105	115	2309
PORTLAND	30	155	174	213	226	268	286	312	294	229	202	146	148	2653
MD. BALTIMORE	30	148	170	211	229	270	295	299	272	238	212	164	145	2653
MASS. BLUE HILL OBS.	10	125	136	165	182	233	248	266	241	211	181	134	135	2257
BOSTON	30	148	168	212	222	263	283	300	280	232	207	152	148	2615
NANTUCKET	22	128	156	214	227	278	284	291	279	242	208	149	129	2585
MICH. ALPENA	24	86	124	198	228	261	303	339	285	204	159	70	67	2324
DETROIT	30	90	128	180	212	263	295	321	284	226	189	98	89	2375
LANSING	30	84	119	175	215	272	305	344	294	228	182	87	73	2378
ESCANABA	30	112	148	204	226	266	283	316	267	198	162	90	94	2366
GRAND RAPIDS	30	74	117	178	218	277	308	349	304	231	188	92	70	2406
MARQUETTE	30	78	113	172	207	248	268	305	251	186	142	68	66	2104
SAULT STE. MARIE	30	83	123	187	217	252	269	309	256	165	133	61	62	2117
MINN. DULUTH	30	125	163	221	235	268	282	328	277	203	166	100	107	2475
MINNEAPOLIS	30	140	166	200	231	272	302	343	296	237	193	115	112	2607
MISS. JACKSON	12	130	147	199	244	280	287	279	287	235	223	185	150	2646
VICKSBURG	30	136	141	199	232	284	304	291	297	254	244	183	140	2705
MO. COLUMBIA	30	147	164	207	232	281	296	341	298	262	225	166	138	2757
KANSAS CITY	30	154	170	211	235	278	313	347	308	266	235	178	151	2846
ST. JOSEPH	23	154	165	211	231	274	301	347	287	260	224	168	144	2766
ST. LOUIS	30	137	152	202	235	283	301	325	289	256	223	166	125	2694
SPRINGFIELD	30	145	164	213	238	278	305	342	310	269	233	183	140	2820
MONT. BILLINGS	21	140	154	208	236	283	301	372	332	258	213	136	129	2762
GREAT FALLS	19	154	176	245	261	299	299	381	342	256	206	132	133	2884
HAVRE	30	136	174	234	268	311	312	384	339	260	202	132	122	2874
HELENA	30	138	168	215	241	292	292	342	336	258	202	137	121	2742
MISSOULA	25	85	109	167	209	261	260	378	328	246	178	90	66	2377
NEBR. LINCOLN	30	173	172	213	244	287	316	356	309	266	237	174	160	2907
NORTH PLATTE	30	181	179	221	246	282	310	343	304	264	242	184	169	2925
OMAHA	30	172	188	222	259	305	332	379	311	270	248	166	145	2997
VALENTINE	30	185	194	229	252	296	323	369	326	275	242	174	172	3037
NEV. ELY	22	186	197	262	260	300	354	359	344	303	255	204	187	3211
LAS VEGAS	8	239	251	314	336	386	411	383	364	345	301	258	250	3838
RENO	30	185	199	267	306	354	376	414	391	336	273	212	170	3483
WINNEMUCCA	30	142	155	207	255	312	346	395	375	316	242	177	139	3061
N. H. CONCORD	23	136	153	192	196	229	261	286	260	214	179	122	126	2354
MT. WASHINGTON OBS.	18	94	98	133	141	162	145	150	143	139	159	89	87	1540
N. J. ATLANTIC CITY	30	151	173	210	233	273	287	298	271	239	218	177	153	2683
TRENTON	30	145	168	203	235	277	294	309	273	239	208	160	142	2653
N. MEX. ALBUQUERQUE	30	221	218	273	299	343	365	340	317	299	279	245	219	3418
ROSWELL	21	218	223	286	306	330	333	341	313	266	266	242	216	3340
N. Y. ALBANY	30	125	151	194	213	266	301	317	286	224	192	115	112	2496
BINGHAMTON	30	94	119	151	170	226	256	266	230	184	158	92	79	2025
BUFFALO	30	110	125	180	212	274	319	338	297	239	183	97	84	2458
NEW YORK	30	154	171	213	237	268	289	302	271	235	213	169	155	2677
ROCHESTER	30	93	123	172	209	274	314	333	294	224	173	97	86	2392
SYRACUSE	30	87	115	165	197	261	295	316	276	211	163	81	74	2241
N. C. ASHEVILLE	30	146	161	211	247	289	292	268	250	235	222	179	146	2646
CAPE HATTERAS	9	152	168	206	259	293	301	286	265	214	202	169	154	2669
CHARLOTTE	30	165	177	230	267	313	316	291	277	247	243	198	167	2891
GREENSBORO	30	157	171	217	231	298	302	287	272	243	236	190	163	2767
RALEIGH	29	154	168	220	255	290	284	277	253	224	215	184	156	2680
WILMINGTON	30	179	180	237	279	314	312	286	273	237	238	206	178	2919
N. DAK. BISMARCK	30	141	170	205	236	279	294	358	307	243	198	130	125	2686
DEVILS LAKE	30	150	177	220	250	291	297	352	302	230	198	123	124	2714
FARGO	30	132	170	210	232	283	288	343	293	222	187	112	114	2586
WILLISTON	29	141	168	215	260	305	312	377	328	247	206	131	129	2819
OHIO CINCINNATI (ABBE)	30	115	137	186	222	273	309	323	295	253	205	138	118	2574
CLEVELAND	30	79	111	167	209	274	301	325	288	235	187	99	77	2352
COLUMBUS	30	112	132	177	215	270	296	323	291	250	210	131	101	2508
DAYTON	10	114	136	195	222	281	313	323	307	268	229	152	124	2664
SANDUSKY	30	100	128	183	229	285	312	343	302	248	201	111	91	2533
TOLEDO	30	93	120	170	203	263	296	331	298	241	196	106	92	2409
OKLA. OKLAHOMA CITY	29	175	182	235	253	290	329	352	331	282	243	201	175	3048
TULSA	18	152	164	200	213	244	287	314	308	281	241	207	172	2783
OREG. BAKER	22	118	143	198	251	302	313	406	368	289	215	132	100	2835
PORTLAND	30	77	97	142	203	246	249	329	275	218	134	87	65	2122
ROSEBURG	30	69	96	148	205	257	278	369	329	255	146	81	50	2283
PA. HARRISBURG	30	132	160	203	230	277	297	319	282	233	200	140	131	2604
PHILADELPHIA	30	142	166	203	231	270	281	288	253	225	205	158	142	2564
PITTSBURGH	25	89	114	163	200	239	260	283	250	234	180	114	76	2202
READING	30	133	151	195	220	259	275	293	259	219	198	144	127	2473
SCRANTON	30	108	138	178	199	251	269	290	249	213	183	120	105	2303
R. I. PROVIDENCE	30	145	168	211	221	271	285	292	267	226	307	153	143	2589
S. C. CHARLESTON	30	188	189	243	284	323	308	297	281	244	239	210	187	2993
COLUMBIA	30	173	183	233	274	312	312	291	283	243	242	202	166	2914
GREENVILLE	26	166	176	227	274	307	300	278	274	239	232	192	157	2822
S. DAK. HURON	30	153	177	213	250	295	321	367	320	260	212	142	134	2844
RAPID CITY	30	164	182	222	245	278	300	348	317	266	228	164	144	2858
TENN. CHATTANOOGA	30	126	146	187	239	290	295	278	266	247	220	169	128	2591
KNOXVILLE	30	124	144	189	237	281	288	277	248	237	213	157	120	2515
MEMPHIS	30	135	152	204	244	296	321	319	314	261	243	180	139	2808
NASHVILLE	30	123	142	196	241	285	308	292	279	250	224	168	126	2634
TEX. ABILENE	13	190	199	250	259	290	347	335	322	276	245	223	201	3137
AMARILLO	30	207	199	258	276	305	338	350	328	288	260	229	205	3243
AUSTIN	30	148	152	207	221	266	302	331	320	261	242	180	160	2790
BROWNSVILLE	30	147	152	187	210	272	297	326	311	246	252	165	151	2716
CORPUS CHRISTI	24	160	165	212	237	295	329	366	341	276	264	194	164	3003
DALLAS	30	155	159	220	238	279	326	341	325	274	240	191	163	2911
DEL RIO	27	173	173	230	237	259	279	331	319	252	240	195	178	2866
EL PASO	30	234	236	299	329	373	369	336	327	300	287	257	236	3583
GALVESTON	30	151	149	203	230	288	322	305	292	257	264	199	151	2811
HOUSTON	30	144	141	193	212	266	298	294	281	238	239	181	146	2633
PORT ARTHUR	30	153	149	209	235	292	317	285	281	252	256	191	148	2768
SAN ANTONIO	30	148	153	213	224	258	292	325	307	261	241	183	160	2765
UTAH SALT LAKE CITY	30	137	155	227	269	329	358	377	346	306	249	171	135	3059
VT. BURLINGTON	30	103	127	184	185	244	270	291	266	199	152	77	80	2178
VA. LYNCHBURG	26	153	169	216	243	288	297	288	264	235	217	177	158	2705
NORFOLK	30	156	174	223	257	304	311	296	282	237	220	182	161	2803
RICHMOND	30	144	166	211	248	280	296	286	263	230	211	176	152	2663
WASH. NORTH HEAD	22	76	97	135	182	221	214	226	186	170	123	87	66	1783
SEATTLE	30	74	99	154	201	247	234	304	248	197	122	77	62	2019
SPOKANE	30	78	120	197	262	308	309	397	350	264	177	86	57	2605
TATOOSH ISLAND	30	70	100	135	182	229	217	235	190	175	129	71	60	1793
WALLA WALLA	30	72	106	194	262	317	335	411	367	280	198	92	51	2685
W. VA. ELKINS	24	110	119	158	198	227	256	225	236	211	186	131	103	2160
PARKERSBURG	30	91	111	155	200	252	277	286	264	230	189	117	93	2265
WIS. GREEN BAY	30	121	148	194	210	251	279	314	266	213	176	110	106	2388
MADISON	30	126	147	196	214	258	285	336	288	230	198	116	108	2502
MILWAUKEE	30	116	134	191	218	267	293	340	292	235	193	125	106	2510
WYO. CHEYENNE	30	191	197	243	237	259	304	318	286	265	242	188	170	2900
LANDER	30	200	208	260	264	301	340	361	326	280	233	186	185	3144
SHERIDAN	30	160	179	226	245	286	303	367	333	266	221	153	145	2884
P. R. SAN JUAN	30	231	229	273	252	240	245	264	257	219	229	217	222	2878

The total solar radiation is the sum of direct, diffuse, and reflected radiation. At present, a statistical approach is the only reliable method of separating out the diffuse component of horizontal insolation. The full detail of this method is contained in an article by Liu and Jordan; we only summarize their results here. First we ascertain the ratio of the daily insolation on a horizontal surface (measured at a particular weather station) to the extraterrestrial radiation on another horizontal surface (outside the atmosphere). This ratio (usually called the *percent of Extraterrestrial radiation*, or % ETR) can be determined from the National Weather Records Center; it is also given in the article by Liu and Jordan. With a knowledge of the % ETR, you can use the accompanying graph to determine the percentage of diffuse radiation of a horizontal surface. For example, 50% ETR corresponds to 38% diffuse radiation and 62% direct radiation.

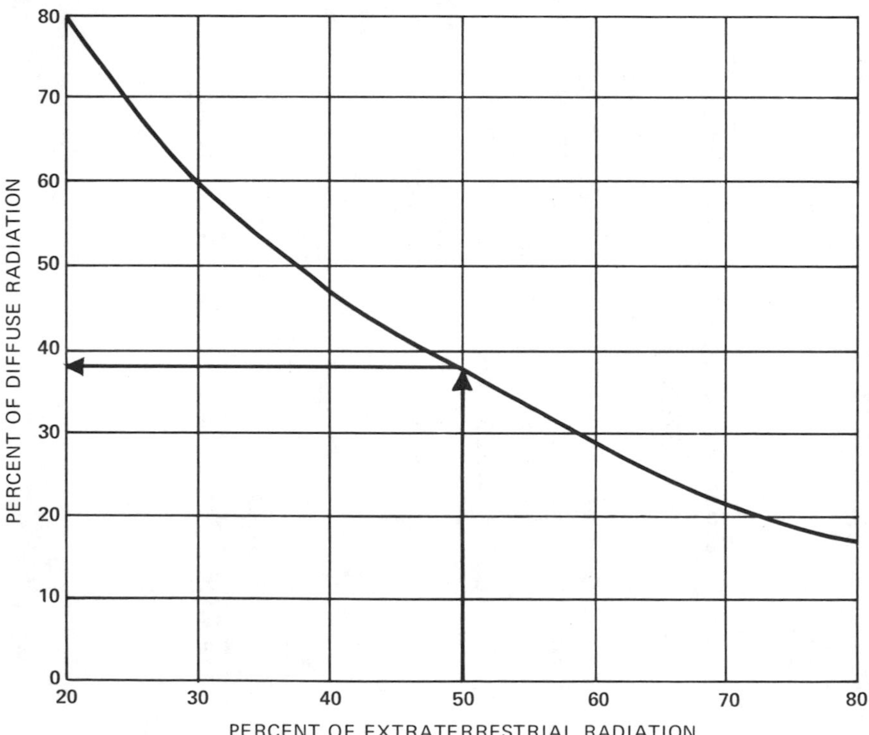

You are now prepared to convert the direct and diffuse components of the horizontal insolation into the daily total insolation on south-facing tilted or vertical surfaces. The conversion factor for the direct component F_D, depends on the latitude, L, the tilt angle of the surface, β, and the *sunset hour angles*, ω and ω', of the horizontal and tilted surfaces:

$$\text{horizontal surface:} \quad \cos\omega = -\tan L \tan\delta$$
$$\text{tilted surface:} \quad \cos\omega' = -\tan(L-\beta)\tan\delta$$

where the declination δ is found from the graph on page 252 and $\beta = 90°$ applies to vertical surfaces. Depending on the value of these two angles ω and ω', the calculation of F_D is slightly different. If ω is less than ω', then

$$F_D = \frac{\cos(L-\beta)}{\cos L} \times \frac{\sin\omega - \omega\cos\omega'}{\sin\omega - \omega\cos\omega}$$

If ω' is smaller than ω, then

$$F_D = \frac{\cos(L-\beta)}{\cos L} \times \frac{\sin\omega' - \omega'\cos\omega'}{\sin\omega - \omega\cos\omega}$$

The direct component of the radiation on a tilted or vertical surface is $I'_D = F_D \times I_D$, where I_D is the direct horizontal insolation.

The treatment of diffuse and reflected radiation is a bit different. The diffuse radiation is assumed to come uniformly from all corners of the sky, so one need only determine the fraction of the sky exposed to a tilted surface and reduce the horizontal diffuse radiation accordingly. The diffuse radiation on a surface tilted at an angle β is

$$I'_d = \frac{1+\cos\beta}{2} \times I_d$$

where I_d is the daily horizontal diffuse radiation. The reflected radiation on a tilted surface is

$$I'_r = \rho \times \frac{1-\cos\beta}{2} \times (I_D + I_d)$$

where ρ is the reflectance of the horizontal surface.

SOURCE: Liu, B.Y.H. and R. C. Jordan, "Availability of Solar Energy for Flat-Plate Solar Heat Collectors." in *Low Temperature Engineering Applications of Solar Energy*, edited by Richard C. Jordan, New York: ASHRAE, 1967.

The hourly, monthly, and yearly heat losses from a house depend on the temperature difference between the indoor and outdoor air, as explained in Chapter 3. To aid in the calculation of these heat losses, ASHRAE publishes the expected winter *design temperatures* and the monthly and yearly total *degree days* for many cities and towns in the United States.

The maximum heat loss rate occurs when the temperature is lowest, and you need some idea of the lowest likely temperature in your locale in order to size a conventional heating unit. The ASHRAE *Handbook of Fundamentals* provides three choices—the "median of annual extremes" and the "99%" and "97½%" design temperatures. The median of annual extremes is the average of the lowest winter temperatures recorded in each of the past 30 to 40 years. The 99% and 97½% design temperatures are the temperatures which are normally *exceeded* during those percentages of the time in December, January and February. We list the 97½% temperatures here together with the average winter temperatures. For example, the temperature will fall below 19°F for 2½% of the time (about 2 days) during a typical Birmingham winter. Consult the ASHRAE *Handbook of Fundamentals* for more detailed listings.

Degree days gauge heating requirements over the long run. One degree day accrues for every day the average outdoor temperature is 1°F below 65°F, which is the base for degree day calculations because most houses don't require any heating until outdoor temperatures fall below this level. For example, if the outdoor air temperature remained constant at 30°F for the entire month of January, then 31 × (65 − 30) = 1085 degree days would result. Both monthly and yearly total degree days are listed in these tables, but only the months from September to May are included here because very little heating is needed in the summer. The yearly total degree days are the sum over *all 12 months.* More complete listings of monthly and yearly degree days can be found in the ASHRAE *Guide and Data Book.*

SOURCES: ASHRAE, *Guide and Data Book*, 1970. Reprinted by permission.
ASHRAE, *Handbook of Fundamentals*, 1972. Reprinted by permission.

State	City	Avg. Winter Temp	Design Temp	Sep	Oct	Nov	Dec	Jan	Feb	Mar	Apr	May	Yearly Total
Ala.	Birmingham	54.2	19	6	93	363	555	592	462	363	108	9	2551
	Huntsville	51.3	13	12	127	426	663	694	557	434	138	19	3070
	Mobile	59.9	26	0	22	213	357	415	300	211	42	0	1560
	Montgomery	55.4	22	0	68	330	527	543	417	316	90	0	2291
Alaska	Anchorage	23.0	−25	516	930	1284	1572	1631	1316	1293	879	592	10864
	Fairbanks	6.7	−53	642	1203	1833	2254	2359	1901	1739	1068	555	14279
	Juneau	32.1	− 7	483	725	921	1135	1237	1070	1073	810	601	9075
	Nome	13.1	−32	693	1094	1455	1820	1879	1666	1770	1314	930	14171
Ariz.	Flagstaff	35.6	0	201	558	867	1073	1169	991	911	651	437	7152
	Phoenix	58.5	31	0	22	234	415	474	328	217	75	0	1765
	Tucson	58.1	29	0	25	231	406	471	344	242	75	6	1800
	Winslow	43.0	9	6	245	711	1008	1054	770	601	291	96	4782
	Yuma	64.2	37	0	0	108	264	307	190	90	15	0	974
Ark.	Fort Smith	50.3	9	12	127	450	704	781	596	456	144	22	3292
	Little Rock	50.5	19	9	127	465	716	756	577	434	126	9	3219
	Texarkana	54.2	22	0	78	345	561	626	468	350	105	0	2533
Calif.	Bakersfield	55.4	31	0	37	282	502	546	364	267	105	19	2122
	Burbank	58.6	36	6	43	177	301	366	277	239	138	81	1646
	Eureka	49.9	32	258	329	414	499	546	470	505	438	372	4643
	Fresno	53.3	28	0	84	354	577	605	426	335	162	62	2611
	Long Beach	57.8	36	9	47	171	316	397	311	264	171	93	1803
	Los Angeles	57.4	41	42	78	180	291	372	302	288	219	158	2061
	Oakland	53.5	35	45	127	309	481	527	400	353	255	180	2870
	Sacramento	53.9	30	0	56	321	546	583	414	332	178	72	2502
	San Diego	59.5	42	21	43	135	236	298	235	214	135	90	1458
	San Francisco	55.1	42	102	118	231	388	443	336	319	279	239	3001
	Santa Maria	54.3	32	96	146	270	391	459	370	363	282	233	2967
Colo.	Alamosa	29.7	−17	279	639	1065	1420	1476	1162	1020	696	440	8529
	Colorado Springs	37.3	− 1	132	456	825	1032	1128	938	893	582	319	6423
	Denver	37.6	− 2	117	428	819	1035	1132	938	887	558	288	6283
	Grand Junction	39.3	8	30	313	786	1113	1209	907	729	387	146	5641
	Pueblo	40.4	− 5	54	326	750	986	1085	871	772	429	174	5462
Conn.	Bridgeport	39.9	4	66	307	615	986	1079	966	853	510	208	5617
	Hartford	37.3	1	117	394	714	1101	1190	1042	908	519	205	6235
	New Haven	39.0	5	87	347	648	1011	1097	991	871	543	245	5897
Del.	Wilmington	42.5	12	51	270	588	927	980	874	735	387	112	4930
D. C.	Washington	45.7	16	33	217	519	834	871	762	626	288	74	4224
Fla.	Daytona Beach	64.5	32	0	0	75	211	248	190	140	15	0	879
	Fort Myers	68.6	38	0	0	24	109	146	101	62	0	0	442
	Jacksonville	61.9	29	0	12	144	310	332	246	174	21	0	1239
	Key West	73.1	55	0	0	0	28	40	31	9	0	0	108
	Lakeland	66.7	35	0	0	57	164	195	146	99	0	0	661
	Miami	71.1	44	0	0	0	65	74	56	19	0	0	214
	Miami Beach	72.5	45	0	0	0	40	56	36	9	0	0	141
	Orlando	65.7	33	0	0	72	198	220	165	105	6	0	766
	Pensacola	60.4	29	0	19	195	353	400	277	183	36	0	1463
	Tallahassee	60.1	25	0	28	198	360	375	286	202	36	0	1485

State	City	Avg. Winter Temp	Design Temp	Sep	Oct	Nov	Dec	Jan	Feb	Mar	Apr	May	Yearly Total
	Tampa	66.4	36	0	0	60	171	202	148	102	0	0	683
	West Palm Beach	68.4	40	0	0	6	65	87	64	31	0	0	253
Ga.	Athens	51.8	17	12	115	405	632	642	529	431	141	22	2929
	Atlanta	51.7	18	18	124	417	648	636	518	428	147	25	2961
	Augusta	54.5	20	0	78	333	552	549	445	350	90	0	2397
	Columbus	54.8	23	0	87	333	543	552	434	338	96	0	2383
	Macon	56.2	23	0	71	297	502	505	403	295	63	0	2136
	Rome	49.9	16	24	161	474	701	710	577	468	177	34	3326
	Savannah	57.8	24	0	47	246	437	437	353	254	45	0	1819
Hawaii	Hilo	71.9	59	0	0	0	0	0	0	0	0	0	0
	Honolulu	74.2	60	0	0	0	0	0	0	0	0	0	0
Idaho	Boise	39.7	4	132	415	792	1017	1113	854	722	438	245	5809
	Lewiston	41.0	6	123	403	756	933	1063	815	694	426	239	5542
	Pocatello	34.8	− 8	172	493	900	1166	1324	1058	905	555	319	7033
Ill.	Chicago	37.5	− 4	81	326	753	1113	1209	1044	890	480	211	6155
	Moline	36.4	− 7	99	335	774	1181	1314	1100	918	450	189	6408
	Peoria	38.1	− 2	87	326	759	1113	1218	1025	849	426	183	6025
	Rockford	34.8	− 7	114	400	837	1221	1333	1137	961	516	236	6830
	Springfield	40.6	− 1	72	291	696	1023	1135	935	769	354	136	5429
Ind.	Evansville	45.0	6	66	220	606	896	955	767	620	237	68	4435
	Fort Wayne	37.3	0	105	378	783	1135	1178	1028	890	471	189	6205
	Indianapolis	39.6	0	90	316	723	1051	1113	949	809	432	177	5699
	South Bend	36.6	− 2	111	372	777	1125	1221	1070	933	525	239	6439
Iowa	Burlington	37.6	− 4	93	322	768	1135	1259	1042	859	426	177	6114
	Des Moines	35.5	− 7	96	363	828	1225	1370	1137	915	438	180	6588
	Dubuque	32.7	−11	156	450	906	1287	1420	1204	1026	546	260	7376
	Sioux City	34.0	−10	108	369	867	1240	1435	1198	989	483	214	6951
	Waterloo	32.6	−12	138	428	909	1296	1460	1221	1023	531	229	7320
Kans.	Dodge City	42.5	3	33	251	666	939	1051	840	719	354	124	4986
	Goodland	37.8	− 2	81	381	810	1073	1166	955	884	507	236	6141
	Topeka	41.7	3	57	270	672	980	1122	893	722	330	124	5182
	Wichita	44.2	5	33	229	618	905	1023	804	645	270	87	4620
Ky.	Covington	41.4	3	75	291	669	983	1035	893	756	390	149	5265
	Lexington	43.8	6	54	239	609	902	946	818	685	325	105	4683
	Louisville	44.0	8	54	248	609	890	930	818	682	315	105	4660
La.	Alexandria	57.5	25	0	56	273	431	471	361	260	69	0	1921
	Baton Rouge	59.8	25	0	31	216	369	409	294	208	33	0	1560
	Lake Charles	60.5	29	0	19	210	341	381	274	195	39	0	1459
	New Orleans	61.0	32	0	19	192	322	363	258	192	39	0	1385
	Shreveport	56.2	22	0	47	297	477	552	426	304	81	0	2184
Me.	Caribou	24.4	−18	336	682	1044	1535	1690	1470	1308	858	468	9767
	Portland	33.0	− 5	195	508	807	1215	1339	1182	1042	675	372	7511
Md.	Baltimore	43.7	12	48	264	585	905	936	820	679	327	90	4654
	Frederick	42.0	7	66	307	624	955	995	876	741	384	127	5087
Mass.	Boston	40.0	6	60	316	603	983	1088	972	846	513	208	5634
	Pittsfield	32.6	− 5	219	524	831	1231	1339	1196	1063	660	326	7578
	Worcester	34.7	− 3	147	450	774	1172	1271	1123	998	612	304	6969

State	City	Avg. Winter Temp	Design Temp	Sep	Oct	Nov	Dec	Jan	Feb	Mar	Apr	May	Yearly Total
Mich.	Alpena	29.7	− 5	273	580	912	1268	1404	1299	1218	777	446	8506
	Detroit	37.2	4	87	360	738	1088	1181	1058	936	522	220	6232
	Escanaba	29.6	− 7	243	539	924	1293	1445	1296	1203	777	456	8481
	Flint	33.1	− 1	159	465	843	1212	1330	1198	1066	639	319	7377
	Grand Rapids	34.9	2	135	434	804	1147	1259	1134	1011	579	279	6894
	Lansing	34.8	2	138	431	813	1163	1262	1142	1011	579	273	6909
	Marquette	30.2	− 8	240	527	936	1268	1411	1268	1187	771	468	8393
	Muskegon	36.0	4	120	400	762	1088	1209	1100	995	594	310	6696
	Sault Ste. Marie	27.7	−12	279	580	951	1367	1525	1380	1277	810	477	9048
Minn.	Duluth	23.4	−19	330	632	1131	1581	1745	1518	1355	840	490	10000
	Minneapolis	28.3	−14	189	505	1014	1454	1631	1380	1166	621	288	8382
	Rochester	28.8	−17	186	474	1005	1438	1593	1366	1150	630	301	8295
Miss.	Jackson	55.7	21	0	65	315	502	546	414	310	87	0	2239
	Meridian	55.4	20	0	81	339	518	543	417	310	81	0	2289
	Vicksburg	56.9	23	0	53	279	462	512	384	282	69	0	2041
Mo.	Columbia	42.3	2	54	251	651	967	1076	874	716	324	121	5046
	Kansas City	43.9	4	39	220	612	905	1032	818	682	294	109	4711
	St. Joseph	40.3	− 1	60	285	708	1039	1172	949	769	348	133	5484
	St. Louis	43.1	4	60	251	627	936	1026	848	704	312	121	4900
	Springfield	44.5	5	45	223	600	877	973	781	660	291	105	4900
Mont.	Billings	34.5	−10	186	487	897	1135	1296	1100	970	570	285	7049
	Glasgow	26.4	−25	270	608	1104	1466	1711	1439	1187	648	335	8996
	Great Falls	32.8	−20	258	543	921	1169	1349	1154	1063	642	384	7750
	Havre	28.1	−22	306	595	1065	1367	1584	1364	1181	657	338	8700
	Helena	31.1	−17	294	601	1002	1265	1438	1170	1042	651	381	8129
	Kalispell	31.4	− 7	321	654	1020	1240	1401	1134	1029	639	397	8191
	Miles City	31.2	−19	174	502	972	1296	1504	1252	1057	579	276	7723
	Missoula	31.5	− 7	303	651	1035	1287	1420	1120	970	621	391	8125
Neb.	Grand Island	36.0	− 6	108	381	834	1172	1314	1089	908	462	211	6530
	Lincoln	38.8	− 4	75	301	726	1066	1237	1016	834	402	171	5864
	Norfolk	34.0	−11	111	397	873	1234	1414	1179	983	498	233	6979
	North Platte	35.5	− 6	123	440	885	1166	1271	1039	930	519	248	6684
	Omaha	35.6	− 5	105	357	828	1175	1355	1126	939	465	208	6612
	Scottsbluff	35.9	− 8	138	459	876	1128	1231	1008	921	552	285	6673
Nev.	Elko	34.0	−13	225	561	924	1197	1314	1036	911	621	409	7433
	Ely	33.1	− 6	234	592	939	1184	1308	1075	977	672	456	7733
	Las Vegas	53.5	23	0	78	387	617	688	487	335	111	6	2709
	Reno	39.3	2	204	490	801	1026	1073	823	729	510	357	6332
	Winnemucca	36.7	1	210	536	876	1091	1172	916	837	573	363	6761
N. H.	Concord	33.0	−11	177	505	822	1240	1358	1184	1032	636	298	7383
N. J.	Atlantic City	43.2	14	39	251	549	880	936	848	741	420	133	4812
	Newark	42.8	11	30	248	573	921	983	876	729	381	118	4589
	Trenton	42.4	12	57	264	576	924	989	885	753	399	121	4980
N. M.	Albuquerque	45.0	14	12	229	642	868	930	703	595	288	81	4348
	Raton	38.1	− 2	126	431	825	1048	1116	904	834	543	301	6228
	Roswell	47.5	16	18	202	573	806	840	641	481	201	31	3793
	Silver City	48.0	14	6	183	525	729	791	605	518	261	87	3705

State	City	Avg. Winter Temp	Design Temp	Sep	Oct	Nov	Dec	Jan	Feb	Mar	Apr	May	Yearly Total
N. Y.	Albany	34.6	−5	138	440	777	1194	1311	1156	992	564	239	6875
	Binghamton	36.6	−2	141	406	732	1107	1190	1081	949	543	229	6451
	Buffalo	34.5	3	141	440	777	1156	1256	1145	1039	645	329	7062
	New York	42.8	11	30	233	540	902	986	885	760	408	118	4871
	Rochester	35.4	2	126	415	747	1125	1234	1123	1014	597	279	6748
	Schenectady	35.4	−5	123	422	756	1159	1283	1131	970	543	211	6650
	Syracuse	35.2	−2	132	415	744	1153	1271	1140	1004	570	248	6756
N. C.	Asheville	46.7	13	48	245	555	775	784	683	592	273	87	4042
	Charlotte	50.4	18	6	124	438	691	691	582	481	156	22	3191
	Greensboro	47.5	14	33	192	513	778	784	672	552	234	47	3805
	Raleigh	49.4	16	21	164	450	716	725	616	487	180	34	3393
	Wilmington	54.6	23	0	74	291	521	546	462	357	96	0	2347
	Winston-Salem	48.4	14	21	171	483	747	753	652	524	207	37	3595
N. D.	Bismarck	26.6	−24	222	577	1083	1463	1708	1442	1203	645	329	8851
	Devils Lake	22.4	−23	273	642	1191	1634	1872	1579	1345	753	381	9901
	Fargo	24.8	−22	219	574	1107	1569	1789	1520	1262	690	332	9226
	Williston	25.2	−21	261	601	1122	1513	1758	1473	1262	681	357	9243
Ohio	Akron-Canton	38.1	1	96	381	726	1070	1138	1016	871	489	202	6037
	Cincinnati	45.1	8	39	208	558	862	915	790	642	294	96	4410
	Cleveland	37.2	2	105	384	738	1088	1159	1047	918	552	260	6351
	Columbus	39.7	2	84	347	714	1039	1088	949	809	426	171	5660
	Dayton	39.8	0	78	310	696	1045	1097	955	809	429	167	5622
	Mansfield	36.9	1	114	397	768	1110	1169	1042	924	543	245	6403
	Toledo	36.4	1	117	406	792	1138	1200	1056	924	543	242	6494
	Youngstown	36.8	1	120	412	771	1104	1169	1047	921	540	248	6417
Okla.	Oklahoma City	48.3	11	15	164	498	766	868	664	527	189	34	3725
	Tulsa	47.7	12	18	158	522	787	893	683	539	213	47	3860
Ore.	Astoria	45.6	27	210	375	561	679	753	622	636	480	363	5186
	Eugene	45.6	22	129	366	585	719	803	627	589	426	279	4726
	Medford	43.2	21	78	372	678	871	918	697	642	432	242	5008
	Pendleton	42.6	3	111	350	711	884	1017	773	617	396	205	5127
	Portland	45.6	21	114	335	597	735	825	644	586	396	245	4635
	Roseburg	46.3	25	105	329	567	713	766	608	570	405	267	4491
	Salem	45.4	21	111	338	594	729	822	647	611	417	273	4754
Pa.	Allentown	38.9	3	90	353	693	1045	1116	1002	849	471	167	5810
	Erie	36.8	7	102	391	714	1063	1169	1081	973	585	288	6451
	Harrisburg	41.2	9	63	298	648	992	1045	907	766	396	124	5251
	Philadelphia	41.8	11	60	297	620	965	1016	889	747	392	118	5144
	Pittsburgh	38.4	5	105	375	726	1063	1119	1002	874	480	195	5987
	Reading	42.4	6	54	257	597	939	1001	885	735	372	105	4945
	Scranton	37.2	2	132	434	762	1104	1156	1028	893	498	195	6254
	Williamsport	38.5	1	111	375	717	1073	1122	1002	856	468	177	5934
R. I.	Providence	38.8	6	96	372	660	1023	1110	988	868	534	236	5954
S. C.	Charleston	57.9	26	0	34	210	425	443	367	273	42	0	1794
	Columbia	54.0	20	0	84	345	577	570	470	357	81	0	2484
	Florence	54.5	21	0	78	315	552	552	459	347	84	0	2387
	Greenville-Spartanburg	51.6	18	6	121	399	651	660	546	446	132	19	2980
S. D.	Huron	28.8	−16	165	508	1014	1432	1628	1355	1125	600	288	8223
	Rapid City	33.4	−9	165	481	897	1172	1333	1145	1051	615	326	7345
	Sioux Falls	30.6	−14	168	462	972	1361	1544	1285	1082	573	270	7839
Tenn.	Bristol	46.2	11	51	236	573	828	828	700	598	261	68	4143
	Chattanooga	50.3	15	18	143	468	698	722	577	453	150	25	3254
	Knoxville	49.2	13	30	171	489	725	732	613	493	198	43	3494
	Memphis	50.5	17	18	130	447	698	729	585	456	147	22	3232
	Nashville	48.9	12	30	158	495	732	778	644	512	189	40	3578
Tex.	Abilene	53.9	17	0	99	366	586	642	470	347	114	0	2624
	Amarillo	47.0	8	18	205	570	797	877	664	546	252	56	3985
	Austin	59.1	25	0	31	225	388	468	325	223	51	0	1711
	Corpus Christi	64.6	32	0	0	120	220	291	174	109	0	0	914
	Dallas	55.3	19	0	62	321	524	601	440	319	90	6	2363
	El Paso	52.9	21	0	84	414	648	685	445	319	105	0	2700
	Galveston	62.2	32	0	6	147	276	360	263	189	33	0	1274
	Houston	61.0	28	0	6	183	307	384	288	192	36	0	1396
	Laredo	66.0	32	0	0	105	217	267	134	74	0	0	797
	Lubbock	48.8	11	18	174	513	744	800	613	484	201	31	3578
	Port Arthur	60.5	29	0	22	207	329	384	274	192	39	0	1447
	San Antonio	60.1	25	0	31	204	363	428	286	195	39	0	1546
	Waco	57.2	21	0	43	270	456	536	389	270	66	0	2030
	Wichita Falls	53.0	15	0	99	381	632	698	518	378	120	6	2832
Utah	Milford	36.5	−1	99	443	867	1141	1252	988	822	519	279	6497
	Salt Lake City	38.4	5	81	419	849	1082	1172	910	763	459	233	6052
Vt.	Burlington	29.4	−12	207	539	891	1349	1513	1333	1187	714	353	8269
Va.	Lynchburg	46.0	15	51	223	540	822	849	731	605	267	78	4166
	Norfolk	49.2	20	0	136	408	698	738	655	533	216	37	3421
	Richmond	47.3	14	36	214	495	784	815	703	546	219	53	3865
	Roanoke	46.1	15	51	229	549	825	834	722	614	261	65	4150
Wash.	Olympia	44.2	21	198	422	636	753	834	675	645	450	307	5236
	Seattle	46.9	28	129	329	543	657	738	599	577	396	242	4424
	Spokane	36.5	−2	168	493	879	1082	1231	980	834	531	288	6655
	Walla Walla	43.8	12	87	310	681	843	986	745	589	342	177	4805
	Yakima	39.1	6	144	450	828	1039	1163	868	713	435	220	5941
W. Va.	Charleston	44.8	9	63	254	591	865	880	770	648	300	96	4476
	Elkins	40.1	1	135	400	729	992	1008	896	791	444	198	5675
	Huntington	45.0	10	63	257	585	856	880	764	636	294	99	4446
	Parkersburg	43.5	8	60	264	606	905	942	826	691	339	115	4754
Wisc.	Green Bay	30.3	−12	174	484	924	1333	1494	1313	1141	654	335	8029
	La Crosse	31.5	−12	153	437	924	1339	1504	1277	1070	540	245	7589
	Madison	30.9	−9	174	474	930	1330	1473	1274	1113	618	310	7863
	Milwaukee	32.6	−6	174	471	876	1252	1376	1193	1054	642	372	7635
Wyo.	Casper	33.4	−11	192	524	942	1169	1290	1084	1020	657	381	7410
	Cheyenne	34.2	−6	219	543	909	1085	1212	1042	1026	702	428	7381
	Lander	31.4	−16	204	555	1020	1299	1417	1145	1017	654	381	7870
	Sheridan	32.5	−12	219	539	948	1200	1355	1154	1051	642	366	7680

The conduction heat flow through a wall, window, door, roof, ceiling, or floor decreases as more *resistance* is placed in the path of the flow. All materials have some resistance to conduction heat flow. Those that have high resistance are called insulators; those with low resistance are called conductors.

Insulators are compared to one another according to their R-values, which are a measure of their resistance, as discussed in Chapter 3. The R-value of a material increases with its thickness—a 2 inch thick sheet of styrofoam has twice the resistance of a 1 inch sheet. And two similar building materials that differ in density will also differ in R-value. Generally, though not always, the lighter material will have a higher R-value because it has more pockets of air trapped in it. Finally, the average temperature of a material also affects its R-value. The colder it gets, the better most materials retard the flow of heat.

Knowledge of the R-values of insulators and other components permits us to calculate the heat transmission through a wall or other building surface, as explained in Chapter 3. Toward this end, we list the R-values of many common building materials in the first table. R-values are given per inch of thickness and for standard thicknesses. If you have some odd size not listed in the table, use the R-value per inch thickness and multiply by its thickness. Unless otherwise noted, the R-values are quoted for a temperature of 75° F.

Further tables list R-values for surface air films and for air spaces, both of which have insulating value. These R-values vary markedly with the reflectance of the surfaces facing the air film or space. Radiation heat flow is very slow across an air space with aluminum foil on one side, for example, and the R-value of such an air space is correspondingly high. This is why fiberglass batt insulation is often coated with an aluminized surface. In the tables we have used three categories of surface: non-reflective (such as painted wood or metal), fairly reflective (such as aluminum-coated paper), and highly reflective (such as metallic foil). The R-value of an air film or air space also depends on the orientation of the surface and the direction of heat flow that we are trying to retard. These differences are reflected in the tables.

The total resistance R_t of a wall or other building surface is just the sum of the R-values of all its components—including air films and spaces. The coefficient of heat transmission, or U-value, is the inverse of the total resistance $U = 1/R_t$. To get the rate of heat loss through a wall, for example, you multiply its U-value by the total surface area of the wall and by the temperature difference between the indoor and outdoor air. The next table in this section lists U-values for windows and skylights. Here again, the U-value depends upon the surface orientation, the direction of heat flow, and the season of the year. The U-values in these tables apply only to the glazing surfaces; to include the effects of a wood sash, multiply these U-values by about 80 to 90 percent, depending upon the area of the wood.

For the avid reader seeking more detailed information about the insulating values of building materials, we recommend Chapter 20 of the 1972 ASHRAE *Handbook of Fundamentals,* from which most of the present data was taken. Tables on pages 360-363 list resistances *and* conductances of many more materials than are given here. Sample calculations of the U-values of typical frame and masonry walls, roofs, and floors are also provided there.

In all of this discussion, no mention has been made of the relative costs of all the various building alternatives. To a large extent, these depend upon the local building materials suppliers. But charts in the next two sections of this Appendix will help you to assess the savings in fuel costs that can be expected from adding insulation.

R-VALUES OF BUILDING MATERIALS

Material and Description		Density (lb/ft³)	R-value* per inch thickness	R-value* for listed thickness
Building Boards, Panels, Flooring				
Asbestos-cement board		120	0.25	–
Asbestos-cement board	1/8"	120	–	0.03
Gypsum or plaster board	3/8"	50	–	0.32
Gypsum or plaster board	1/2"	50	–	0.45
Plywood (see Siding Materials)		34	1.25	–
Sheathing, wood fiber (impregnated or coated)	25/32"	20	–	2.06
Wood fiber board (laminated or homogenous)		26	2.38	–
Wood fiber, hardboard type		65	0.72	–
Wood fiber, hardboard type	1/4"	65	–	0.18
Wood subfloor	25/32"	–	–	0.98
Wood, hardwood finish	3/4"	–	–	0.68
Building Paper				
Vapor-permeable felt		–	–	0.06
Vapor-seal, 2 layers of mopped 15 lb felt		–	–	0.12
Vapor-seal plastic film		–	–	negl.
Finish Materials				
Carpet and fibrous pad		–	–	2.08
Carpet and rubber pad		–	–	1.23
Cork tile	1/8"	–	–	0.28
Terrazzo	1"	–	–	0.08
Tile (asphalt, linoleum, vinyl, rubber)		–	–	0.05
Gypsumboard	1/2"	–	–	0.45
Gypsumboard	5/8"	–	–	0.56
Hardwood flooring	25/32"	–	–	0.68
Insulating Materials				
Blankets and Batts:				
Mineral wool, fibrous form (from rock, slag or glass)		0.5	3.12	–
		1.5-4.0	3.70	–
Wood fiber		3.2-3.6	4.00	–
Boards and Slabs:				
Cellular glass	90°F	9	2.44	–
	60°F		2.56	–
	30°F		2.70	–
	0°F		2.86	–
Corkboard	90°F	6.5-8.0	3.57	–
	60°F		3.70	–
	30°F		3.85	–
	0°F		4.00	–
	90°F	12	3.22	–
	60°F		3.33	–
	30°F		3.45	–
	0°F		3.57	–
Glass fiber	90°F	4.0-9.0	3.85	–
	60°F		4.17	–
	30°F		4.55	–
	0°F		4.76	–
Expanded rubber (rigid)	75°F	4.5	4.55	–
Expanded polyurethane (R-11 blown; 1" thickness or more)	100°F	1.5-2.5	5.56	–
	75°F		5.88	–
	50°F		6.25	–
	25°F		5.88	–
	0°F		5.88	–
Expanded polystyrene, extruded	75°F	1.9	3.85	–
	60°F		4.00	–
	30°F		4.17	–
	0°F		4.55	–
Expanded polystyrene, molded beads	75°F	1.0	3.57	–
	30°F		3.85	–
	0°F		4.17	–
Mineral fiberboard, felted				
core or roof insulation		16-17	2.94	–
acoustical tile[1]		18	2.86	–
acoustical tile[1]		21	2.73	–
Mineral fiberboard, molded acoustical tile[1]		23	2.38	–
Wood or cane fiberboard				
acoustical tile	1/2"	–	–	1.19
acoustical tile	3/4"	–	–	1.78
interior finish		15	2.86	–

R-VALUES OF BUILDING MATERIALS

Material and Description		Density (lb/ft^3)	R-value* per inch thickness	R-value* for listed thickness
Insulating roof deck[2]	1"	—	—	2.78
	2"	—	—	5.56
	3"	—	—	8.33
Shredded wood (cemented, preformed slabs)		22	1.67	—
Loose Fills:				
Macerated paper or pulp		2.5-3.5	3.57	—
Mineral wool	90°F	2.0-5.0	3.33	—
	60°F		3.70	—
	30°F		4.00	—
	0°F		4.35	—
Perlite (expanded)	90°F	5.0-8.0	2.63	—
	60°F		2.78	—
	30°F		2.94	—
	0°F		3.12	—
Vermiculite (expanded)	90°F	7.0-8.2	2.08	—
	60°F		2.18	—
	30°F		2.27	—
	0°F		2.38	—
Sawdust or shavings		0.8-15	2.22	—
Masonry Materials—Concretes				
Cement mortar		116	0.20	—
Gypsum-fiber concrete (87½% gypsum, 12½% concrete)		51	0.60	—
Lightweight aggregates (expanded shale, clay or slate; expanded slags, or cinders; pumice; perlite or vermiculite; cellular concretes)		120	0.19	—
		100	0.28	—
		80	0.40	—
		60	0.59	—
		40	0.86	—
		20	1.43	—
Sand and gravel or stone aggregate (oven dried)		140	0.11	—
Sand and gravel or stone aggregate (not dried)		140	0.08	—
Stucco		116	0.20	—
Masonry Units				
Brick, common[3]		120	0.20	—
Brick, face[3]		130	0.11	—

R-VALUES OF BUILDING MATERIALS

Material and Description		Density (lb/ft^3)	R-value* per inch thickness	R-value* for listed thickness
Clay tile, hollow				
1 cell deep	3"	—	—	0.80
1 cell deep	4"	—	—	1.11
2 cells deep	6"	—	—	1.52
2 cells deep	8"	—	—	1.85
3 cells deep	10"	—	—	2.22
3 cells deep	12"	—	—	2.50
Concrete block, 3 oval core				
Sand and gravel aggregate	4"	—	—	0.71
	8"	—	—	1.11
	12"	—	—	1.28
Cinder aggregate	3"	—	—	0.86
	4"	—	—	1.11
	8"	—	—	1.72
	12"	—	—	1.89
Lightweight aggregate	3"	—	—	1.27
(expanded shale, clay	4"	—	—	1.50
slate or slag; pumice)	8"	—	—	2.00
	12"	—	—	2.72
Concrete blocks, rectangular core				
Sand and gravel aggregate				
2 core, 36 lb[4]	8"	—	—	1.04
same, filled cores[5]		—	—	1.93
Lightweight aggregates				
3 core, 19 lb[4]	6"	—	—	1.65
same, filled cores[5]		—	—	2.99
2 core, 24 lb[4]	8"	—	—	2.18
same, filled cores[5]		—	—	5.03
3 core, 38 lb[4]	12"	—	—	2.48
same, filled cores[5]		—	—	5.82
Stone, lime or sand		—	0.08	—
Granite, marble		150-175	0.05	—
Plastering Materials				
Cement plaster, sand aggregate		116	0.20	—
Gypsum plaster				
Lightweight aggregate	1/2"	45	—	0.32
Lightweight aggregate	3/8"	45	—	0.39
Same, on metal lath	3/4"	—	—	0.47
Perlite aggregate		45	0.67	—
Sand aggregate		105	0.18	—

R-VALUES OF BUILDING MATERIALS

Material and Description		Density (lb/ft³)	R-value* per inch thickness	R-value* for listed thickness
Same, on metal lath	3/4"	–	–	0.10
Same, on wood lath	3/4"	–	–	0.40
Vermiculite aggregate		45	0.59	–
Roofing Materials				
Asbestos-cement shingles		120	–	0.21
Asphalt roll roofing		70	–	0.15
Built-up roofing	3/8"	70	–	0.44
Slate roofing	1/2"	–	–	0.05
Wood shingles		–	–	0.94
Siding Materials				
Shingles				
Asbestos-cement		120	–	0.21
Wood, 16" with 7½" exposure		–	–	0.80
Wood, double 16" with 12" exposure		–	–	1.19
Wood, plus insulating backer board	5/16"	–	–	1.40
Siding				
Asbestos-cement lapped	1/4"	–	–	0.21
Asphalt roll siding		–	–	0.15
Asphalt insulating siding	1/2"	–	–	1.46
Wood, drop (1" X 8")		–	–	0.79
Wood, drop (½" X 8" lapped)		–	–	0.81
Wood, bevel (¾" X 10", lapped)		–	–	1.05
Plywood, lapped	3/8"	–	–	0.59
Plywood	1/4"	–	–	0.31
	3/8"	–	–	0.47
	1/2"	–	–	0.62
	5/8"	–	–	0.78
	3/4"	–	–	0.94
Stucco		116	0.20	–
Sheathing, insulating board	1/2"	–	–	1.32
(regular density)	25/32"	–	–	2.04
Woods				
Hardwoods (maple, oak)		45	0.91	–
Softwoods (fir, pine)		32	1.25	–

R-VALUES OF BUILDING MATERIALS

Material and Description		Density (lb/ft³)	R-value* per inch thickness	R-value* for listed thickness
	25/32"	32	–	0.98
	1–5/8"	32	–	2.03
	2–5/8"	32	–	3.28
	3–5/8"	32	–	4.55
Wood Doors				
Solid core	1"	–	–	1.56
	1–1/4"	–	–	1.82
	1–1/2"	–	–	2.04
	2"	–	–	2.33

*Representative values intended for use as design values of dry building materials in normal use.
[1] R-values of acoustical tile depend upon the board and the type, size and depth of perforations; these are average values.
[2] Roof deck insulation is made in thicknesses to meet these standards; thickness may vary somewhat with manufacturer.
[3] Face brick and common brick do not always have these densities and R-values.
[4] Weights of blocks approximately 7–5/8" high by 15–3/8" long.
[5] Vermiculite, perlite, or mineral wool insulation.
SOURCE: ASHRAE *Handbook of Fundamentals.* 1967. Reprinted by permission.

R-VALUES OF AIR FILMS

Type and Orientation of Air Film	Direction of Heat Flow	R-value for Air Film On: Non-reflective surface	Fairly reflective surface	Highly reflective surface
Still air:				
Horizontal	up	0.61	1.10	1.32
Horizontal	down	0.92	2.70	4.55
45° slope	up	0.62	1.14	1.37
45° slope	down	0.76	1.67	2.22
Vertical	across	0.68	1.35	1.70
Moving air:				
15 mph wind	any*	0.17	–	–
7½ mph wind	any†	0.25	–	–

*Winter conditions.
†Summer conditions.
SOURCE: ASHRAE, *Handbook of Fundamentals,* 1972. Reprinted by permission.

R-VALUES OF AIR SPACES

Orientation & Thickness of Air Space		Direction of Heat Flow	R-value for Air Space Facing:‡		
			Non-reflective surface	Fairly reflective surface	Highly reflective surface
Horizontal	¾"	up*	0.87	1.71	2.23
	4"		0.94	1.99	2.73
	¾"	up†	0.76	1.63	2.26
	4"		0.80	1.87	2.75
	¾"	down*	1.02	2.39	3.55
	1½"		1.14	3.21	5.74
	4"		1.23	4.02	8.94
	¾	down†	0.84	2.08	3.25
	1½"		0.93	2.76	5.24
	4"		0.99	3.38	8.03
45° slope	¾"	up*	0.94	2.02	2.78
	4"		0.96	2.13	3.00
	¾"	up†	0.81	1.90	2.81
	4"		0.82	1.98	3.00
	¾"	down*	1.02	2.40	3.57
	4"		1.08	2.75	4.41
	¾"	down†	0.84	2.09	3.34
	4"		0.90	2.50	4.36
Vertical	¾"	across*	1.01	2.36	3.48
	4"		1.01	2.34	3.45
	¾"	across†	0.84	2.10	3.28
	4"		0.91	2.16	3.44

‡One side of the air space is a non-reflective surface.
*Winter conditions.
†Summer conditions.
SOURCE: ASHRAE, *Handbook of Fundamentals*, 1972. Reprinted by permission.

U-VALUES OF WINDOWS AND SKYLIGHTS

Description	U-values[1]	
	Winter	Summer
Vertical panels:		
Single pane flat glass	1.13	1.06
Insulating glass—double[2]		
3/16" air space	0.69	0.64
1/4" air space	0.65	0.61
1/2" air space	0.58	0.56
Insulating glass—triple[2]		
1/4" air spaces	0.47	0.45
1/2" air spaces	0.36	0.35
Storm windows		
1-4" air space	0.56	0.54
Glass blocks[3]		
6 X 6 X 4" thick	0.60	0.57
8 X 8 X 4" thick	0.56	0.54
same, with cavity divider	0.48	0.46
Single plastic sheet	1.09	1.00
Horizontal panels:[4]		
Single pane flat glass	1.22	0.83
Insulating glass—double[2]		
3/16" air space	0.75	0.49
1/4" air space	0.70	0.46
1/2" air space	0.66	0.44
Glass blocks[3]		
11 X 11 X 3" thick, with cavity divider	0.53	0.35
12 X 12 X 4" thick, with cavity divider	0.51	0.34
Plastic bubbles[5]		
single-walled	1.15	0.80
double-walled	0.70	0.46

[1] in units of $Btu/hr/ft^2/°F$
[2] double and triple refer to the number of lights of glass.
[3] nominal dimensions.
[4] U-values for horizontal panels are for heat flow *up* in winter and *down* in summer.
[5] based on area of opening, not surface.
SOURCE: ASHRAE, *Handbook of Fundamentals*, 1972. Reprinted by permission.

By adding insulating materials to a wall or other building surface, you can lower its U-value, or heat transmission coefficient. But it takes a much greater amount of insulation to lower a small U-value than it does to lower a large U-value. For example, adding 2 inches of polyurethane insulation (R = 12) to a solid 8-inch concrete wall reduces the U-value from 0.66 to 0.07, or almost a factor of 10. Adding the same insulation to a good exterior stud wall reduces the U-value from 0.069 to 0.038, or less than a factor of 2. Mathematically, if U_i is the initial U-value of a building surface, and R is the resistance of the added insulation, the final U-value, U_f, is:

$$U_f = \frac{U_i}{1 + RU_i}$$

If you don't have a pocket calculator handy, the following chart will help you to tell at a glance the effects of adding insulation to a wall or other building surface. The example shows you how to use this chart.

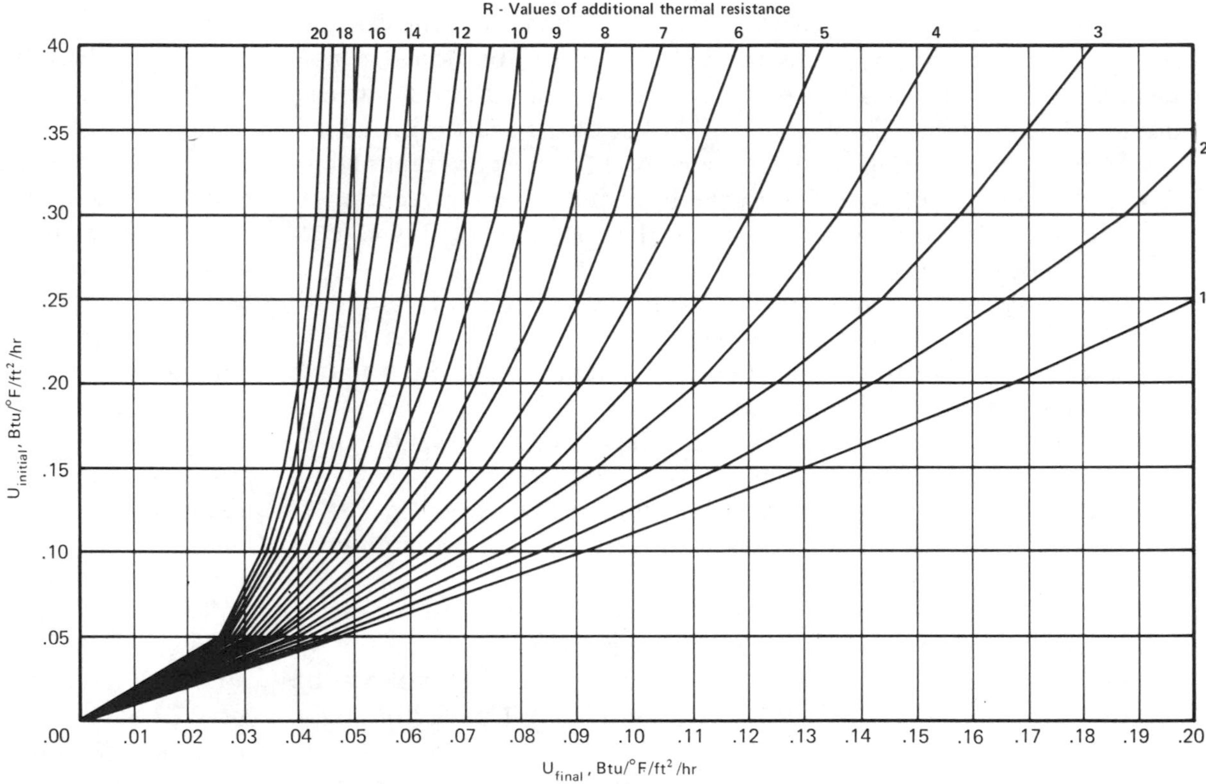

NOTES: Final U value obtained by adding an amount of thermal resistance, R, to an exterior skin having a coefficient of heat transmission, $U_{initial}$.

Example: If 3½ inches of fiberglass insulation (R = 11) is added to an uninsulated stud wall having a U-value of 0.23, what is the final U-value? When adding the insulation, you remove the insulating value of the air space (R = 1.01) inside the wall, so the net increase in resistance is R = 10. To use the chart, begin at U_i = 0.23 on the left-hand scale. Move horizontally to intersect the curve numbered R = 10. Drop down from this point ot the bottom scale to find the final U-value, U_f = 0.069, which agrees with calculations in Chapter 3. With this information, you can now use the Heat Conduction Cost Chart in the next section to find the fuel savings resulting from the added insulation.

The Heat Conduction Cost Chart provided here simplifies the calculation of total seasonal heat loss through wall, roof, or floor. It also facilitates the study of alternative constructions and possible savings due to added insulation. An example of the use of the chart is included:

- For a building surface with a U-value of 0.58, start at point (1);
- follow up the oblique line to the horizontal line representing the total heating degree days for the location, in this case 7,000 degree days (2);
- move vertically from this point to find a heat loss of 95,000 Btu/ft^2 per season for the surface (3);
- continue vertically to the oblique line representing the total area of the surface, 100 square feet (4);
- moving horizontally from this point, the total heat loss through the entire surface for the season is 9,500,000 Btu (5);
- continue horizontally to the oblique line representing the cost per million Btu of heat energy, in this case $9 per million Btu (6);
- moving vertically down from this point, the total cost for the season of the heat through that surface is $86 (7).

The lower right graph converts the apparent cost to a "real cost of energy" through the use of a multiplication factor. This factor might reflect:

(1) Estimated future cost of energy—design decisions based on present energy costs make little sense as costs soar.

(2) Real environmental cost of using fossil fuels—this particularly includes pollution and the depletion of natural resources, both directly as fuels burn and indirectly as they are brought to the consumer from the source.

(3) Initial investment cost—use of the proper multiplication factor would give the quantity of increased investment made possible by resultant yearly fuel savings.

For example, heat costs may increase by a factor of 10. Continue down from the last point until you intersect the oblique line representing the multiplication factor 10 (8). Then move horizontally left to arrive at the adjusted seasonal heating cost through the building surface of $862 (9).

The numerical values of the chart can be changed by a factor of ten. For example, to determine the heat transfer through a really good exterior wall, U = 0.05, use U = 0.5 on the chart and divide the final answer by ten. Each of the graphs can be used independently of one another. For example, knowing a quantity of energy and its price, the upper right graph gives the total cost of that energy.

The Heat Conduction Cost Chart can help you compare the energy costs of two different methods of insulation. For example, an insulated stud wall has a U-value of 0.07 and an uninsulated wall has a U-value of 0.23. The *difference* 0.23 − 0.07 = 0.16 can be run through the chart in the same way as done for a single U-value. Assuming 5000 degree days, 100 square feet of wall, and $9 per million Btu, the savings in heating costs for one year is about $21, or more than the cost of insulation.

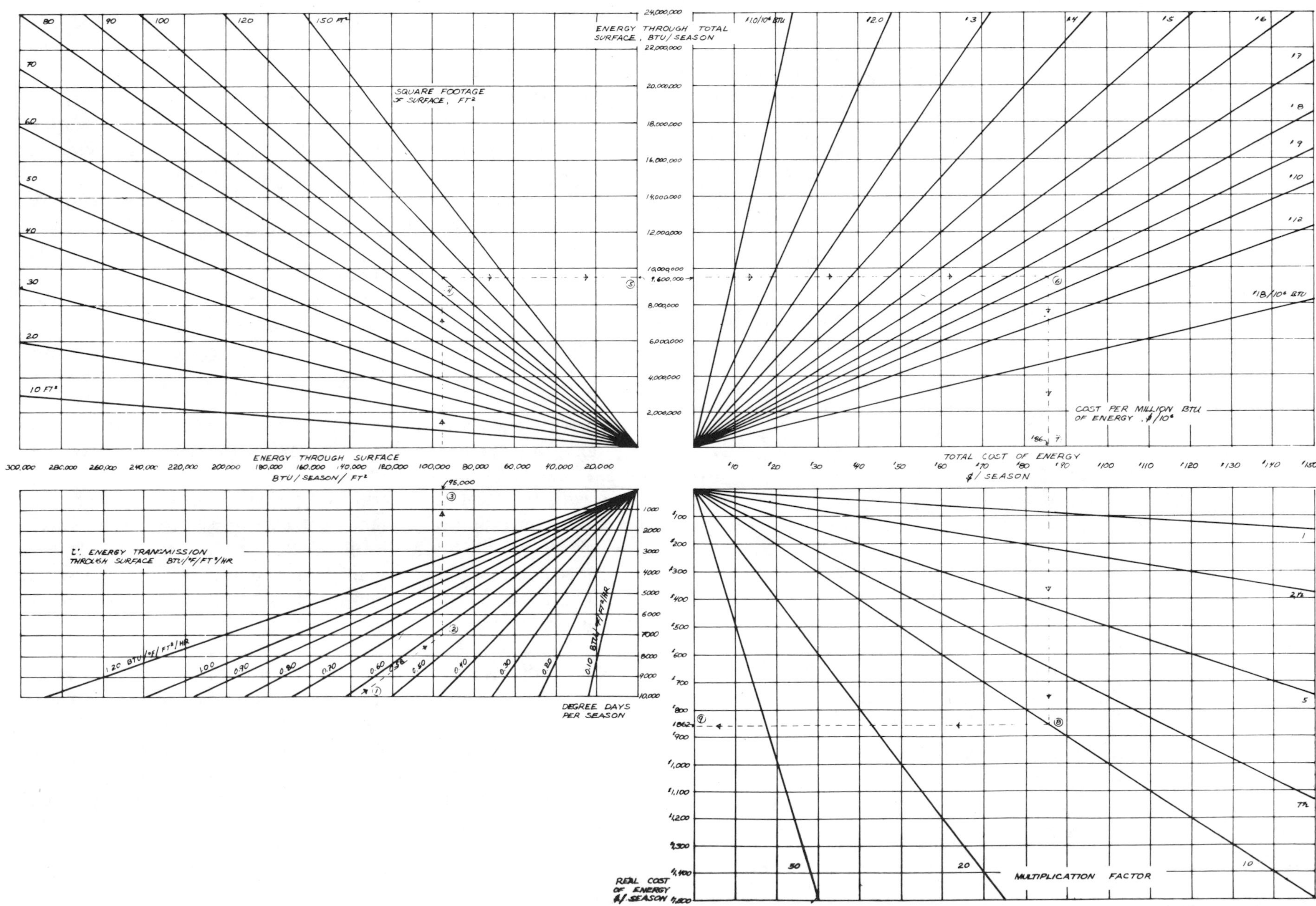

SQUARE FOOTAGE OF SURFACE, FT²
ENERGY THROUGH TOTAL SURFACE, BTU/SEASON
ENERGY THROUGH SURFACE BTU/SEASON/FT²
TOTAL COST OF ENERGY $/SEASON
COST PER MILLION BTU OF ENERGY, $/10⁶
"U" ENERGY TRANSMISSION THROUGH SURFACE BTU/°F/FT²/HR
DEGREE DAYS PER SEASON
REAL COST OF ENERGY $/SEASON
MULTIPLICATION FACTOR

The use of this chart to calculate the costs of heat loss through air infiltration is similar to the use of the Heat Conduction Cost Chart. An example of its use is included:

- Start at point (1) for an air infiltration rate of 45 ft^3/hr/ft;
- follow up the oblique line to the horizontal line representing the total heating degree days for the location, in this case 7000 degree days (2);
- move vertically from this point to find that 146,000 Btu are consumed each heating season per crack foot (3);
- continue vertically to the oblique line corresponding to the total crack length, in this case 30 feet (4);
- move horizontally from this point to the total seasonal heat loss through the window crack—4,400,000 Btu (5);
- continue horizontally to the oblique line representing the cost per million Btu of heat energy, in this case $6 per million Btu (6);
- move vertically down from this point to the total cost for heat lost through the crack during an entire heating season, or $26.75 (7).

As with the Heat Conduction Cost Chart, the bottom right graph permits a conversion of this apparent cost to a "real cost of energy" through the use of a multiplication factor. In this example, a factor of 10 is used. Continue down from the last point until you intersect the oblique line representing a multiplication factor 10 (8). Then move horizontally left to arrive at an adjusted heating cost of about $270 per heating season (9).

As with the previous chart, you can use this Air Infiltration Cost Chart to make quick evaluations of the savings resulting from changes in the rate of air infiltration. For example, if a wood-sash, double-hung window is weatherstripped, the air infiltration rate will drop from 39 to 24 cubic feet per hour per crack foot. By moving through the chart from a starting point of 15 ft^3/hr/ft, you arrive ultimately at the savings resulting from weatherstripping. Assuming 5000 degree days, 15 feet of crack, and $9 per million Btu, we get an immediate savings of about $6 in the first heating season. Since weatherstripping costs a few cents per foot, it can pay for itself in fuel savings within a few weeks.

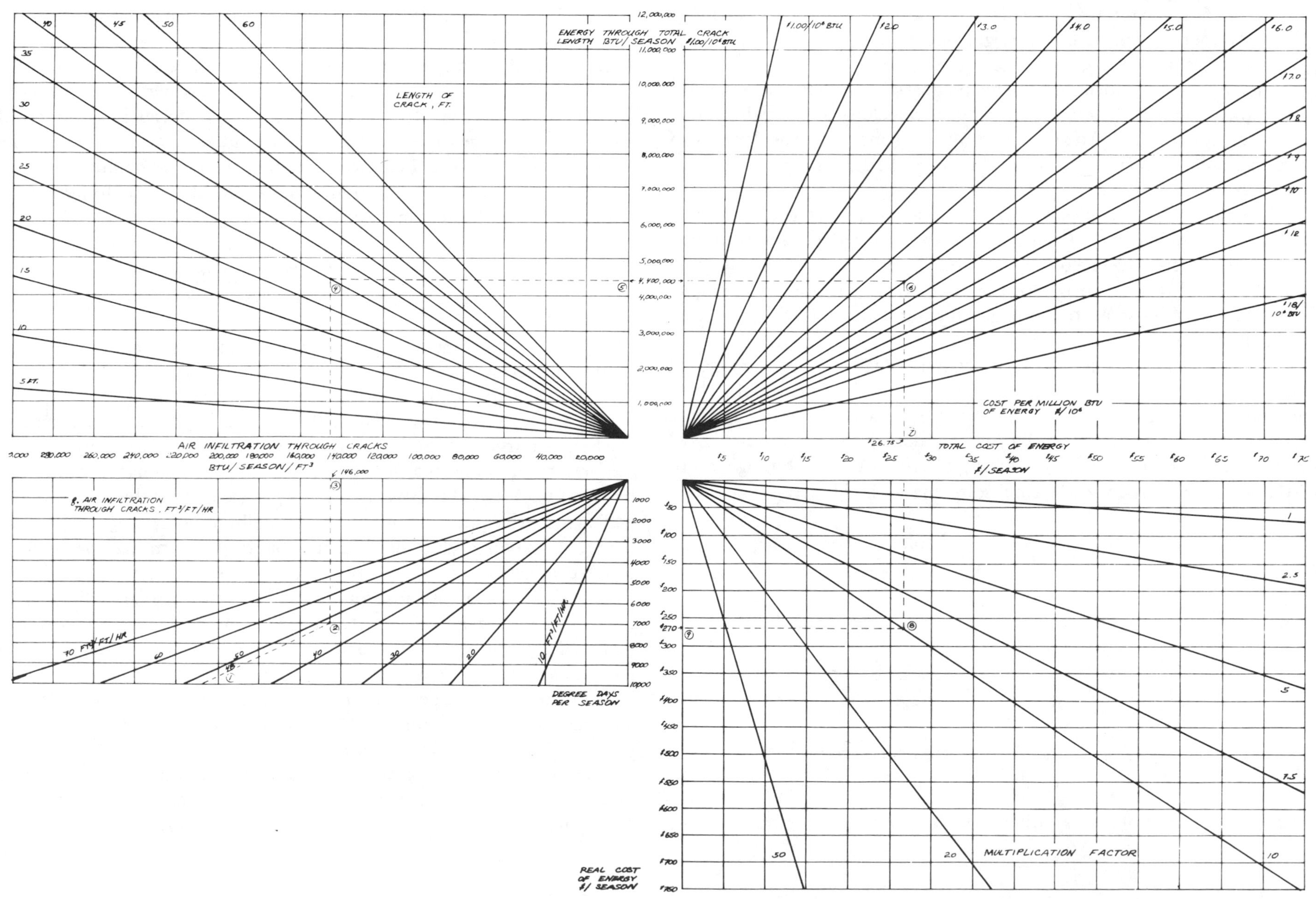
ENERGY THROUGH TOTAL CRACK LENGTH BTU/SEASON
LENGTH OF CRACK, FT.
AIR INFILTRATION THROUGH CRACKS BTU/SEASON/FT³
COST PER MILLION BTU OF ENERGY $/10⁶
TOTAL COST OF ENERGY $/SEASON
8. AIR INFILTRATION THROUGH CRACKS, FT³/FT/HR
DEGREE DAYS PER SEASON
MULTIPLICATION FACTOR
REAL COST OF ENERGY $/SEASON

Radiation is an important method of heat transfer between two surfaces. As explained in Chapter 3, *any* warm body emits energy in the form of electromagnetic radiation. We might say every warm object has an "aura" that is invisible to our eyes.

Sunlight is one form of electromagnetic radiation and thermal radiation is another. They differ only in wavelength. Sunlight comes in wavelengths ranging from 0.3 to 3.0 microns. The wavelengths of thermal radiation from warm bodies (say 100°F) range from 3 to 50 microns.

When radiation strikes a surface of any material, it is either absorbed, reflected or transmitted. Each material absorbs, reflects and transmits radiation differently—according to its physical and chemical characteristics and the wavelength of the incoming radiation. For example, glass transmits most of the sunlight hitting it but absorbs almost all thermal radiation.

We can assign numerical ratings that gauge the percentage of radiation absorbed, reflected or transmitted by a material. These numbers depend upon the temperature of the material and the wavelength of the radiation. We usually define the *absorptance* α of a material as the ratio of solar energy (in the wavelength range 0.3 to 3.0 microns) absorbed to the total solar energy incident:

$$\alpha = \frac{I_a}{I} = \frac{\text{absorbed solar energy}}{\text{incident solar energy}}$$

The *reflectance* ρ and *transmittance* τ are similarly defined ratios:

$$\rho = \frac{I_r}{I} = \frac{\text{reflected solar energy}}{\text{incident solar energy}}$$

$$\tau = \frac{I_t}{I} = \frac{\text{transmitted solar energy}}{\text{incident solar energy}}$$

Because all the sunlight is either absorbed, reflected or transmitted, $\alpha + \rho + \tau = 1$. For opaque solids, no energy is transmitted, so that $\alpha + \rho = 1$ or $\rho = 1 - \alpha$. If we know the absorptance of an opaque material, we also know its reflectance.

Once absorbed, this radiant energy is transformed into heat energy—the motion of molecules. The body becomes warmer and emits more radiation of its own. The *emittance* ϵ of a material is a numerical indicator of that material's propensity to radiate away its energy. The emittance is defined as the ratio of the thermal radiation emitted from a material to the thermal radiation emitted by a hypothetical "blackbody" with the same shape and temperature:

$$\epsilon = \frac{R_m}{R_b} = \frac{\text{radiation from material}}{\text{radiation from blackbody}}$$

With $\epsilon = 1$, the blackbody is a theoretically "perfect" emitter of thermal radiation.

A knowledge of the absorptances and emittances of materials helps us to evaluate their relative thermal performance. For example, brick, masonry and concrete have emittances around 0.9—so they are better heat radiators than galvanized iron, which has an emittance between 0.13 and 0.28. With an absorptance greater than 0.9, asphalt paving absorbs much more of the sunlight than sand (α = 0.60 to 0.75), as anyone who has walked barefoot from parking lot to beach can testify.

The ratio α/ϵ of the absorptance (of shortwave solar radiation) to the emittance (of longwave thermal radiation) has special importance in the design of solar collectors. In general, you want materials with high values of α/ϵ for the absorber coating. Then a large percentage of solar radiation is absorbed, but only a small amount lost by re-radiation. Materials with high values of both α and α/ϵ are called "selective surfaces," as explained in Chapter 6.

The ensuing tables list absorptances and emittances of many common and some uncommon materials. They are grouped into two categories according to whether α/ϵ is less than or greater than 1.0.

CLASS I SUBSTANCES: Absorptance to Emittance Ratios (α/ϵ) Less than 1.0

Substance	α	ϵ	α/ϵ
White plaster	0.07	0.91	0.08
Snow, fine particles, fresh	0.13	0.82	0.16
White paint on aluminum	0.20	0.91	0.22
Whitewash on galvanized iron	0.22	0.90	0.24
White paper	0.25-0.28	0.95	0.26-0.29
White enamel on iron	0.25-0.45	0.90	0.28-0.50
Ice, with sparse snow cover	0.31	0.96-0.97	0.32
Snow, ice granules	0.33	0.89	0.37
Aluminum oil base paint	0.45	0.90	0.50
Asbestos felt	0.25	0.50	0.50
White powdered sand	0.45	0.84	0.54
Green oil base paint	0.50	0.90	0.56
Bricks, red	0.55	0.92	0.60
Asbestos cement board, white	0.59	0.96	0.61
Marble, polished	0.5-0.6	0.90	0.61
Rough concrete	0.60	0.97	0.62
Concrete	0.60	0.88	0.68
Grass, wet	0.67	0.98	0.68
Grass, dry	0.67-0.69	0.90	0.76
Vegetable fields and shrubs, wilted	0.70	0.90	0.78
Oak leaves	0.71-0.78	0.91-0.95	0.78-0.82
Grey paint	0.75	0.95	0.79
Desert surface	0.75	0.90	0.83
Common vegetable fields and shrubs	0.72-0.76	0.90	0.82
Red oil base paint	0.74	0.90	0.82
Asbestos, slate	0.81	0.96	0.84
Ground, dry plowed	0.75-0.80	0.70-0.96	0.83-0.89
Linoleum, red-brown	0.84	0.92	0.91
Dry sand	0.82	0.90	0.91
Green roll roofing	0.88	0.91-0.97	0.93
Slate, dark grey	0.89	—	—
Bare moist ground	0.90	0.95	0.95
Wet sand	0.91	0.95	0.96
Water	0.94	0.95-0.96	0.98
Black tar paper	0.93	0.93	1.0
Black gloss paint	0.90	0.90	1.0

CLASS I SUBSTANCES: Absorptance to Emittance Ratios (α/ϵ) (Continued) Less than 1.0

Substance	α	ϵ	α/ϵ
Small hole in large box, furnace or enclosure	0.99	0.99	1.0
"Hohlraum," theoretically perfect black body	1.00	1.0	1.0

CLASS II SUBSTANCES: Absorptance to Emittance Ratios (α/ϵ) Greater than 1.0

Substance	α	ϵ	α/ϵ
Black silk velvet	0.99	0.97	1.02
Alfalfa, dark green	0.97	0.95	1.02
Lamp black	0.98	0.95	1.03
Black paint on aluminum	0.94-0.98	0.88	1.07-1.11
Granite	0.55	0.44	1.25
Dull brass, copper, lead	0.2-0.4	0.4-0.65	1.63-2.0
Graphite	0.78	0.41	1.90
Stainless steel wire mesh	0.63-0.86	0.23-0.28	2.70-3.0
Galvanized sheet iron, oxidized	0.80	0.28	2.86
Galvanized iron, clean, new	0.65	0.13	5.00
Aluminum foil	0.15	0.05	3.00
Cobalt oxide on polished nickel*	0.93-0.94	0.24-0.40	3.9
Magnesium	0.30	0.07	4.3
Chromium	0.49	0.08	6.13
Nickel black on galvanized iron*	0.89	0.12	7.42
Cupric oxide on sheet aluminum*	0.85	0.11	7.73
Nickel black on polished nickel*	0.91-0.94	0.11	8.27-8.55
Polished zinc	0.46	0.02	23.0

*Selective surfaces

SOURCES: ASHRAE, *Handbook of Fundamentals*, 1972.
Bowden, *Alternative Sources of Energy*, July 1973.
Duffie and Beckman, *Solar Energy Thermal Processes*, 1974.
McAdams, *Heat Transmission*, 1954.
Severns and Fellows, *Air Conditioning and Refrigeration*, 1966.
Sounders, *The Engineer's Companion*, 1966.

Different materials absorb different amounts of heat while undergoing the same temperature rise. Ten pounds of water will absorb 100 Btu during a 10° F temperature rise, but 10 pounds of cast iron will absorb only 12 Btu over the same range. There are two common measures of the ability of a material to absorb and store heat—its specific heat and its heat capacity.

The *specific heat* of a material is the number of Btu's absorbed by a pound of that material as its temperature rises 1° F. All specific heats vary with temperature and a distinction must be made between the *true* and the *mean* specific heat. The true specific heat is the number of Btu's absorbed per pound per ° F temperature rise at a fixed temperature. Over a wider temperature range, the mean specific heat is the average number of Btu's absorbed per pound per ° F temperature rise. In the following table, only true specific heats are given—for room temperature unless otherwise noted.

The *heat capacity* of a material is the amount of heat absorbed by one cubic foot of that material during a 1° F temperature rise. The heat capacity is just the product of the density of the material (in lb/ft^3) times its specific heat. Specific heats, heat capacities, and densities of common building materials and other substances are given in the following table.

Material	Specific Heat (Btu/lb/°F)	Density (lb/ft^3)	Heat Capacity (Btu/ft^3/°F)
Air (at 1 atmosphere)	0.24 [75]	0.075	0.018
Aluminum (alloy 1100)	0.214	171	36.6
Asbestos fiber	0.25	150	37.5
Asbestos insulation	0.20	36	7.2
Ashes, wood	0.20	40	8.0
Asphalt	0.22	132	29.0
Bakelite	0.35	81	28.4
Brick, building	0.2	123	24.6
Brass, red (85% Cu, 15% Zn)	0.09	548	49.3
Brass, yellow (65% Cu, 35% Zn)	0.09	519	46.7
Bronze	0.104	530	55.1
Cellulose	0.32	3.4	1.1
Cement (Portland clinker)	0.16	120	19.2
Chalk	0.215	143	30.8
Charcoal (wood)	0.20	15	3.0
Clay	0.22	63	13.9
Coal	0.3	90	27.0
Concrete (stone)	0.22	144	31.7
Copper (electrolytic)	0.092	556	51.2
Cork (granulated)	0.485	5.4	2.6
Cotton (fiber)	0.319	95	30.3
Ethyl alcohol	0.68	49.3	33.5
Fireclay brick	0.198 [212]	112	22.2
Glass, crown (soda-lime)	0.18	154	27.7
Glass, flint (lead)	0.117	267	31.2
Glass, pyrex	0.20	139	27.8
Glass, "wool"	0.157	3.25	0.5
Gypsum	0.259	78	20.2
Hemp (fiber)	0.323	93	30.0
Ice	0.487 [32]	57.5	28.0
Iron, cast	0.12 [212]	450	54.0
Lead	0.031	707	21.8
Limestone	0.217	103	22.4
Magnesium	0.241	108	26.0
Marble	0.21	162	34.0
Nickel	0.105	555	58.3
Octane	0.51	43.9	22.4
Paper	0.32	58	18.6
Paraffin	0.69	56	38.6
Porcelain	0.18	162	29.2
Rock salt	0.219	136	29.8
Salt water	0.75	72	54.0
Sand	0.191	94.6	18.1
Silica	0.316	140	44.2
Silver	0.056	654	36.6
Steel (mild)	0.12	489	58.7
Stone (quarried)	0.2	95	19.0
Tin	0.056	455	25.5
Tungsten	0.032	1210	38.7
Water	1.0 [39]	62.4	62.4
Wood, white oak	0.570	47	26.8
Wood, white fir	0.65	27	17.6
Wood, white pine	0.67	27	18.1
Zinc	0.092	445	40.9

*Values are for room temperature unless otherwise noted in brackets.

CONVERSION FACTORS

Multiply:	By:	To obtain:
Acres	43,560	Square feet
Acre-feet	1,233.5	Cubic meters
Barrels, oil (crude)	5.8×10^6	Btu
Barrels, oil	5.615	Cubic feet
Btu	777.48	Foot-pounds
Btu	1,055	Joules
Btu	0.29305	Watt-hours
Btu/hr/ft^2/°F	5.682×10^4	Watts/cm^2/°C
Btu per square foot	0.271	Langleys (cal/cm^2)
Calories	3.9685×10^{-3}	Btu
Calories	4.184	Joules
Cords	128	Cubic feet
Cubic feet	0.037037	Cubic yards
Cubic feet	7.48	Gallons
Cubic feet per second	448.83	Gallons per minute
Feet of water (39.2°F)	0.4335	Pounds per square inch
Feet of water	0.88265	Inches of mercury at 32°F
Gallons	0.1337	Cubic feet
Gallons of water at 60°F	8.3453	Pounds
Horsepower	33,000	Foot-pounds per minute
Horsepower	42.42	Btu per minute
Horsepower	2,546	Btu per hour
Horsepower	1.014	Metric horsepower

CONVERSION FACTORS

Multiply:	By:	To obtain:
Horsepower	0.7457	Kilowatts
Inches of mercury at 32°F	0.4912	Pounds per square inch
Kilowatts	56.90	Btu per minute
Kilowatts	1.341	Horsepower
Kilowatt-hours	3,413	Btu
Kilowatt-hours	2.66×10^6	Foot-pounds
Langleys (cal/cm^2)	3.69	Btu per square foot
Langleys per minute	0.0698	Watts per square centimeter
Microns	1×10^{-4}	Centimeters
Months (mean calendar)	730.1	Hours
Newtons	0.22481	Pounds (force)
Pounds of water	0.1198	Gallons
Pounds per square inch	0.068046	Standard atmospheres
Pounds per square inch	51.715	Millimeters of mercury at 0°C
Standard atmospheres	14.696	Pounds per square inch
Tons (short)	2,000	Pounds
Tons (short)	0.907185	Metric tons
Tons (metric)	2,204.62	Pounds
Tons of refrigeration	12,000	Btu per hour
Therms	1×10^5	Btu
Watts	3.413	Btu per hour
Watts	0.00134	Horsepower

METRIC / ENGLISH EQUIVALENTS

English Measure	Metric Equivalent
inch	2.54 centimeters
foot	30.50 centimeters
yard	0.91 meter
mile (statute)	1.60 kilometers
square inch	6.45 square centimeters
square foot	929.00 square centimeters
square yard	0.84 square meter
square mile	2.60 square kilometers
ounce	28.30 grams
pound (mass)	0.45 kilogram
short ton	907.00 kilograms
fluid ounce	29.60 milliliters
pint	0.47 liter
quart	0.95 liter
gallon	3.78 liters
cubic foot	0.03 cubic meter
cubic yard	0.76 cubic meter
Btu	251.98 calories
pound (force)	4.45 newtons

Metric Measure	English Equivalent
millimeter	0.04 inch
centimeter	0.39 inch
meter	3.28 feet
meter	1.09 yards
kilometer	0.62 miles
square centimeter	0.16 square inch
square meter	1.19 square yards
square kilometer	0.38 square mile
gram	0.035 ounces
kilogram	2.20 pounds
ton (1,000 kg)	1.10 short tons
milliliters	0.03 fluid ounce
liter	1.06 quarts
liter	0.26 gallon
cubic meter	35.3 cubic feet
cubic meter	1.3 cubic yards
calorie	0.004 Btu
newton	0.225 pound (force)

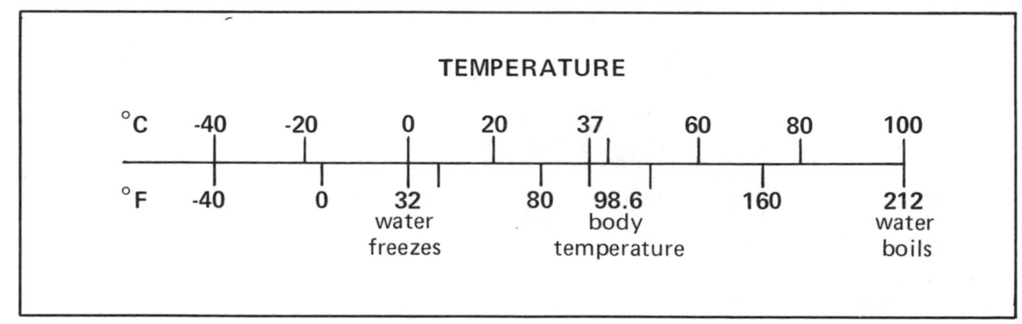

Many readers still need additional design or engineering assistance before they can proceed confidently with the construction of a solar home. Others prefer to leave the whole enterprise in the hands of qualified people. Most of the following firms and individuals have an established record of competence in the field of solar heating and cooling. You can contact them for help on a specific project, but try not to bother them with inquiries for free information—they are probably very overworked right now! This list is by no means complete and it does not constitute a warranty of their services.

ABR Partnership, 1200 Walnut Street, Denver, CO 80204 (John D. Anderson)

Alten Associates, 3080 Olcott Street, Suite 200D, Santa Clara, CA 95051 (Klause Heinemann, President).

The Architects Taos, Box 1884, Taos, NM (Keith W. Haggard and William Mingenbach).

Banwell, White, and Arnold, Architects, Inc., 2 Wheelock Street, Hanover, NH 03755 (Stuart C. White).

Berkeley Solar Group, 3026 Shattuck Avenue, Berkeley, CA 94703 (Marshall F. Merriam, Douglas C. Daniels, Bruce A. Wilcox).

Bridgers and Paxton, Consulting Engineers, Inc., 213 Truman Street NE, Albuquerque, NM 87108 (Frank H. Bridgers and D. D. Paxton).

Burt, Hill, and Associates, 610 Mellon Bank Building, Butler, PA 16001 (Richard P. Rittleman).

Butler, Lee Porter, 3375 Clay Street, San Francisco, CA 94118 (architect and planner).

Colorado Sunworks, P.O. Box 455, Boulder, CO 80302 (Paul Shippee).

Crowther, Kruse, and McWilliams, Architects, 2830 East Third Avenue, Denver, CO 80206 (Richard L. Crowther).

D.A.W.N. Associates, 1220 North Progress Street, Apt. 3, Blacksburg, VA 24060 (Eugene Eccli).

Donovan and Bliss Co., Chocorua, NH 03817 (Mary Donovan and Raymond Bliss).

Dresser, Peter van, 634 Garcia Street, Santa Fe, NM 87501 (ecologist and designer).

Dubin-Bloome Associates, Consulting Engineers, 42 West 39th Street, New York, NY 10018 (Fred S. Dubin and David L. Hartman).

Environmental Consulting Services, Inc., 1929 Walnut Street, Boulder, CO 80302 (Dr. Jan D. Kreider, President)

Glave, Neman, Anderson and Associates, Inc., 209 West Franklin Street, Richmond, VA 23220 (James M. Glave and Richard B. Prud'homme).

Hankins and Anderson, Inc., 2117 North Hamilton Street, Richmond, VA 23230 (Richard P. Hankins).

Harrison Fraker Architects, 245 Nassau Street, Princeton, NJ.

Interactive Resources, Inc., 39 Washington Avenue, Point Richmond, CA 94801 (T. K. Butt, John E. Clinton, Dale A. Sartor).

John Yellott Engineering Associates, Inc., 901 West El Caminito Drive, Phoenix, AZ 85021 (John I. Yellott).

Joint Venture, Inc., 1406 Pearl Street, Boulder, CO 80302 (Alan Brown, President).

Kelbaugh, Douglas, 9 Pine Street, Princeton, NJ (designer).

Living Systems, Route 1, Box 170, Winters, CA 95694 (Jonathan Hammond).

Low Energy Design, 6639 West Windsor Street, Phoenix, AZ 85035 (Daniel Aiello and Guilford A. Rand).

Lumpkins, William, 535 Camino del Monte Sol, Santa Fe, NM 87501.

Mass Design Architects and Planners, Inc., 18 Brattle Street, Cambridge, MA 02138 (Gordon F. Tully, President).

Materials Consultants, Inc., 2150 South Josephine Street, Denver, CO 80210 (Jerry D. Plunkett).

Mathew, Henry, Route 3, Box 768, Coos Bay, OR 97420.

Moore, Grover, Harper, Architects, Essex, CT 06426 (Charles W. Moore).

People Space Co., 259 Marlboro Street, Boston, MA 02116 (Robert F. Shannon).

Prowler, Donald, 2216 Saint James Street, Philadelphia, PA (consultant).

Saunders, Norman B. 15 Ellis Road, Weston, MA 02193 (engineer and consultant).

Shore, Ron, P.O. Box 238, Snowmass, CO 81654 (designer).

Sky Therm Processes and Engineering Co., 2424 Wilshire Boulevard, Los Angeles, CA 90057 (Harold R. Hay).

Solar Group, Inc., 2830 East Third Avenue, Denver, CO 80206 (Richard L. Crowther).

Solar Service Corp., 306 Cranford Road, Cherry Hill, NJ 08003 (Irwin Spetgang and Malcolm Wells).

Sunray Energy Corp., 107 Cienega Street, Santa Fe, NM 87501 (Herman G. Barkmann).

Sunstructures, Inc., 225 East Liberty Street, Ann Arbor, MI 48104 (Wayne Appleyard, Edward J. Kelley, Jr. and Richard McMath).

Thomason Solar Homes, Inc., 6802 Walker Mill Road SE, Washington, DC 20027 (Harry E. Thomason, President).

Total Environmental Action, Inc., Church Hill, Harrisville, NH 03450 (Bruce Anderson, Charles J. Michal, Daniel Scully).

Watson, Donald R. P.O. Box 401, Guilford, CT 06437 (architect).

Wright, David. 960 Camino Santander, Santa Fe, NM 87501 (architect).

Zomeworks Corp., P.O. Box 712, Albuquerque, NM 87103 (Steve Baer).

The following firms are marketing collectors and other components that can be used in solar heating and cooling systems. Addresses and telephone numbers are provided together with a short description of the products offered. You should include a stamped, self-addressed envelope if you write them for further information.

This list is by no means exhaustive, and it does not constitute a warranty of equipment performance. More detailed listings can be found in the *Energy Primer* (available for $5.50 prepaid from Whole Earth Truck Store, 558 Santa Cruz Avenue, Menlo Park, CA 94025). This excellent book contains descriptions and evaluations of the solar equipment offered by these and many more companies.

Contemporary Systems in Housing Design
Route L, Box 66
Ashland, NH 03217
Phone: (603) 968-7841
Attn: John C. Christopher

Developing air-type collector constructed of vinyl and aluminum in a molded, foam-insulated case, with fiberglass cover. Also making electronic control circuitry.

Daystar
41 Second Avenue
Burlington, MA 01803
Phone: (617) 272-8460
Attn: Clifton Smith

Marketing "Daystar 20" liquid-type collector. Claim good efficiency at low and high temperature differentials. Also developing other solar components.

Deko Labs
P.O. Box 12841
Gainesville, FL 32604
Phone: (904) 372-6009
Attn: Donald DeKold

Marketing differential temperature comparator controls for the circulation pumps between the solar collector and the storage. Uses solid-state sensors and has a "freeze control" option.

Energex Corporation
481 Tropicana
Las Vegas, NV 89119
Phone: (702) 736-2994
Attn: Alfred Jenkins

Marketing "Energex 751 Series Solar Energy Absorbers." Copper tube-in-plate absorber for solar heating and domestic water heating.

FAFCO
138 Jefferson Drive
Menlo Park, CA 94025
Phone: (415) 364-6772
Attn: Freeman A. Ford, President

Marketing black, extruded plastic collectors for solar swimming pool heating.

Falbel Energy Systems Corporation
472 Westover Road
Stamford, CT 06092
Phone: (203) 323-7477
Attn: Gerlad Falbel

Marketing "FES Delta" concentrating collector and hot water heater. Collector uses both front and rear surfaces of absorber for solar collection.

Filon
12333 South Van Ness Avenue
Hawthorne, CA 90250
Phone: (213) 757-5141

Marketing "Filon" panels–Tedlar coated, fiberglass-reinforced polyester sheets–good for greenhouses and low-temperature solar collectors.

Grumman Aerospace
Bethpage, NY 11714
Phone: (516) 575-7261
Attn: Gregory K. Knowles

Marketing "Sunstream 50A" liquid-type collectors with copper-tube fluid passages; also "Model 50" Solar Collector Water Heating System."

Kalwall Corporation
P.O. Box 237
Manchester, NH 03105
Phone: (603) 727-3861
Attn: Keith Harrison, President
Scott F. Keller, Solar Components

Marketing "Sun-Lite" fiberglass-reinforced polyester solar collector cover panels.

Natural Power, Inc.
New Boston, NH 03070
Phone: (603) 487-2456
Attn: Richard L. Katzenberg

Specializes in instrumentation and controls for the alternate energy industry. Marketing temperature monitors and solar heating controls.

Northrup, Inc.
302 Nichols Drive
Hutchins, TX 75141
Phone: (214) 225-4291
Attn: Lynn L. Northrup, Jr.

Developing a concentrating collector that uses a Fresnel lens to focus the sun's rays.

Olin Brass Corporation
East Alton, IL 62024
Phone: (618) 258-2443
Attn: John Barton

Providing Roll-Bond tube-in-plate absorbers of aluminum and copper.

PPG Industries, Inc.
One Gateway Center
Pittsburgh, PA 15222
Phone: (412) 434-2645
Attn: Richard Lewchuk

Marketing liquid-type collectors with aluminum or copper Roll-Bond absorbers and two cover plates of tempered Herculite glass.

Revere Copper and Brass, Inc.
Research and Development Center
P.O. Box 151
Rome, NY 13440
Phone: (315) 338-2022
Attn: William J. Heidrich

Marketing copper-plate collector for integration into a roof. Copper tubes are clipped to plate of this liquid-type collector.

Rho Sigma Unlimited
5108 Melvin Avenue
Tarzana, CA 91356
Phone: (213) 342-4376
Attn: R. J. Schlesinger, President

Offering solid-state differential thermostat controls for solar heating and domestic water heaters. Catalog available.

Solaron Corporation
4850 Olive Street
Denver, CO 80022
Attn: William Barker

Sells air-type collectors and complete solar heating systems.

Sunearth, Inc.
Route 1, Box 337
Green Lane, PA 18054
Attn: Howard Katz, President

Marketing liquid-type collectors with steel absorbers and copper tubes. Subsidiary Sunearth Construction Co. specializes in installing solar heating equipment.

Sunworks, Inc.
669 Boston Post Road
Guilford, CT 06437
Phone: (203) 453-6191
Attn: Everett Barber

Marketing liquid-type and air-type collectors with copper absorbers and selective surface by Enthone.

Thomason Solaris Systems
6802 Walker Mill Road
Washington, DC 20027
Attn: Harry E. Thomason

Specializing in solar house design and "Solaris" collectors with water trickling over corrugated aluminum absorber.

Wormser Scientific Co.
88 Foxwood Road
Stamford, CT 06903
Phone: (203) 322-1981
Attn: Eric Wormser

Offering "Pyramidal Optics" concentrating solar collector system that can be built into the attic of a house.

Zomeworks Corporation
P.O. Box 712
Albuquerque, NM 87103
Phone: (505) 242-5354
Attn: Stephen C. Baer

Offering Skylid passive insulating shutters and Beadwall systems. Also plans for Drumwall, Bread Box solar water heater. Sells magnetic clips for Nitewall and thermosiphoning solar water heaters that can be used in freezing climates.

Glossary

absorbent – the less volatile of the two working fluids used in an absorption cooling device.
absorber – the blackened surface in a collector that absorbs the solar radiation and converts it to heat energy.
absorptance – the ratio of solar energy absorbed by a surface to the solar energy striking it.
active system – a solar heating or cooling system that requires external mechanical power to move the collected heat.
air-type collector – a collector with air as the heat transfer fluid.
altitude – the angular distance from the horizon to the sun.
ASHRAE – abbreviation for the American Society of Heating, Air-conditioning and Refrigerating Engineers.
auxiliary heat – the extra heat provided by a conventional heating system for periods of cloudiness or intense cold, when a solar heating system cannot provide enough.
azimuth – the angular distance between true south and the point on the horizon directly below the sun.

British thermal unit, or **Btu** – the quantity of heat needed to raise the temperature of 1 pound of water 1°F.

calorie – the quantity of heat needed to raise the temperature of 1 gram of water 1°C.
coefficient of heat transmission, or **U-value** – the rate of heat loss in Btu per hour through a square foot of a wall or other building surface when the difference between indoor and outdoor air temperatures is 1°F.
collector – any of a wide variety of devices used to collect solar energy and convert it to heat.
collector efficiency – the ratio of heat energy extracted from a collector to the solar energy striking the cover, expressed in percent.
concentrating collector – a device which uses reflective surfaces to concentrate the sun's rays onto a smaller area, where they are absorbed and converted to heat energy.
conductance – a property of a slab of material equal to the quantity of heat in Btu per hour that flows through one square foot of the slab when a 1°F temperature difference is maintained between the two sides.
conduction – the transfer of heat energy through a material by the motion of adjacent atoms and molecules.
conductivity – a measure of the ability of a material to permit conduction heat flow through it.
convection – the transfer of heat energy from one location to another by the motion of fluids which carry the heat.
cover plate – a sheet of glass or transparent plastic that sits above the absorber in a flat-plate collector.

degree-day – a unit that represents a 1°F deviation from some fixed reference point (usually 65°F) in the mean daily outdoor temperature.
design heat load – the total heat loss from a house under the most severe winter conditions likely to occur.
design temperature – a temperature close to the lowest expected for a location, used to determine the design heat load.
diffuse radiation – sunlight that is scattered from air molecules, dust, and water vapor and comes from the entire sky vault.
direct methods – techniques of solar heating in which sunlight enters a house through the windows and is absorbed inside.
direct radiation – solar radiation that comes straight from the sun, casting shadows on a clear day.
double-glazed – covered by two panes of glass or other transparent material.

emittance – a measure of the propensity of a material to emit thermal radiation.
eutectic salts – a group of materials that melt at low temperatures, absorbing large quantities of heat.

flat-plate collector – a solar collection device in which sunlight is converted to heat on a plane surface, without the aid of reflecting surfaces to concentrate the rays.
forced convection – the transfer of heat by the flow of warm fluids, driven by fans, blowers, or pumps.

Glaubers salt – sodium sulfate ($Na_2SO_4 \cdot 10H_2O$), a eutectic salt that melts at 90°F and absorbs about 104 Btu per pound as it does so.
gravity convection – the natural movement of heat through a body of fluid that occurs when a warm fluid rises and cool fluid sinks under the influence of gravity.

header – the pipe that runs across the top (or bottom) of an absorber plate, gathering (or distributing) the heat transfer fluid from (or to) the grid of pipes that run across the absorber surface.
heat capacity – a property of a material, defined as the quantity of heat needed to raise one cubic foot of the material 1°F.
heat exchanger – a device, such as a coiled copper tube immersed in a tank of water, that is used to transfer heat from one fluid to another through an intervening metal surface.
heating season – the period from about October 1 to about May 1, during which additional heat is needed to keep a house warm.
heat pump – a mechanical device that transfers heat from one medium (called the heat source) to another (the heat sink), thereby cooling the first and warming the second.
heat sink – a medium or container to which heat flows (see **heat pump**).
heat source – a medium or container from which heat flows (see **heat pump**).
heat storage – a device or medium that absorbs

collected solar heat and stores it for periods of inclement or cold weather.

heat storage capacity – the ability of a material to store heat as its temperature increases.

indirect system – a solar heating or cooling system in which the solar heat is collected exterior to the building and transferred inside using ducts or piping and, usually, fans or pumps.

infiltration – the movement of outdoor air into the interior of a building through cracks around windows and doors or in walls, roofs, and floors.

infrared radiation – electromagnetic radiation, whether from the sun or a warm body, that has wavelengths longer than visible light.

insolation – the total amount of solar radiation–direct, diffuse and reflected–striking a surface exposed to the sky.

insulation – a material with high resistance or R-value that is used to retard heat flow.

integrated system – a solar heating or cooling system in which the solar heat is absorbed in the walls or roof of a dwelling and flows to the rooms without the aid of complex piping or ducts.

langley – a measure of solar radiation, equal to one calorie per square centimeter.

life-cycle costing – an estimating method in which the long-term costs such as energy consumption, maintenance, and repair can be included in the comparison of several system alternatives.

liquid-type collector – a collector with a liquid as the heat transfer fluid.

natural convection – see **gravity convection.**

nocturnal cooling – the cooling of a building or heat storage device by the radiation of excess heat into the night sky.

passive system – a solar heating or cooling system that uses no external mechanical power to move the collected solar heat.

percentage of possible sunshine – the percentage of daytime hours during which there is enough direct solar radiation to cast a shadow.

photosynthesis – the conversion of solar energy to chemical energy by the action of chlorophyll in plants and algae.

photovoltaic cells – semi-conductor devices that convert solar energy into electricity.

radiant panels – panels with integral passages for the flow of warm fluids, either air or liquids. Heat from the fluid is conducted through the metal and transferred to the rooms by thermal radiation.

radiation – the flow of energy across open space via electromagnetic waves, such as visible light.

reflected radiation – sunlight that is reflected from surrounding trees, terrain or buildings onto a surface exposed to the sky.

refrigerant – a liquid such as freon that is used in cooling devices to absorb heat from surrounding air or liquids as it evaporates.

resistance, or **R-value** – the tendency of a material to retard the flow of heat.

retrofitting – the application of a solar heating or cooling system to an existing building.

risers – the flow channels or pipes that distribute the heat transfer liquid across the face of an absorber.

R-value – see **resistance.**

seasonal efficiency – the ratio of solar energy collected and used to that striking the collector, over an entire heating season.

selective surface – an absorber coating that absorbs most of the sunlight hitting it but emits very little thermal radiation.

shading coefficient – the ratio of the solar heat gain through a specific glazing system to the total solar heat gain through a single layer of clear, double-strength glass.

shading mask – a section of a circle that is characteristic of a particular shading device. This mask is superimposed on a circular sun path diagram to determine the time of day and the months of the year when a window will be shaded by the device.

solar house, or **solar tempered house** – a dwelling that obtains a large part, though not necessarily all, of its heat from the sun.

solar radiation – electromagnetic radiation emitted by the sun.

specific heat – the quality of heat, in Btu, needed to raise the temperature of 1 pound of a material 1°F.

sun path diagram – a circular projection of the sky vault, similar to a map, that can be used to determine solar positions and to calculate shading.

thermal capacity – the quantity of heat needed to warm a collector up to its operating temperature.

thermal mass, or **thermal inertia** – the tendency of a building with large quantities of heavy materials to remain at the same temperature or to fluctuate only very slowly; also, the overall heat storage capacity of a building.

thermal radiation – electromagnetic radiation emitted by a warm body.

thermosiphoning – see **gravity convection.**

tilt angle – the angle that a flat collector surface forms with the horizontal.

trickle-type collector – a collector in which the heat transfer liquid flows down channels in the front face of the absorber.

tube-in-plate absorber – an aluminum or copper sheet metal absorber plate in which the heat transfer fluid flows through passages formed in the plate itself.

tube-type collector – a collector in which the heat transfer liquid flows through metal tubes that are wired, soldered, or clamped to the absorber plate.

ultraviolet radiation – electromagnetic radiation, usually from the sun, with wavelengths shorter than visible light.

unglazed collector – a collector with no transparent cover plate.

U-value – see **coefficient of heat transmission.**

Index

Page numbers printed in **bold face** indicate tables, photographs, or illustrations.

The authors are grateful to the following people for their help and encouragement.

Michelle Artese	Hoyt Hottel
Steve Baer	James Janecek
David Bainbridge	Lee Johnson
Jay Baldwin	Henry Mathew
Everett Barber	Scott Matthews
Danielle Bar Gadda	Charles Michal
Richard Blazej	Lea Poisson
Raymond Bliss	Norman Saunders
Lee Porter Butler	Daniel Scully
Douglas Coonley	Ken Smith
Martha Coulton	Thyra Stevenson
Doug Daniels	Susan Swanson
Paul Davis	Maria Telkes
Albert Deitz	Harry Thomason
Mary Donovan	Karen Tolman
Fred Dubin	Peter Voorhees
William Edmondson	J.D. Walton
Erich Farber	Donald Watson
David French	Malcolm Wells
Susan A. Garrett	Hilda Wetherbee
Richard Gordon	Austin Whillier
Jonathan Hammond	Bill Wilson
Harold Hay	Bill Yanda
Dennis Holloway	John Yellott

Photo credits:

Page 25, courtesy of Maria Telkes; p. 29, top, Raymond Bliss; p. 29, bottom, courtesy of Bridgers and Paxton; p. 42, courtesy of Living Systems; pp. 44, 49, 124, courtesy of Total Environmental Action, Inc.; p. 46, courtesy of Dennis Holloway; p. 48, courtesy of Solar Energy Laboratory, University of Florida; p. 51, courtesy of the Institute of Energy Conversion, University of Delaware; p. 73, courtesy of Pacific Gas and Electric Company; pp. 101, 213, courtesy of Zomeworks Corporation; pp. 102, 130, 131, 231, Peter Voorhees; pp. 108, 140, 144, 240, Michael Riordan; p. 110, Lee Porter Butler; pp. 117, 122, 123, courtesy of J. D. Walton; p. 125, Charles J. Michal; pp. 136, 137, 138, courtesy of Harold Hay; p. 182, courtesy of Wormser Scientific Corp.; p. 215, Lynn Nelson; p. 219, courtesy of Lee Johnson; p. 221, Linda Goodman; p. 236, William F. Yanda

Biographical Notes

Bruce Anderson was born in 1947 in Portland, Maine, and grew up in Mankato, Minnesota. He attended the Massachusetts Institute of Technology where he received a B.S. degree in Art and Design and an M.S. degree in Architecture. Moved by the need for fresh architectural approaches that take dwindling energy resources into account, Bruce wrote *Solar Energy and Shelter Design* as his Master's Thesis. The published manuscript has been used as a textbook at many colleges and universities. Following work with the New York firms of I. M. Pei and Dubin-Mindell-Bloome Associates, Bruce moved to Harrisville, New Hampshire in 1974 to found Total Environmental Action, Inc., a research, education, and design firm dedicated to pursuing energy and lifestyle alternatives. In addition to his work as president of TEA, Bruce is currently the executive editor of *Solar Age* magazine and a director of the American section of the International Solar Energy Society. He lectures extensively and has testified before Congress on the uses of solar energy in architecture. He has also written *Solar Energy: Fundamentals in Building Design,* a McGraw-Hill textbook.

Michael Riordan, born in 1946 in Springfield, Massachusetts, was raised in the northern Connecticut village of Hazardville. He received both his B.S. and Ph.D. degrees in Physics from the Massachusetts Institute of Technology. Michael moved to California in 1973 and worked as a Research Associate in high-energy physics at the Stanford Linear Accelerator Center in Menlo Park. When approached by Bruce, an old MIT roommate and crew partner, with the idea of publishing this book Michael left that position to co-found Cheshire Books. In addition to directing the affairs of Cheshire Books, he is currently the architecture editor for the revised *Energy Primer* and is working on a book of his own.